D0480538

STUDIES IN PHYSICAL GEOGRAPHY

Historical Change in the Physical Environment

STUDIES IN PHYSICAL GEOGRAPHY
Edited by K. J. Gregory

Also published in the series

Applied Climatology
J. E. HOBBS

Ecology and Environmental Management
C. C. PARK

Geomorphological Processes
E. DERBYSHIRE, K. J. GREGORY AND J. R. HAILS

Man and Environmental Processes
Edited by
K. J. GREGORY AND D. E. WALLING

STUDIES IN
PHYSICAL GEOGRAPHY

Historical Change in the Physical Environment: a guide to sources and techniques

J. M. HOOKE
Senior Lecturer in Geography, Portsmouth Polytechnic

R. J. P. KAIN
Senior Lecturer in Geography, University of Exeter

BUTTERWORTH SCIENTIFIC
London Boston
Durban Singapore Sydney Toronto Wellington

First published 1982

British Library Cataloguing in Publication Data

Historical change in the physical environment: a guide to sources and techniques.—(Studies in physical growth)
1. Physical geography—Great Britain—Bibliography
I. Hooke, J. M. II. Kain, R. J. P. III. Series
914.Z GB181

ISBN 0-408-10743-X

Typeset by Tunbridge Wells Typesetting Services Ltd
Printed and bound by Page Bros (Norwich) Ltd

Preface

Human geographers frequently look to the past in their search for explanations of present conditions but historical sources have been used less for interpreting features of the physical environment. We contend that evidence from the past has considerable potential for providing longer-term perspectives for studies of current physical processes, for understanding the nature and causes of change and, above all, for understanding the magnitude of the impact of human activities on the physical environment. The main purpose of this book is to provide guidance to the use of historical sources in analyses of the physical environment; it is aimed particularly at physical geographers and other earth scientists with no formal training in the use of historical material.

It is not possible to discuss all extant sources fully, but in Chapter 1 the content and availability of major British sources is reviewed. Comparative international sources are indicated in Chapter 2, where the complementary role of field evidence is also discussed. Some techniques of accuracy assessment and source analysis are treated in Chapter 3. The rest of the book is devoted to an examination of applications of historical evidence in various fields. We review the purpose, methods and results of studies in which historical sources have been used in climatology, geomorphology, biogeography, ecology, hydrology and geophysics. This review is not necessarily comprehensive but rather demonstrates the range of applications that have been made.

Many people have provided help and advice in the preparation of this book and we are very grateful to all with whom we have discussed our work. We would like to thank especially Professor K. J. Gregory for his help and encouragement throughout, Mr T. Bacon who took great care with the diagrams and Mr A. Teed AIIP for the photographs.

Janet Hooke and Roger Kain
Manchester and Exeter

Contents

Illustrations

Tables

Introduction

The reconstruction of past conditions and events and the analysis of changes over time are important objectives in all branches of the earth sciences, though the relevant timescales differ from one field to another. Some physical features are very stable, so that study of change is only appropriate and instructive on a timescale of thousands of years. By contrast, others vary considerably in the short term so that timescales of a few decades can provide realistic frameworks.

This book is concerned mainly with a review of studies and sources which fall into what can be broadly termed the 'historical' period. During this time, man has had a major influence on the nature of physical systems and has recorded information about them in written, graphical and statistical forms. We propose to adopt a catholic definition of what constitutes an 'historical' source so that we include written documents, both published and unpublished, and also a wide range of pictorial, map and numerical sources.

The depth of the historical period and the length of time for which records are available varies in different parts of the world. Documentary sources from the ancient civilizations of the Old World date back several thousands of years, but in areas of more recent European settlement, such as North America, there are generally few records available older than 400 years.

Within the different fields of earth science the extent of use and indeed the general applicability of historical sources varies considerably, but some initial generalizations can perhaps be made. Historical sources have long been used in the more applied areas of physical geography as in coastal studies, for example, where an historical approach was fashionable in 'pre-quantitative' days. Quantification brought with it a general overshadowing of methods involving 'subjective' judgements and 'intuitive' assessments which the use of much historical source material necessarily involves. Associated with this trend towards a more scientific method was a change in emphasis from evolutionary approaches aimed at elucidating chronology to intensive process studies measuring system dynamics. There is now, however, a growing awareness of the need to put the results of such process studies into longer time-perspectives. Two further developments in physical geography have highlighted the value of historical sources. One is the increasing emphasis ascribed to the influence of man on the physical environment. Often process studies are too short in timescale to reveal the full extent of the impact of human activities, for which purpose historical evidence can be very valuable, not least for calibrating predictive models. A second impetus to historical study in recent years has been that process studies have shown that many changes are much more rapid and extensive than formerly thought, and thus that change can be effectively detected within the span of historical sources. Coates and Vitek (1980) sum up the position well: 'There is no substitute for the compilation of long and accurate records regarding historical developments of the phenomenon in question.'

THE USES OF HISTORICAL SOURCES

The major uses of historical sources in the earth sciences are to reconstruct situations or events at a particular date, to analyse changes over time and to explain these changes. Generally speaking, historical sources are more accurate as well as more abundant close to the present, and their value and applicability tend to increase commensurately. Their value also depends on the relative availability of other types of information such as field-derived evidence. Historical documents may be used simply to identify the presence or absence of features at a particular time or their general characteristics, such as shape. Alternatively, they may be employed in highly quantitative analyses, as when planimetric data is taken from maps and subjected to statistical analysis. Techniques have also been developed, primarily in climatology, for quantifying the written content of historical sources.

Historical sources can be usefully employed in conjunction with field-derived evidence from deposits, morphology, human artefacts, pollen and tree-ring analysis. In some applications, historical sources may provide the prime information and field evidence be used to corroborate it; in other instances these roles may be reversed and the historical sources used to corroborate field evidence. In yet other cases, historical sources may be used simply to date a feature and to establish a datum from which interpretation can proceed independently of documentary material.

TYPES OF SOURCES

Historical sources can be divided into graphical, written and statistical categories and each type may be found in manuscript or published form. These subdivisions have been used in this book to provide a convenient basis for discussion.

In the extreme view, any source material used by an earth scientist can be considered 'historical' even if it dates from only a few years ago. However, it is not our intention to discuss material of similar nature to the current data sources with which earth scientists are familiar. Modern maps and aerial photographs are examples—they do not entail the techniques of search and analysis required for historical sources proper. It is further suggested that documents written by our contemporaries can hardly be considered historical, and thus a generation back from the present provides a useful working baseline. This said, it must be acknowledged that some documents and research reports produced in recent years have already been used in analyses of change over time.

The nature of the historical record varies from country to country according to particular cultural, political and economic histories. The emphasis in this book, as noted in the Preface, is on British sources, but these also provide a general indication of the range of material available in other Old World and especially European countries. The other main

English-speaking countries are those of the New World, where there are rich archives of sources associated with exploration and settlement.

Early maps are probably the most important source of all, particularly for the analysis of conditions at particular dates and for monitoring changes in physical features. They do, though, have the crucial limitation for analysis of change in that they only indicate the situation at often widely separated dates perhaps quite unrelated to significant events in the natural environment. Care must thus be taken in drawing inferences from comparisons of map evidence through time. Other *graphical sources* include a variety of building and industrial plans and drawings, engravings, paintings and photographs. Aerial photographs are very valuable but are mostly post-Second World War in date. *Written sources* are an heterogeneous group but certain general types can be recognized such as manorial and estate, legal, ecclesiastical and personal papers, newspapers and journals. Published histories and descriptions can be very valuable, though care has to be taken in the use of all secondary compilations. Published articles vary from the professional and technical to the lay and impressionistic. Oral evidence should not be overlooked but can only cover a limited time span. The most useful historical *statistics* for the earth scientist relate to climate, hydrology and land use. All these types of sources may be found in public archives, museums and libraries, though some papers still remain in private collections.

AREAS OF APPLICATION

The extent to which historical sources have been employed varies among the different branches of the earth sciences. Climatology is the field in which the application of historical evidence is most highly developed. This can be partly explained by the magnitude of climatic variations, the extent to which measurements have been made, the importance of climate to man's activities and also by the value of historical analysis for prediction and forecasting. The study of fluctuations in glaciers was originally closely associated with analysis of climatic change, but a separate field has now been developed concerned with understanding glacier régimes and processes. The rich historical sources relating to areas such as Iceland and Scandinavia also fostered a strong historical emphasis at an early stage of scientific glacier studies.

In geomorphology, the earliest applications were mostly related to practical needs, particularly on coasts where change is rapid and has caused problems for navigation, the livelihood of ports and the safety of buildings and other structures. Geomorphologists have long been aware of some spectacularly rapid river movements but the widespread and systematic study of channel changes is comparatively recent. In other geomorphological fields, the main use of historical sources has been in relation to the reconstruction of catastrophic events such as landslides and floods. Historical analysis has also been used in the study of processes, such as soil erosion, which have increased in rate significantly in the historical period. In the fields of geology and geophysics, the major applications are in relation to events such as earthquakes and volcanic eruptions, though more gradual changes such as those associated with relative sea-level movements have been studied with historical documents.

In biogeography, historical sources have been used extensively in those parts of the world where the change from natural to cultivated vegetation has taken place within the documentary period. Many of these studies also have the more practical aim of understanding the type of vegetation the land will support and how it changes under different conditions. In settled areas, study of vegetation is inevitably connected with the history of land use and human activities, though a more ecological approach is now apparent. Historical study of soils is rather limited by their nature and relatively slow development.

HISTORICAL APPROACH AND TECHNIQUES

The use of historical records does not necessarily entail the acquisition of any particular 'skills' which an earth scientist does not normally possess except in rare instances for deciphering very old documents and translating dead languages. It does, however, require a particular attitude of mind. It can be difficult and time-consuming to locate useful records, especially if the researcher has little prior guidance on the date of an event. Once records have been located they must be read or examined meticulously and this may involve sifting through much unwanted material. Evidence must be corroborated by other sources and the accuracy, reliability and dating should be carefully checked. Often individual pieces of information, especially qualitative descriptions, will appear of little use; generally the earth scientist desires quantifiable information and preferably at several dates. This differs from the approach of some historians whose prime interest might be the individual case and for whom short descriptive statements are valuable. However, even in physical geography there is a place for this type of information particularly in relation to intermittent events such as floods and landslides. Watson (1962) emphasized this point when he wrote in relation to the analysis of historical sources in climatology that, 'all data, even the most quantitative and precise, requires critical interpretation, but not even the most qualitative and indirect evidence should be ignored'.

The main purpose of this book is to demonstrate how historical sources can be used to help to understand the character of changes within the physical environment. The sequential stages in the use of sources for these purposes can be summarized as follows:

(1) Define the purpose and scope of the study.
(2) Identify possible sources and investigate their location, availability, characteristics and content.
(3) Search for relevant material within potentially suitable documents.
(4) Check general accuracy and reliability of documents.
(5) Make a pilot study to test the proposed framework for data abstraction.
(6) Abstract information from data sources.
(7) Corroborate all information obtained from historical documents.
(8) Process data using procedures such as standardization and quantification.
(9) Analyse and interpret the data.

In Chapter 1 information is provided on the major types of historical sources likely to be useful in analyses of the physical environment. It is impossible to provide details on all sources, and so our aim is to give some general guidance and then to refer the reader to publications which provide a starting point for searches. The bulk of this discussion concerns British sources but in Chapter 2 some limited international perspective is presented and the linking of documentary sources with field evidence is also examined. In Chapter 3 techniques for assessing accuracy and reliability of sources are outlined and data manipulation and analysis are discussed. The remaining part of the book reviews published studies to demonstrate how historical material has been utilized. Applications are discussed under the broad themes of the reconstruction of past conditions (Chapter 4), the analysis of processes and events (Chapter 5) and the explanation of changes (Chapter 6). Illustrative examples are drawn from a number of major areas of study within the earth sciences.

1
Sources of Historical Evidence

The primary purpose of this chapter is to introduce some source material which can add an historical dimension to physical geography studies in Britain. In the following chapter some international perspective is provided by reviewing the comparative availability of selected sources overseas. For the purposes of this review, three main categories of source material are used. These are discrete in themselves and they also relate to different types of environmental problems. The first group includes a wide variety of *graphical material,* particularly manuscript and printed maps. Maps are probably the most important source of historical evidence for physical geographers, and so a large proportion of this chapter is devoted to their discussion. Topographical maps can provide evidence of form and pattern and, when available in a series, can assist with monitoring change through time, albeit often at irregular and widely spaced intervals. Pictorial representations of the physical landscape such as paintings, engravings and photographs are also discussed in this section. These can be very difficult to interpret but where a series is available they can be of great value when used in conjunction with written documents.

The second category of evidence includes a wide variety of *written material*—those documentary sources which have a bearing on the natural landscape or physical processes. These range from published topographical accounts and technical articles in scientific periodicals and newspapers to manuscript diaries and letters.

Although techniques exist for extracting quantitative data from written sources, *statistical series* can prove invaluable if available for the historical period. Many systematic data series are too recent to be considered here but three groups of data sources, which are fundamental to physical geography and for which there are sources dating back over 100 years, are considered. These are meteorological, hydrological and land-use statistics.

Historical sources relating to environmental conditions in the past are becoming much more accessible as more privately owned documentary material is entrusted to public care in archive offices. The concurrent sophistication of storage and retrieval systems like the computerized document processing used at the English Public Record Office's new repository at Kew in Surrey are also helping in this respect.

Few historical studies in the earth sciences can rely entirely on locally available materials. The catalogues of national collections must also be consulted, in particular the official archives at the PRO and the great collections in the British Library. There are general guides to working in the PRO (PRO 1963–1968, 1967) which can be consulted prior to a visit to London. The List and Index Society has published photographic reproductions of many

PRO search-room indexes, such as the one which lists every parish tithe map for England and Wales (List and Index Society, 1971, 1972). Catalogues of manuscripts and maps held in other important collections such as the Bodleian Library and the British Library are indexed and printed so that they, too, are available for prior consultation (e.g. British Museum, 1844, 1967, 1978; Adams, 1966, 1970; Warren, 1965).

The second tier of public archives in England and Wales is the system of county record offices. These originated in Bedfordshire and Bristol about fifty years ago and have since been established in all counties. They are staffed by qualified archivists who have now classified, and in many cases indexed, the official records relating to the local area. Addresses and brief particulars of all these offices (as well as national and other local record-holding institutions) are listed in a recently revised publication, *Record Repositories in Great Britain* (Royal Commission on Historical Manuscripts, 1979). This is organized county by county and, most valuably, notes the principal published guides which describe the contents of each record office and the reprographic and similar facilities available. In addition, many offices produce mimeographed guides to matters concerning research on their premises and to particular categories of material. The British Library Science Reference Library publishes a guide to government department and other 'science-oriented' libraries and information bureaux. This provides details of hours of opening, facilities, and contents (British Library Science Reference Library, 1976). In Britain a number of large landowners still retain their estate papers but may allow access on application. Information can be obtained from the National Register of Archives, Chancery Lane, London.

In short, the pattern of record repositories in Britain has evolved in an 'informal' manner. Long before the establishment of county record offices in the 1920s most public libraries and a few museums had acquired historical documents relating to the local area. The City Librarian of Exeter, for example, secured a wealth of material relating to both Devon and Cornwall before the Devon Record Office was established. The British Museum and the University Libraries of Oxford and Cambridge have also absorbed large numbers of documents and other old-established institutions such as the John Rylands Library, Manchester, have done the same (Dopson, 1955). Local libraries, museums and institutions with archives which are open to the public are listed in *Record Repositories in Great Britain* (1979). The local studies rooms of principal public libraries may also have useful maps and printed sources. They are listed in telephone directories.

Having found some historical evidence to illuminate a particular problem, earth scientists should not encounter many general difficulties when deciphering and transcribing the kind of material likely to be found most useful. There may, though, be problems of interpretation, as noted in subsequent sections of this chapter. Most documents encountered will date from the 'modern' period and will be written in English, not Latin. If difficulty is encountered with transcription, then one can turn to guides on palaeography and reading local records (Grieve, 1959; Hector, 1980), while archival staff will also be on hand to help resolve queries. Further to these matters of the logistics of working with historical documents one should perhaps note that the use of pens is banned in some offices and that there are often restrictions on tracing maps. To help on these and similar points a glance through one of the number of guides for researchers in local history can provide much helpful advice (West, 1962; Emmison, 1966; Stephens, 1973).

1.1 GRAPHICAL SOURCES

Estate, enclosure and tithe maps

The three categories of estate, enclosure and tithe maps belong to one and the same 'family' of large-scale and usually manuscript plans produced by privately employed surveyors. Estate maps, usually in manuscript until the late nineteenth century, are surveys of the land belonging to a particular individual or institution and were commissioned by them. Some are maps of just a single farm, perhaps as small as 40 acres, while others are surveys of extensive landed estates of upwards of 10 000 acres and extending over several parishes.

Enclosure and tithe maps are similar to estate maps in that they were privately commissioned (all three categories of work were also often undertaken by the same surveyors) and portray a small area, usually a township or parish, at a large scale. However, they differ from private estate plans in that they were instruments by which legislation was implemented. Most enclosure awards with their accompanying maps date from the period 1770–1850, while the vast majority of tithe maps were completed around the year 1840.

Estate maps

These are maps of estates of various kinds such as manors, farms, parks and mines. Their surveyors were not necessarily concerned with a 'scientific geodesic survey' but rather with indicating to their patrons what land they owned, how the boundaries lay, what estate buildings there were and perhaps how the farmland was used (Baker, 1962a; Emmison, 1963; Hull, 1973). Matters such as the relief of the land were of minor concern, though occasionally hill-shading and, on later maps, hachuring might be used. In fact, a first impression might be that the body of estate maps comprises a singularly unhelpful source of evidence for the earth scientist. Though produced for purposes a world apart from the needs of the student of, say, landform development, there are a number of redeeming factors associated with the nature and devêlopment of estate plans which can render them quite useful.

Estate maps are very numerous. A rough estimate made in 1966 put the total number of estate maps dated before 1850 and available for consultation in libraries and local record offices at over 20 000 (Emmison, 1966). This averages almost two maps for every parish of England and Wales, and private owners are continually depositing their plans in record offices to which the public has access. A great majority of estate plans show the land of small estates, but perhaps one in ten covers an entire parish while a small minority encompasses a number of adjacent parishes. Seldom will a complete blank be drawn and some counties, Kent for example, have particularly rich collections (Baker, 1962b, 1965; Hull, 1973). Scotland is also rather well endowed in this respect (Third, 1957; Adams, 1967).

A second redeeming factor is that estate maps provide a source of quite reasonable topographic information as far back as the late sixteenth century. A few survive from the late Middle Ages, their number increases in the Tudor and Stuart periods and they are plentiful from about 1700 onwards. Indeed, manuscript estate maps flourished during the whole of the period from *c.* 1575 to the mid-nineteenth century (Redstone and Steer, 1953). The years around 1550, in fact, marked a revolution in local cartography (Harley, 1972).

Until this time, the word 'surveyor' was really synonymous with 'valuer'. Estates were described by metes and bounds and any maps were produced with the most elementary knowledge of mensuration (Taylor, 1947, 1954; Price, 1955; Lynam in Emmison, 1947).

Improvement came about for two main reasons. First, there was an increase in the demand for surveys from the late sixteenth century onwards (Darby, 1933; Steer, 1962):

> The re-distribution of church lands . . ., the consolidation and exchange of openfield strips, the improvement of estates and the changing ownership of land during the Commonwealth and the Restoration, are all reflected in the output of estate maps. (Harley, 1972)

Second, this increase in demand coincided with technical advances in surveying. University mathematicians such as John Norden were attracted to the profession (Walters, 1968; Ravenhill, 1971, 1972). John Aldgate, for example, who drew plans of the Lennard estates at West Wickham in Kent in the mid-seventeenth century, referred to himself as 'welwiller unto the Mathematticks' and the Stuart surveyor, George Russell, was a master at the Sir Joseph Williamson Mathematical School at Rochester (Hull, 1973). Phillips (1980) has discussed the background of William Fowler, who mapped estates in the West Midlands in the seventeenth century. Instruments and methods were also changing to come in line with the best of Continental practice (Kiely, 1979).

Some sixteenth-century surveys produced with a plane table are far from accurate, but from about 1570 simple theodolites came into more common use (Hull, 1973). Leonard Digges, writing in 1571, praised the plane table as the best method, but in 1596 Ralph Agas strongly advised the theodolite (Emmison, 1947; Crone, 1966; Ravenhill, 1973). The plane table certainly provided a rapid, if not very accurate, method of plotting straight on to paper placed on the plotter board.

The theodolite is an instrument used to measure the horizontal angles between fixed points on the ground. With the aid of a measured base line a triangulated framework of survey lines can be built up. Measurements of the distances at right-angles from these lines to the objects to be plotted are then measured with a chain. Though methods improved with time, it must not be assumed that every surveyor was equally competent. Indeed, Emmison refers to some eighteenth-century maps which bear little resemblance to the actual properties described (Emmison, 1947).

A major advantage of estate maps is that they are usually drawn at a very large scale because of the relatively small area surveyed. The basis of all scales was the 22-yard chain, and usually the surveyor selected a convenient number of chains to the inch for his maps without concern for the number of inches to the mile. The commonest scale is 3 chains to an inch or 26.7 inches to a mile (1:2376). Larger scales of 32 (1:1950), 35.5 (1:1785), 40 (1:1584), 45.7 (1:1386) and 53.3 inches to a mile (1:1189) are quite common, while the 6-chain or 13.3 inches to a mile scale (1:4752) is the smallest scale normally used for estate mapping.

Estate maps were usually inventories of particular farming landscapes and so their prime concern was to communicate information about fields, farms, buildings and land use (Thomas, 1966) (Figure 1.1). They also provide some information on the physical form of the land. Rivers and streams are carefully noted because they were commonly used as boundaries, either of estates themselves or of parishes and townships. Such other details as

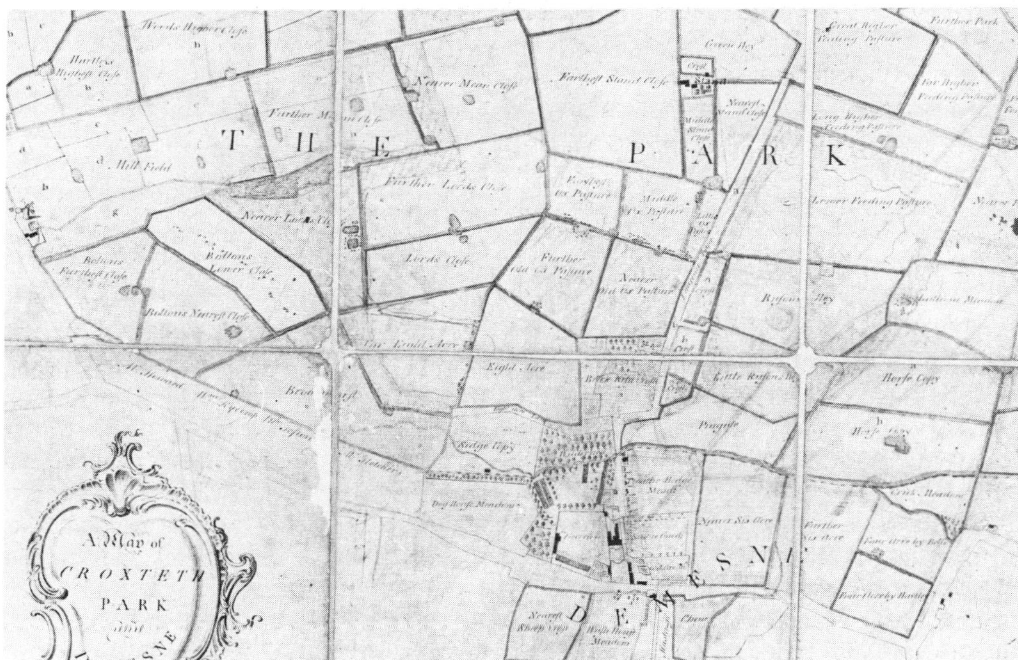

Figure 1.1. AN EXTRACT FROM THE ESTATE PLAN OF CROXTETH PARK AND DEMESNE, 1769. (Source: Lancashire Record Office DDM 14/24)

are shown depend largely on local circumstances. Maps of maritime parishes quite commonly locate cliffs, marshes, sandspits and beaches, while those of upland parishes show moors, combes, quarries and valley scars (Yates, 1964; Emmison, 1966; Norden reprint 1966; de Boer, 1969; Carr, 1969).

Estate maps have been deposited in all the main national archives. The PRO at Kew and the Scottish Record Office have collections as have national libraries like the Bodleian at Oxford, the British Library in London and the National Library of Wales. Estate maps are relatively easily traced in the published catalogues of these institutions (British Museum, 1962 reprint; Adams, 1966 and 1970; PRO, 1967; and see Harley, 1972).

The bulk of estate maps, however, are held in collections of estate papers deposited with county record offices. There are only a few published catalogues of these. One of the first to appear, and the model for all subsequent catalogues, was for Essex (Emmison, 1947). Catalogues for Buckinghamshire (Bucks Record Office, 1962 and 1964), Huntingdon (Dickinson, 1968), Bedfordshire (Emmison and Fowler, 1930), Kent (Hull, 1973), Sussex (Steer, 1962 and 1968) and Hertfordshire (Walne, 1969) have also been published. However, a majority of *Guides* published by county record offices list maps briefly and in most offices estate maps have been catalogued and indexed under parishes in typescript or on cards. Long searches should not be necessary providing the parishes in which features lie are known. A residue of estate maps is still in private hands. These are usually the property of remaining large landed estates and it is occasionally possible to obtain permission to inspect the family papers on written application to the estate office.

Enclosure maps

Enclosure maps record the local administration of the parliamentary enclosure movement which by private act of parliament or by General Enclosure Acts reallocated common fields, pastures and wastes, and so created the pattern of enclosed countryside familiar to us today. The enclosure movement was a long and gradual process, and only in the last accelerated period of parliamentary enclosure in the eighteenth and nineteenth centuries were maps normally produced. Parishes which had been gradually enclosed by private agreement over the course of several centuries may have no parliamentary enclosure award or perhaps just one relating to some very small remaining portion of commons and waste (West, 1962). One or more commissioners were appointed under the acts and, after carefully considering the claims of all the various parties, the commissioners made an award reallocating the relevant lands which were illustrated on a map (Turner, 1980) (Figure 1.2). The nature of physical–topographical detail shown on enclosure maps is similar to that on estate maps; rivers, estuaries, lakes and ponds, the coast and similar features will be noted, often with the greatest care where they form part of the parish boundary. As with estate maps (discussed above) and the body of tithe maps (noted below), enclosure maps vary from highly detailed and accurate surveys to the crudest of sketches, perhaps showing only a few fields or a small piece of common or waste. As a general rule, planimetric accuracy should be considered suspect until tested using one of the methods noted in Chapter 3. Enclosure maps are also a good record of the extent of commons, heaths and moors at the time of their compilation, as very many enclosure awards dealt not with common arable fields but with common pasture or unimproved waste. Such maps are not only to be found in upland Britain but also record the positions of commons and waste in lowland counties like Essex (Emmison, 1969).

An original drawing and a copy were required and both of these frequently survive, either in the local county record office or with perhaps one remaining in the care of churchwardens in the parish chest. A few have found their way into private hands or into local libraries, while the Public Record Office also has a set relating to both Crown and privately held property. The listing of enclosure acts and awards owes much to the lifetime's work of the late W. E. Tate. His handlists of enclosures published county by county in the Proceedings of local history and archaeology societies have been brought together and edited with an introduction by M. E. Turner (Tate, 1978). Many county record offices have produced lists of enclosures in their custody, some on the excellent model of the *Catalogue of Maps in the Essex Record Office* which provides details of the maps as well as the awards (Emmison, 1969). The availability of published lists can be judged from those noted by West (1962), Harley (1972) and those listed by the *Royal Commission on Historical Manuscripts* (1979).

Tithe maps

When parliament passed the Tithe Commutation Act in 1836 it put an end to a long and bitter period of dispute over payment of tithes for the support of the church (Evans, 1976, 1978). The settlement was effected by a minute field-by-field survey of every piece of titheable land in the country. The tithe surveys cover about 29.5 million acres in all, about three-quarters of England and Wales.

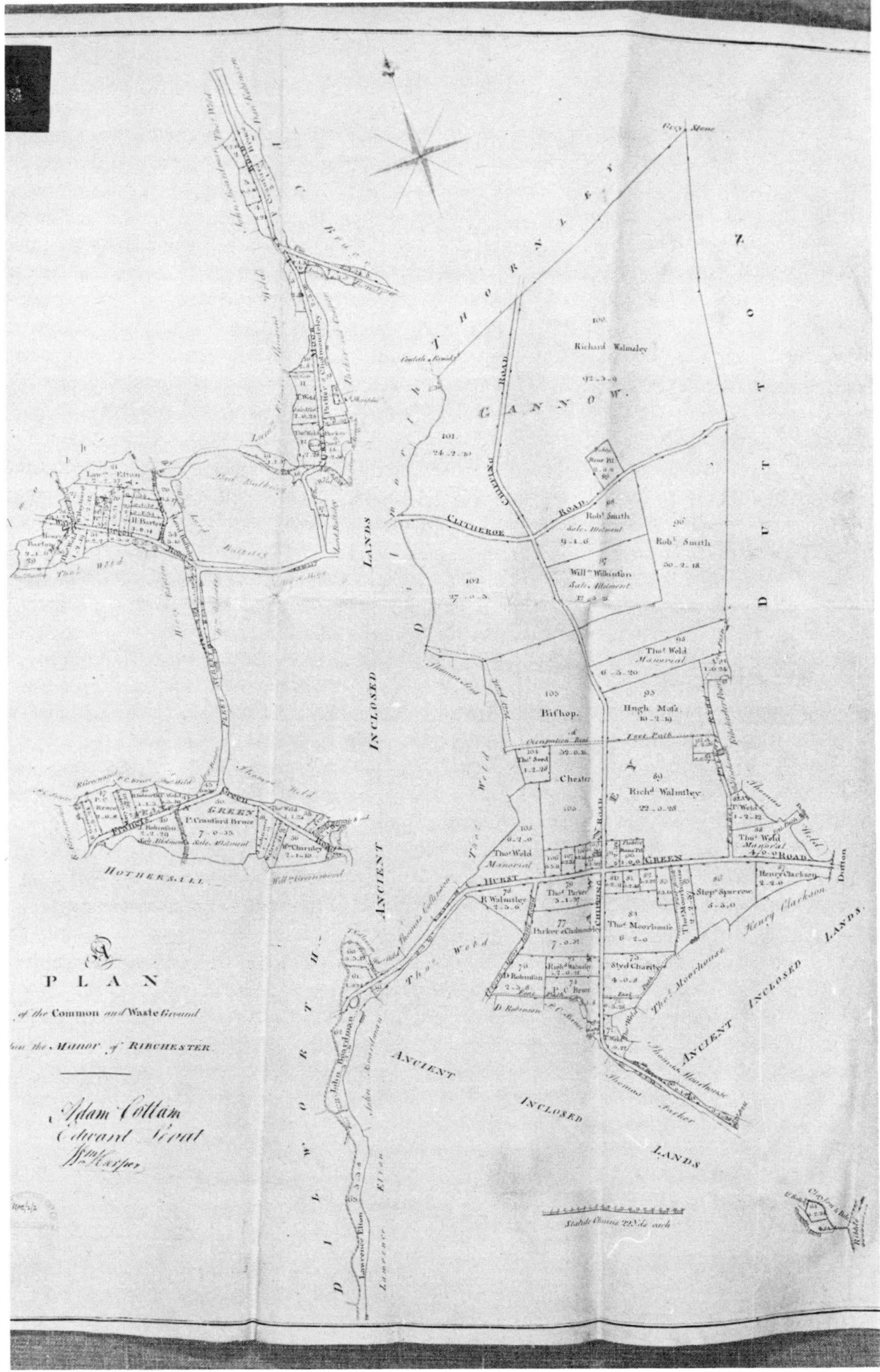

Figure 1.2. PART OF THE ENCLOSURE MAP OF CHIPPING, MILTON AND RIBCHESTER, 1812, SHOWING SOME OF THE COMMON AND WASTE WHICH WAS ENCLOSED BY ACT OF PARLIAMENT IN 1808.
(Source: Lancashire Record Office AE 2/2)

Though differing in the span of dates which they cover, enclosure and tithe maps are in other ways complementary. It is quite rare to find both an enclosure map and a tithe map for the same parish; rather their distributions are approximate mirror images of each other as tithe was usually commuted at the same time as a parish was enclosed. However, not all parishes with a tithe map lack an enclosure map nor was every place without a tithe survey enclosed by act of parliament from the eighteenth century onwards.

Tithe maps are usually drawn at a scale of 3 chains to an inch (about 26.7 inches to a mile, 1:2376) (Figure 1.3). Some very large parishes were surveyed at smaller 4- or 6-chain scales but, even so, many maps are 10 square metres or more in size. The specifications for tithe maps were drawn up by Lieutenant R. K. Dawson of the Royal Engineers, although the actual survey work was entrusted to private, usually local, land surveyors appointed by parish landowners (Kain, 1975). The maps drawn for each district usually show the boundaries of fields, woods, roads, streams and the position of buildings. On some plans conventional symbols are used to identify different types of land. For example, commons, heaths, moors and marshes are often differentiated (Figure 1.4). Each map is accompanied by a written document known as a tithe apportionment in which further data for each parcel of land is recorded, including its agricultural land use (Prince, 1959; Kain, 1974). Thus the tithe surveys represent the first comprehensive record of the pattern of land use at a national level. The maps and apportionments can be related because each parcel of land on the map is allotted a reference number which is used to identify it in the apportionment schedule and so the land use of any area can be identified. For a large area this can be a tedious and time-consuming operation, so for areas extending over several parishes it is likely that the more easily obtained parish totals discussed under section 1.3 of this chapter might be just as effective. It should be noted also that historical geographers have already transcribed land use from the tithe apportionments for quite considerable areas of the country, as indicated in Kain (1979).

In many parishes landowners were able to save themselves some of the cost of a new survey by using maps drawn originally for other purposes, notably estate plans or enclosure maps. Sometimes they cobbled together a map from several earlier surveys and some modern work. In consequence, some tithe maps are potentially inaccurate. All, however, were inspected by Lieutenant Dawson's staff and subjected to a series of checks. In evidence given in 1844 before the *Select Committee Inquiring into the State of Large Towns. . .*, he explained how his system of checks worked:

> Any discrepancy between the map and field notes leads to the rejection of the map; and when doubts arise as to the originality of the field notes . . . recourse is had to testing upon the ground.

Those maps attaining a standard of accuracy admissable in courts of law were affixed with the Tithe Commissioners' seal and became known as 'first class maps' (Figure 1.5). The remainder, if sufficiently accurate for the immediate purposes of commutation but falling short of the highest standards of planimetric accuracy, were known as 'second class maps'. Contemporary nineteenth-century opinion derided this category. Much criticism was, however, impressionistic; few commentators attempted to quantify the amount of error they encountered. A modern study of the planimetric accuracy of tithe maps suggests that, as a body, the 'second class' tithe maps are tolerably accurate (Hooke and Perry, 1976).

An original and two copies of each map and apportionment were produced and certified

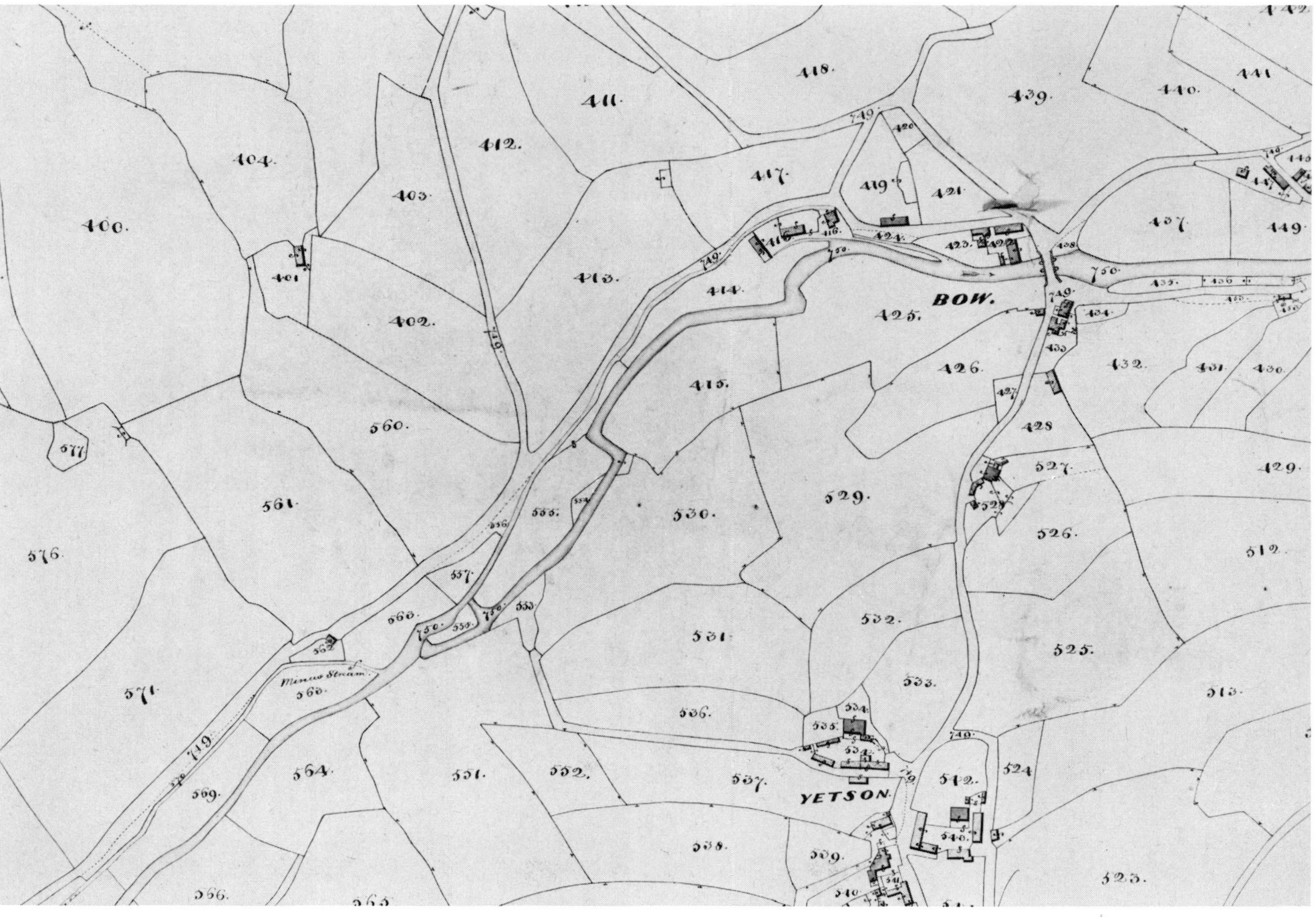

Figure 1.3. AN EXTRACT FROM THE 'FIRST CLASS' TITHE MAP OF ASHPRINGTON, DEVON.
Drawn at a scale of 3 chains to an inch in 1843 showing the course of the Harbourne river. (Source: Devon Record Office)

Figure 1.4. PART OF THE 'SECOND CLASS' TITHE MAP OF GITTISHAM, DEVON.
Drawn at a scale of 6 chains to an inch on which arable and pasture fields are distinguished by colour washes and woodland, etc. by conventional symbols. (Source: Devon Record Office)

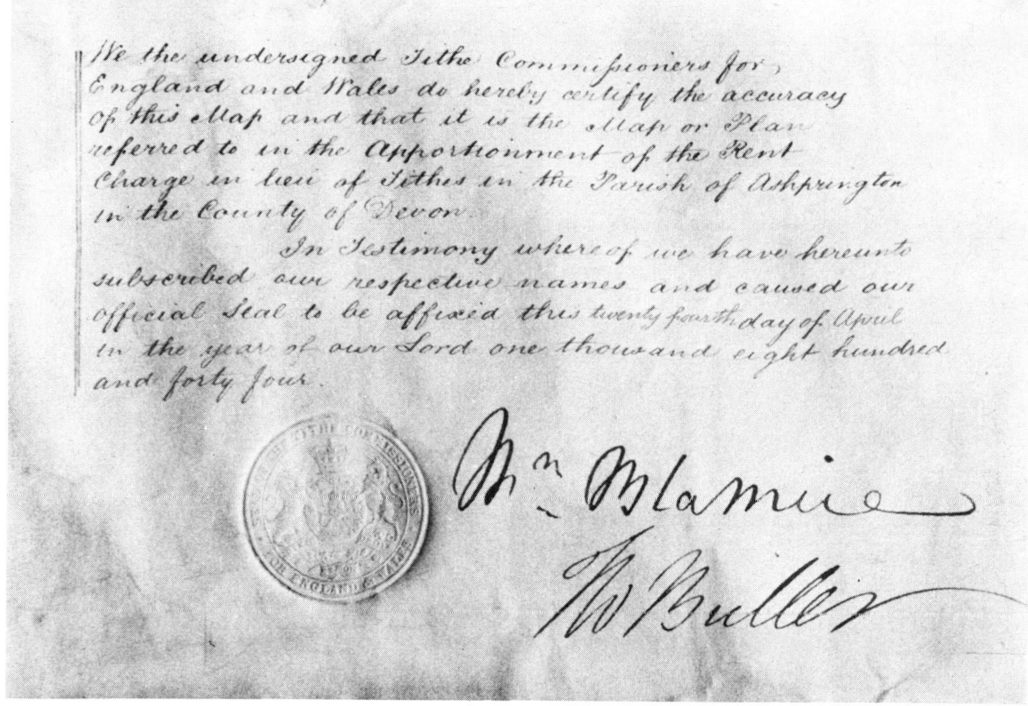

We the undersigned Tithe Commissioners for
England and Wales do hereby certify the accuracy
of this Map and that it is the Map or Plan
referred to in the Apportionment of the Rent
Charge in lieu of Tithes in the Parish of Ashprington
in the County of Devon.

In Testimony whereof we have hereunto
subscribed our respective names and caused our
Official Seal to be affixed this twenty fourth day of April
in the year of our Lord one thousand eight hundred
and forty four.

Wm Blamire
T W Buller

Figure 1.5. THE TITHE COMMISSIONERS' SEAL WHICH IDENTIFIES A 'FIRST CLASS' MAP. (Source: Devon Record Office)

by the Tithe Commissioners. The only advantage to be gained from consulting the original is if information on the survey network is required—construction lines are marked in red ink. The originals are now in the custody of the PRO at their Kew repository. A copy of the survey was deposited with the incumbent and churchwardens of each district to be kept in the parish chest. A second copy was deposited with the relevant diocesan registry. The copies have sometimes been lost or damaged but more usually transferred to county record offices. About 10 000 tithe surveys are available for local consultation in county record offices, although the actual number of parishes to which they refer is somewhat fewer than this because of some duplication of diocesan and parish copies for the same place. A number of county record offices have supplemented their collections with photocopies of the originals where nineteenth-century copies have been lost or are badly damaged. The PRO has microfilm copies of all the original tithe maps and apportionments and positive copies of these can be purchased. The definitive catalogue of tithe surveys is that prepared by the PRO from Tithe Redemption Commission lists. It has been published in xerox form by the List and Index Society (1971 and 1972).

Deposited plans

After 1794 there was a statutory requirement for promoters of canals and river navigations to deposit plans, sections and books of reference with parliament and with clerks of the peace (later clerks to the county councils) for public inspection. Similar deposits were required for docks and harbours (from 1800), railways (from 1802) and turnpike roads and bridges (from 1814). The land adjoining or close to the course of the proposed undertaking was surveyed in plan, and from 1812 vertical sections showing gradients were also deposited.

As might be expected, road, canal and river navigation plans figure prominently among early deposited plans. In the mid-nineteenth century there is a great quantity of railway plans; works in connection with the public utilities of gas, water and electricity come down into the present century. Schemes for docks, piers and harbour improvements are interspersed throughout the period (Steer, 1968). The plans relate to proposed schemes for which private acts of parliament were being sought; only a small proportion were actually carried out (Emmison, 1969).

Deposited plans vary widely in amount of recorded detail, standards of draughtmanship and accuracy (Figures 1.6 and 1.7). Some are little more than outline sketches of a proposed route, others are detailed drawings executed by the country's leading civil engineers like John Rennie, Robert Stephenson and William Cubitt. After 1807, all deposited plans were required to be at a minimum scale of 4 inches to a mile (1:15 840). In a physical geography context, the river navigation plans are particularly important because of the detail with which they portray plan profiles, often detailing gravel bars, shoals, saltings and adjacent marshes.

The House of Lords Records Office preserves very many deposited plans (Bond,

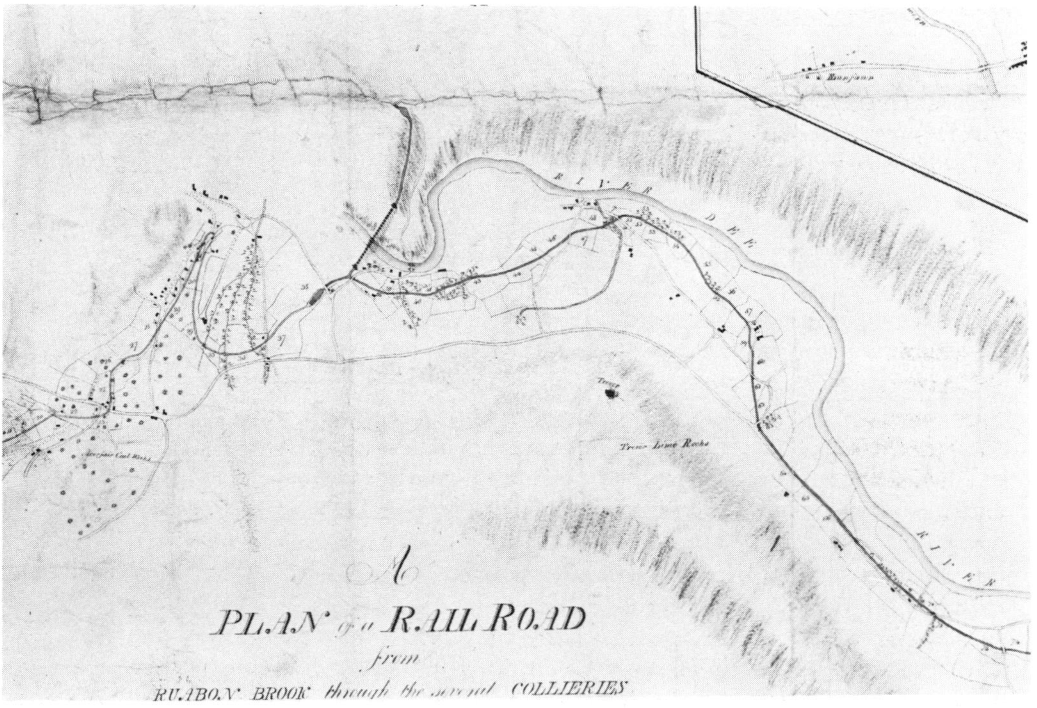

Figure 1.6. PLAN OF THE PROPOSED 'RAILROAD' FROM RUABON BROOK TO THE ELLESMERE CANAL, 1803. (Source: Cheshire Record Office QDP 17)

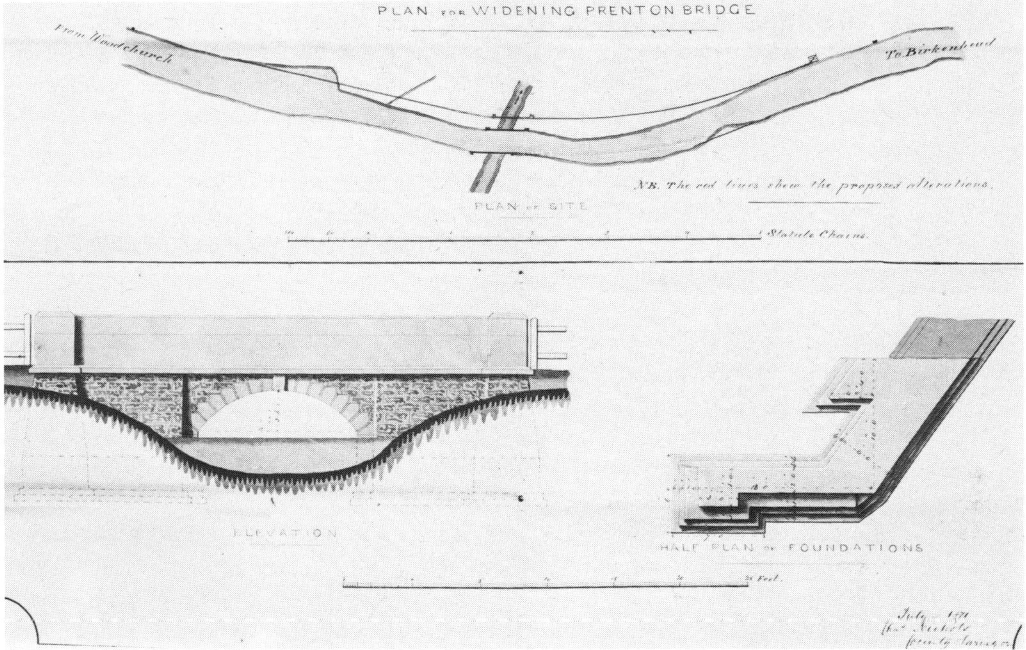

Figure 1.7. PLAN AND CROSS-SECTIONS FOR THE WIDENING OF PRENTON BRIDGE BY CHARLES NICHOLS, 1871. (Source: Cheshire Record Office QAR 95)

1959–1960, 1971). Copies left for public inspection with clerks of the peace will now be found in the respective county record office. A number of other record offices have good collections of transport plans and excellent guidance to these is provided in *Maps for the Local Historian* (Harley, 1972).

County maps

This category includes a variety of maps drawn on a county basis, dating from Elizabethan times to the nineteenth century. The most useful for environmental studies are those topographic maps drawn at scales of 1 or 2 inches to a mile (1:63 360 or 1:31 680) in the eighteenth and early nineteenth centuries. The mid-eighteenth century was perhaps the high point of making county maps. New surveys were started about this time and rapidly published, so that for broad areas of Britain there is fairly comparable map evidence (Harley, 1965, 1972). The period 1765–1780 was characterized by the most intense activity during which 65 per cent of England (32 900 square miles) was portrayed on twenty-five maps. A second generation of some sixty surveys, many executed by Christopher Greenwood, were carried out after the Napoleonic Wars between 1817 and 1839 (Laxton, 1976). This chronology is illustrated on Figure 1.8.

As sources of data these maps are a variable quantity. Most provide details of the pattern of rivers, estuaries, coasts and other landforms as well as distinguishing settlements and roads (Grigg, 1967) (Figure 1.9). A considerable number of them show land use. On some of them, like Andrew Armstrong's map of Northumberland (1769), there is little more than a

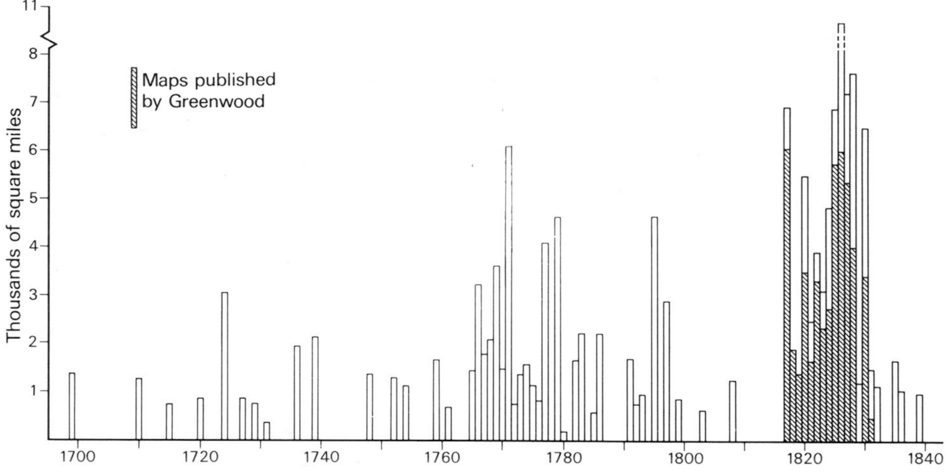

Figure 1.8. THE CHRONOLOGY OF COUNTY MAPPING IN ENGLAND, *C.*1700 TO *C.*1840. (Source: Laxton (1976))

generalized indication of woodland but at the other extreme are, for example, the remarkably comprehensive surveys of land use carried out by John Rocque in the middle of the eighteenth century (Darby, 1973). Following the publication of his map of London at 5 inches to the mile (1:12 672) in 1746, Rocque produced a similar, virtual 'land use map' of the whole of Middlesex at 2 inches to a mile (1:31 680) in 1754. The pattern of land use

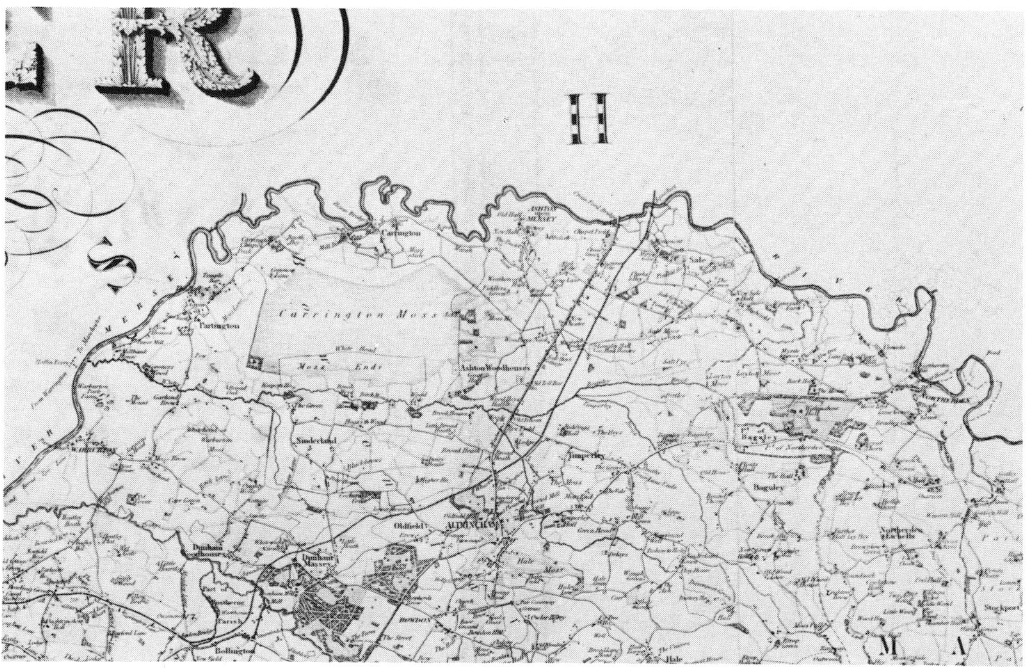

Figure 1.9. THE NORTHERN PART OF BRYANT'S MAP OF THE COUNTY OF CHESTER, 1831

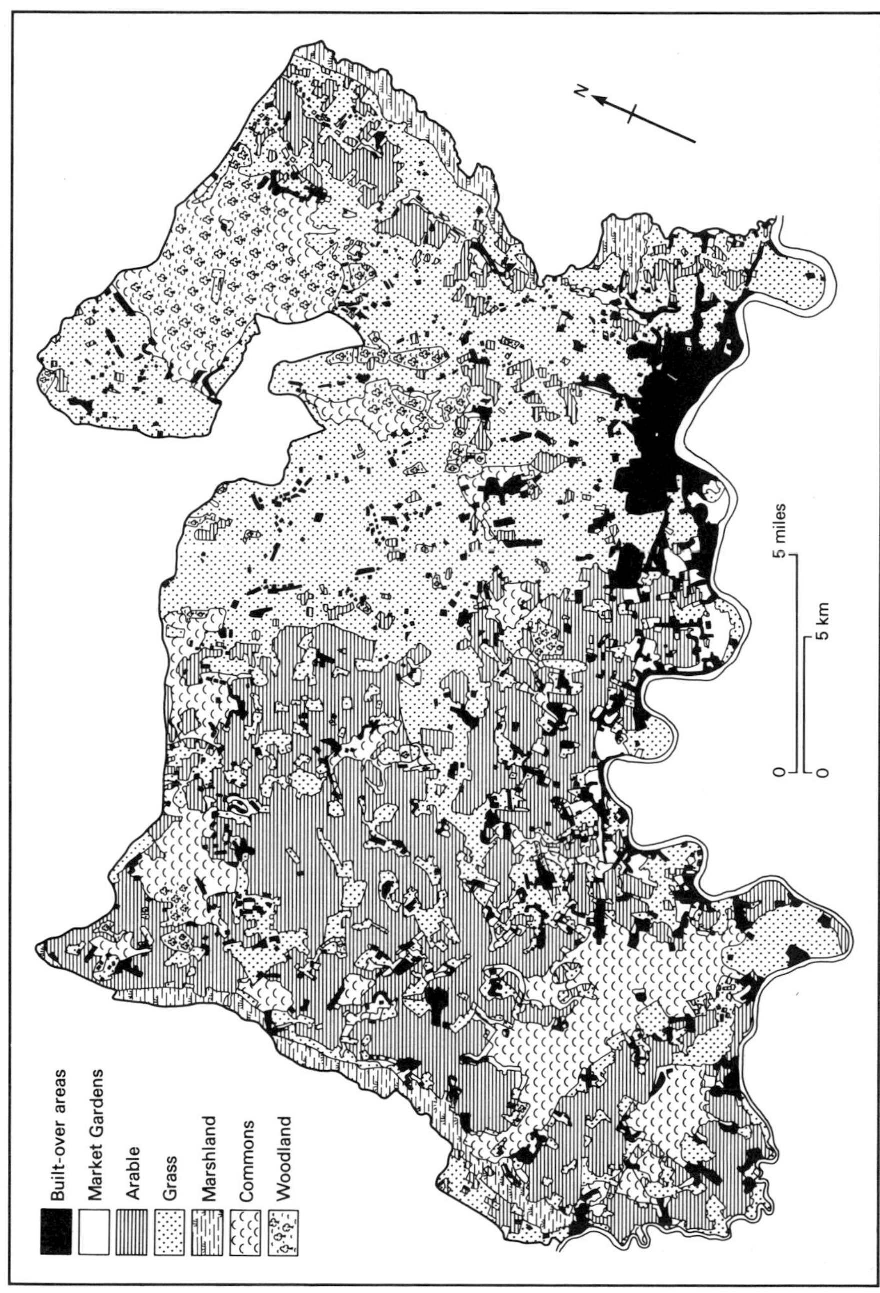

Figure 1.10. LAND USE IN MIDDLESEX AT THE TIME OF JOHN ROCQUE'S SURVEY IN 1754. (Source: Willatts (1937))

recorded on this map is reproduced on Figure 1.10 and indicates the potential of large-scale county maps for providing early land-use information which could be of interest directly or in relation to the analysis of environmental change. Careful examination of county maps suggests that often the representation of fields is purely diagrammatic, since they are all of extremely large size. Although comparison of Rocque's map with others

> suggests that there may also be some occasional inaccuracies in the representation of utilisation, there is no doubt whatever that on the whole the map gives an excellent representation of the surface utilisation of Middlesex in the middle of the eighteenth century. (Willatts, 1937)

Many other writers of county reports of the First Land Utilisation Survey of Britain made use of county maps to provide an historical perspective to their accounts of twentieth-century land use (e.g. East in Stephenson and East, 1936; Mosby, 1938; Cameron, 1941; Howell, 1941; Smith and Stamp, 1941; Stamp and Willatts, 1941). In short, the analysis of county maps can provide a broad picture of land use some two centuries ago, but the information should not be accepted uncritically or much regard placed on the accuracy of recording at the individual field scale.

Recent cartometric studies, and especially the important work of Laxton (1976), have assessed the geodetic and planimetric qualities of county maps and the accuracy of the topographic information portrayed on them. Laxton compared first, the displacement of fixed points like churches on three county maps from their 'true' positions established by the Ordnance Survey and second, the representation of boundaries on two county maps with those on modern maps. All these tests underscore the presence of error, but its nature and magnitude do not vary in any consistent way. Laxton wisely refrains from advancing definite conclusions on the relative geodetic merits of the various county map-makers not least because:

> of the tenacious way in which the county maps display their individuality, even when they are the work of the same hands.

His sample results do suggest that direct comparisons of sections of rivers or details of coastal features should not be undertaken without some attempt at evaluating error. An observation from just one part of one county map should counsel caution. On Andrews, Dury and Herbert's map of Kent (1769) the drainage pattern of what is now south-east London is incorrectly shown. A number of confluences are misplaced and the courses of streams like the Ravensbourne are incorrectly plotted.

Contemporaneous with these privately produced county maps in England was the *Military Survey of Scotland* carried out between 1747 and 1755 under the direction of William Roy. Fair copies of the maps of this survey were produced at a scale of 1 inch to 1000 yards with colour wash used to distinguish water, woodland, cultivated land and moorland. Evidence in Arrowsmith's 'oral history' of the survey, compiled from recollections of the field surveyors, suggests that the courses of rivers, lines of roads, banks of lakes and a number of intermediate points were surveyed (Arrowsmith, 1809). According to Skelton (1967):

> field boundaries in the map are merely conventional, while the limits of cultivation and waste, of inclosed and uninclosed land, of drained and undrained, of woodland and open country, are carefully and accurately defined.

M. L. Parry (1975) has checked the accuracy of the Military Survey's cultivation mapping by reference to field evidence of the relict landscape in the area of the Lammermuir Hills in south-east Scotland. Correspondence between the Military Survey and air photograph evidence on reverted moorland suggests that, except in a small number of cases, the content of the Military Survey is a reasonably accurate record of the distribution of cultivation in the mid-eighteenth century.

Access to county maps has been improved in recent years by the publication of a number of excellent facsimile editions. Some of these are listed by Harley (1972). A fundamental national bibliography of county maps is R. A. Skelton's *County Atlases of the British Isles 1579–1850* (1970). A basic handlist is E. M. Rodger's *The Large Scale Maps of the British Isles 1596–1850*, the second edition of which was published by the Bodleian Library, Oxford (1972). Laxton (1976) has compiled a list of large-scale surveys of the counties of England and Wales 1699–1839. Maps of Scotland are detailed in the third edition of *The Early Maps of Scotland*, edited for the Royal Scottish Geographical Society by D. G. Moir in 1973.

Ordnance Survey maps

The Ordnance Survey, formally established in 1791, is the official mapping and survey organization in Britain. Two comprehensive handbooks to Ordnance Survey maps have been published (Harley and Phillips, 1964; Harley, 1975) and a *History of the Ordnance Survey* has been edited by Seymour (1980). Users are strongly recommended to refer to these for details. Reference has already been made in this chapter to the way in which published Ordnance Survey maps can be used as a datum against which the reliability of other, usually earlier, manuscript maps and plans can be judged. However, the various editions of the small-scale Ordnance Survey maps (from the beginning of the nineteenth century onwards) and the large-scale maps (from the second half of the nineteenth century) are, in their own right, sources of landform, vegetation and land-use data for the earth scientist concerned with providing a time dimension in the more recent past. Many references to the use of Ordnance Survey maps are also to be found in the studies reviewed in later chapters of this book.

The main editions, revisions and dates of publication of early Ordnance Survey maps are summarized in Table 1.1. The original 'basic' edition on a Cassini projection with a number of meridians was published at the 1:63 360 scale by reduction from manuscript drawings constructed at scales from 2 inches (1:31 680) to 6 inches to a mile (1:10 560). From the 1820s, publication of the Irish Survey had been at 6 inches to a mile (1:10 560) and from 1840 this scale was adopted for the as then unsurveyed areas of northern England with the county, usually each with a separate meridian for its Cassini projection, used as the basis for publication. Thereafter, as Harley (1975) has put it:

> A long controversy, the so-called 'Battle of the Scales' lasting for about twenty years, was to surround the choice of a suitable base scale for the maps of Great Britain. A series of Royal Commissions and Parliamentary Committees debated the alternatives and took voluminous evidence but not until 1858 was a final decision taken to adopt the 1:2500 scale for cultivated areas, the 1:10,560 for uncultivated areas, and 1:500 (c. 10ft to the mile) for towns with a population of 4000 and over. Map series on smaller scales, including the 1:63,360, were to be derived from these basic large scale maps.

Table 1.1
EARLY 1-INCH, 6-INCH AND 25-INCH ORDNANCE SURVEY MAPPING OF ENGLAND
(after Harley and Phillips, 1964; Harley, 1975)

1-inch maps

Old Series or First Edition
Survey began 1795 (Kent)
Publication complete 1873 (Isle of Man)

New Series or Second Edition
Survey began 1840 (north of Hull–Preston line)
Publication complete 1893 (revised 1893–1898)

Third Edition
Survey began 1901
Publication complete 1913

Fourth Edition
Survey began 1913
Publication complete 1926

6-inch and 25-inch maps

County Series First Edition
Survey began 1840 (on 6-inch scale, on 25-inch scale after 1853)
Survey ended 1893 (with resurvey of areas previously on 6-inch scale)
Revision
First 1891–1914
Second (incomplete) 1904–1923

As a result of the 1938 Davidson Committee's report, most of whose recommendations were implemented after the war, the 1:2500 survey was recast on national instead of county sheet lines on a national projection and a National Grid was superimposed on all plans and maps (Harley, 1975).

In the two short sections below, some notes are provided on two sets of documents associated with published Ordnance Survey maps which can provide valuable data additional to that obtainable from the map sheets themselves. These are, first, the unpublished manuscript drawings from which the Old Series One-inch maps (1:63 360) were produced, and second, the published Books of Reference to the early 1:2500 surveys.

Manuscript drawings for the Old Series One-inch maps

In 1792 a triangulation programme of England and Wales was begun by the Trigonometrical (later Ordnance) Survey and the first numbered sheets of the 'One-inch' series appeared in 1805. The full set of 110 sheets was not completed until 1873, but by 1820 maps of the country south of a line from Pembrokeshire to Essex had been published (Harley and Phillips, 1964). Although essentially a military survey, these maps can supplement published county maps particularly for the study of vegetation. It is best to consult the manuscript drawings at scales of 2 inches, 3 inches and 6 inches to a mile which have been deposited in the British Library map library (Seymour, 1980). Though their precise status in the map-making process is not entirely clear, their larger scale ensures rather more accurate transfer of data or measurement of areas. However, the results of a number of studies reviewed later in this book, notably Carr (1962) and Hooke (1977a), demonstrate the existence of significant planimetric error on published Old Series maps. Generally, their use is not to be recommended for the study of change in topographic features. Most valuable are the 6-inch (1:10 560) manuscript drawings and, although these cover only a very small part of southern England, they portray a full range of land uses. The 3-inch (1:21 120) drawings cover a larger area but are more varied in character. Some again are a comprehensive record of land use, but others show little additional information beyond that recorded on the printed maps (Wallis, 1981). A great majority of the drawings are at the smallest 2-inch scale and do not, for example, distinguish grassland separately. As a whole, the Ordnance Survey surveyors' drawings are an underused and undertested source of vegetation evidence (Coppock, 1968). A number of agricultural geographers have used this information in studies of farming and land use which indicate the potential and also some of the pitfalls of this source (Coppock, 1958; Thomas, 1963).

Books of reference to the First Edition 25-inch plans

In 1853 the Ordnance Survey began the mapping of County Durham at the 1:2500 scale and in 1855 it was decided to make this series countrywide except for uncultivated areas, which were to be mapped at the smaller 6-inch scale (Harley and Phillips, 1964). A characteristic of the 25-inch series is that every field is allotted a reference number on the map and that in the early years of the survey its area and land use were recorded and printed in an 'Area Book' (later Book of Reference) which accompanied the map sheets of each parish. The publication of land use in the Books of Reference was discontinued in 1880, and after 1888 the acreages but not the land use were printed on the maps themselves (Harley, 1979). Books of Reference containing land-use data are available for about a quarter of England and Wales, i.e. for those areas where the 1:2500 sheets were published as parish maps. South-east England is especially well covered. An approximate picture of the distribution of coverage can be obtained from Figure 1.11, which is a map of the area covered by 1:2500 sheets in 1880 (after Coppock, 1968) while the various published formats are discussed by Harley (1979).

Relatively little use has been made of this source to reconstruct land-use patterns. This is despite the fact that the Books of Reference relate to a key period in British land-use history near the time of maximum arable cultivation in most of England (Henderson, 1936). However, some land-use maps based on the data do figure in some of the county reports of the First Land Utilisation Survey of Great Britain, and these can be consulted for further information on the potentials of this source (Green and Moon, 1940; Stamp and Willatts, 1941; Briault, 1942). Harley (1979) has reviewed many of the problems of interpretation which arise from the Ordnance Survey's mapping of land-use information at the 1:2500 scale.

First Land Utilisation Survey of Great Britain

On 1 January 1933 the first 1-inch maps of the First Land Utilisation Survey of Great Britain were published; by the end of 1935 the bulk of the field work was accomplished and in 1946 the last of the county reports accompanying the 170 sheets of the survey was published (Stamp, 1962). The results of the field survey, plotted initially by volunteers using Ordnance Survey 6-inch sheets as base maps, is a record of land use prior to the great

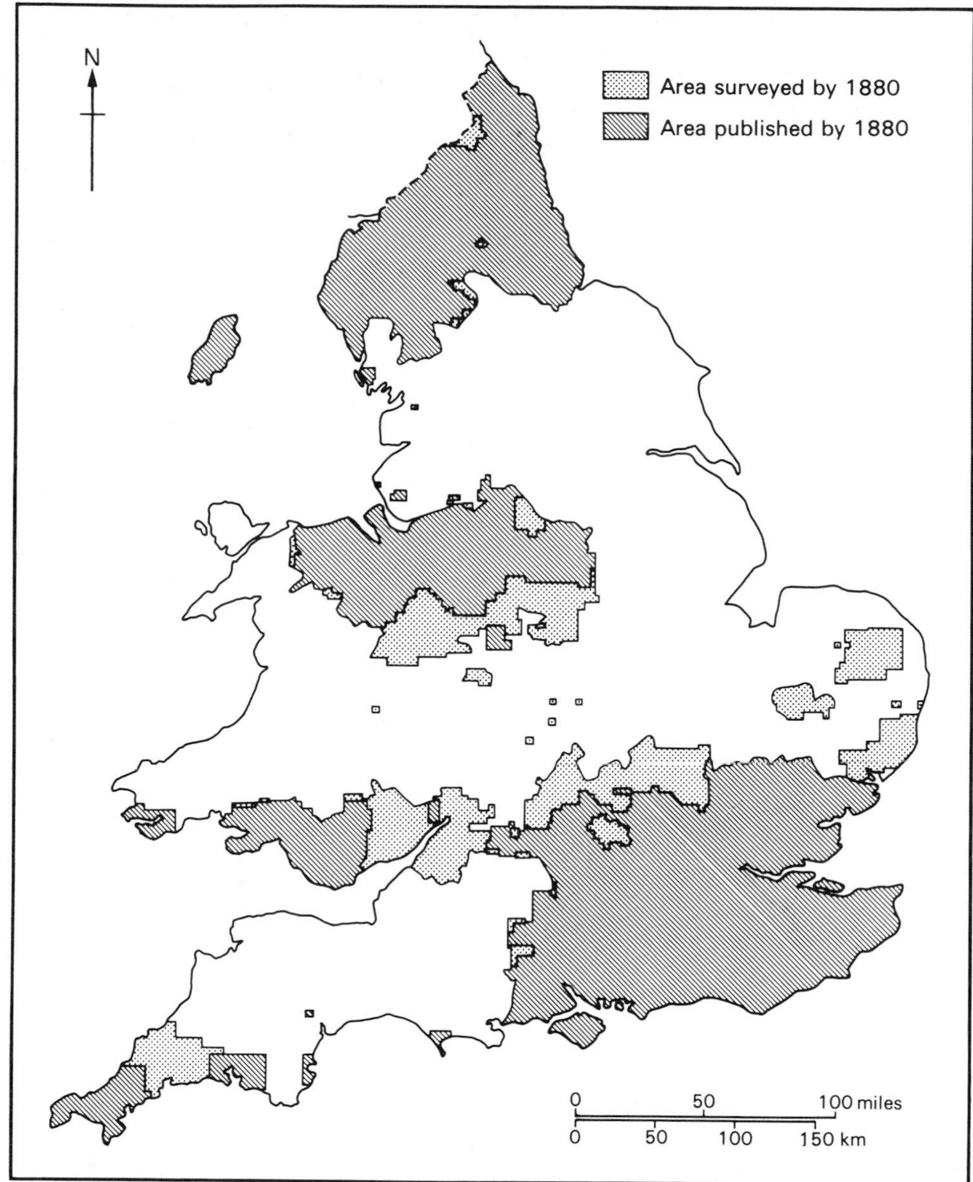

Figure 1.11. AREA COVERED BY 1:2500 ORDNANCE SURVEY MAPS IN 1880.
(Source: Coppock (1968))

changes induced by the Second World War, when in many counties, in Kent for example, the arable acreage rose close to its all-time peak of the 1870s (Garrad, 1954; Coppock, 1954). Though not by any means an 'instantaneous' cross-section of land use, the maps of the First Land Utilisation Survey approach this ideal more closely than any produced before or since. The land-use categories are perhaps rather broad and some of the amateur field surveyors did encounter problems of definition, but their 20 000 or so 6-inch maps were subjected to as rigorous a system of checks as the budget of the Survey permitted (Stamp, 1962). The published 1-inch sheets may be satisfactory for some purposes, but where greater detail is required the extant manuscript 6-inch maps may be consulted by prior arrangement with the Department of Geography, King's College, London.

Marine charts

The needs of national defence on the one hand and the awakening of maritime enterprise on the other stimulated the compilation of marine charts in the sixteenth century (Robinson, 1962). The main effort in this early period was concentrated along the southern Channel coastlands with much of the work being done by Continental engineers. By the end of the sixteenth century the work of Elizabethan hydrographers not only increased in output but also showed a distinct improvement in quality. In 1580, for example, Robert Norman produced the first detailed picture of the intricate pattern of banks and channels in the outer Thames estuary, even if the Kent and Essex coasts were greatly distorted (Robinson, 1962). Methods of running traverse, where accuracy depended on true measurement of the speed of the ship and the fixing of only prominent points by compass intersection with the

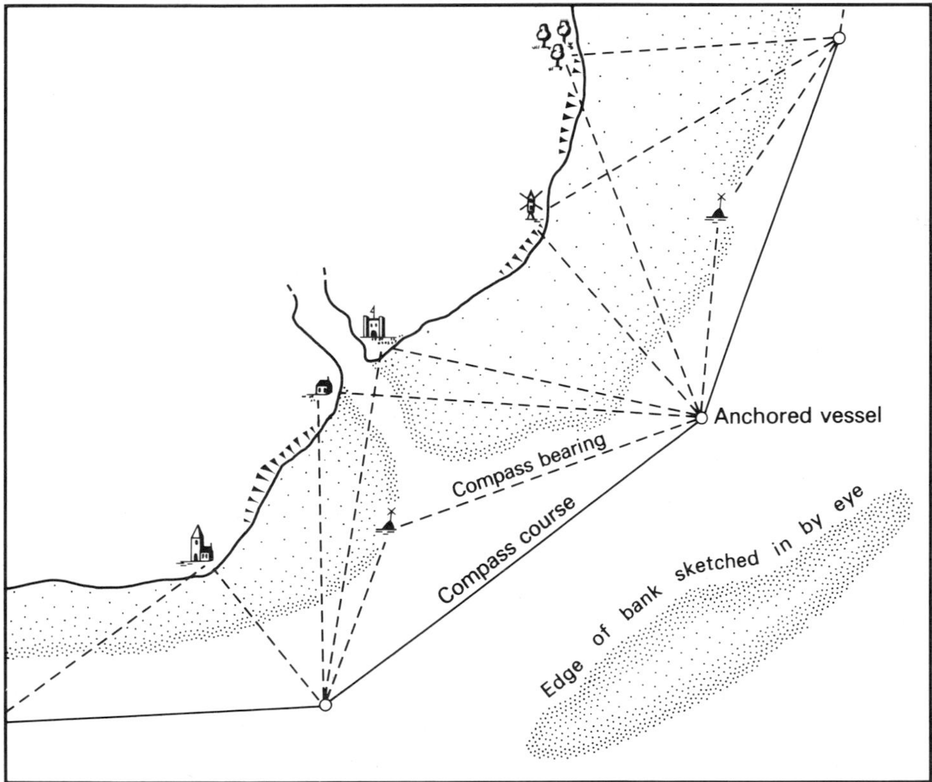

Figure 1.12. THE RUNNING TRAVERSE METHOD OF MARINE SURVEY.
(Source: Robinson (1962))

remaining coastline sketched by eye, meant that errors were common on an intricate coast. This method of working is illustrated on Figure 1.12. Despite these inaccuracies such early charts can still be of value, as de Boer (1969) has demonstrated in his study of the evolution of Spurn Point. The decision about whether or not to accept an 'inaccurate' map, as with all such historical evidence, must depend upon the nature of the environmental question under examination. This most fundamental of all historical canons appears to escape some physical scientists.

As with land surveying, the founding of the Royal Society (charter granted 1662) encouraged the formulation of a firm, theoretical basis for hydrographic surveying. Robert Hooke, the Society's Curator of Experiments, invented an instrument designed to sound water depth to replace the old lead line. It was not particularly successful but it is indicative of the mechanical ingenuity which was being allied to mathematical advances, such as solving the resection problem (Robinson, 1962).

Theory, however, was often far in advance of practice in the seventeenth and eighteenth centuries. Murdoch Mackenzie's (senior) much vaunted mid-eighteenth-century survey of the Orkneys shows many inaccuracies judged by present-day standards. Murdoch Mackenzie (junior) and his assistant Graeme Spence did much to combine theory with sound practice. Part of Mackenzie's 6-inch to a nautical mile chart of Poole Harbour is illustrated in Figure 1.13. As no topographic maps were available a detailed survey of the coastal area was made before sounding could start. This was done by establishing a base line on the sands of Studland Bay from Studland village to South Haven Point. Points on land were then fixed by triangulation from this. In fact, Mackenzie's charts show much more land detail than the Ordnance Survey sheets published almost thirty years later (Robinson, 1962).

In 1795 the Hydrographic Department of the Admiralty was established and by the mid-1820s its chart-makers were working on what became known as the 'Grand Survey of the British Isles' (Day, 1967). Some very accurate surveys were made such as that of the river Thames by Frank Bullock, illustrated in Figure 1.14. Francis Beaufort, appointed Hydrographer in 1829, described it thus:

> A standard survey of the River Thames will be an important epoch in its maritime history; for besides the practical value of a correct plan of its present limits, depths, currents, etc., we shall be able to watch the smallest indications of those changes which nature is always producing through the nature of the stream . . .

Bullock began his survey in 1830 on scales between 5 and 22 inches to a mile (1:12 672 and 1:2880):

> like George Thomas in the Shetlands, he found that accuracy had to be bought with time and it was not until 1836 that he reached the Nore off Southend. Where his surveys overlapped with Spence and Thomas at the mouth of the river, a comparison between the two showed that the banks and channels had changed considerably in the intervening years. (Robinson, 1962)

During the period when Francis Beaufort was Hydrographer the previous total of forty-four Admiralty charts rose to 255, covering virtually all the coast of the United Kingdom. The size and scale of these charts varies a great deal. Harbour charts were usually drawn at 10 to 20 inches (1:63360 to 1:3168) to a mile, estuaries, bays and harbour approaches at usually 1 to 10 inches to a mile (1:63 360 to 1:6336) (Harley, 1972). Early printings of charts often pre-date the Ordnance Survey large-scale plans of an area and, once published, were subject to almost continuous revision. Dating the content of these can be difficult and, if not carefully done, can present problems of interpretation (Harley, 1972; Carr, 1980).

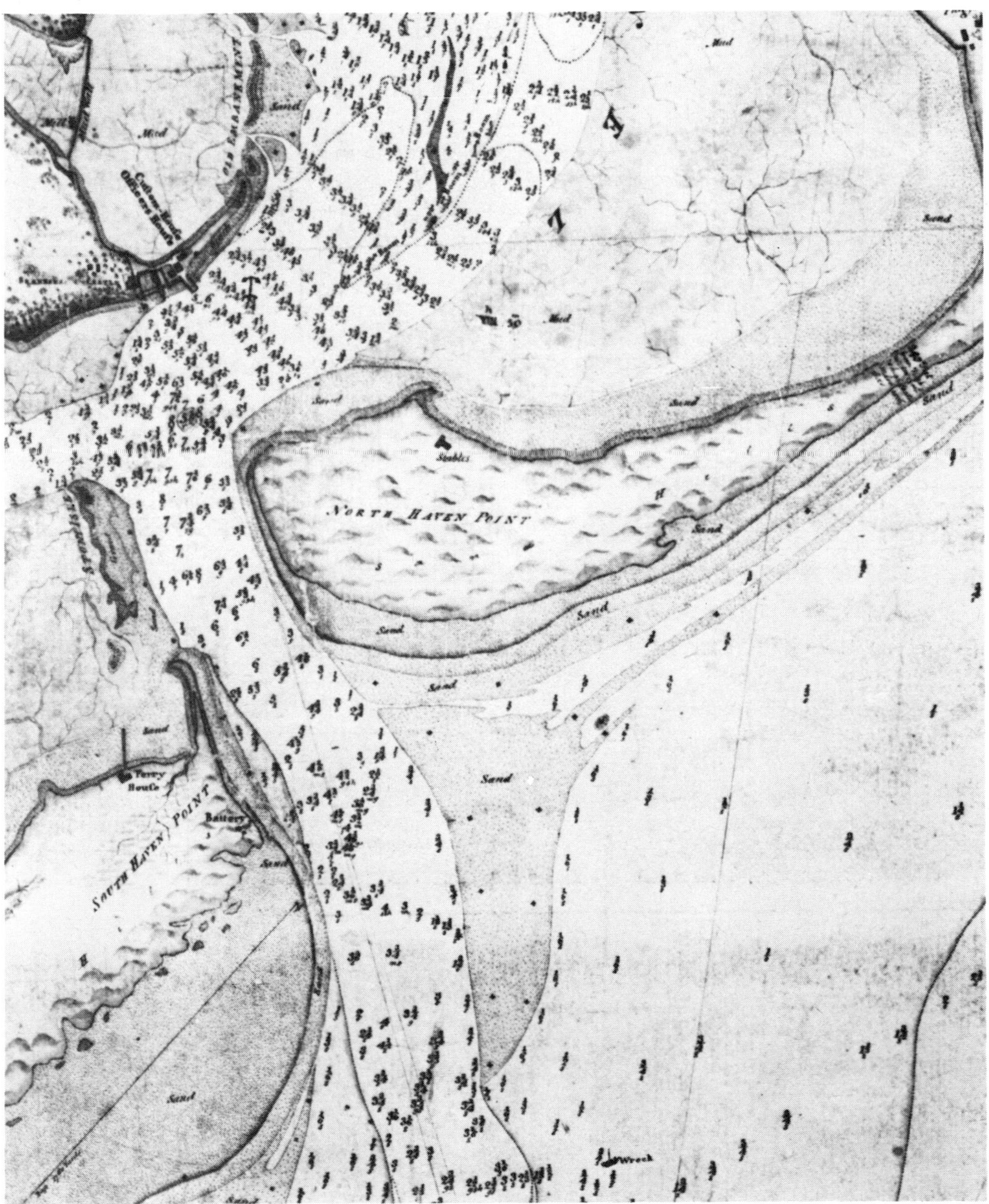

Figure 1.13. PART OF MURDOCH MACKENZIE JUNIOR'S 6-INCH TO A MILE CHART OF POOLE HARBOUR.
(Source: photograph by courtesy of the Hydrographer of the Navy and Controller of HM Stationery Office)

In his *Marine Cartography in Britain*, A. H. W. Robinson (1962) has listed a large number of marine charts from Tudor times to the nineteenth century, and indicated where a copy of each can be consulted. Harley (1972) in a chapter on 'Marine Charts', in his *Maps for the Local Historian A Guide to the British sources*, has listed many repositories where

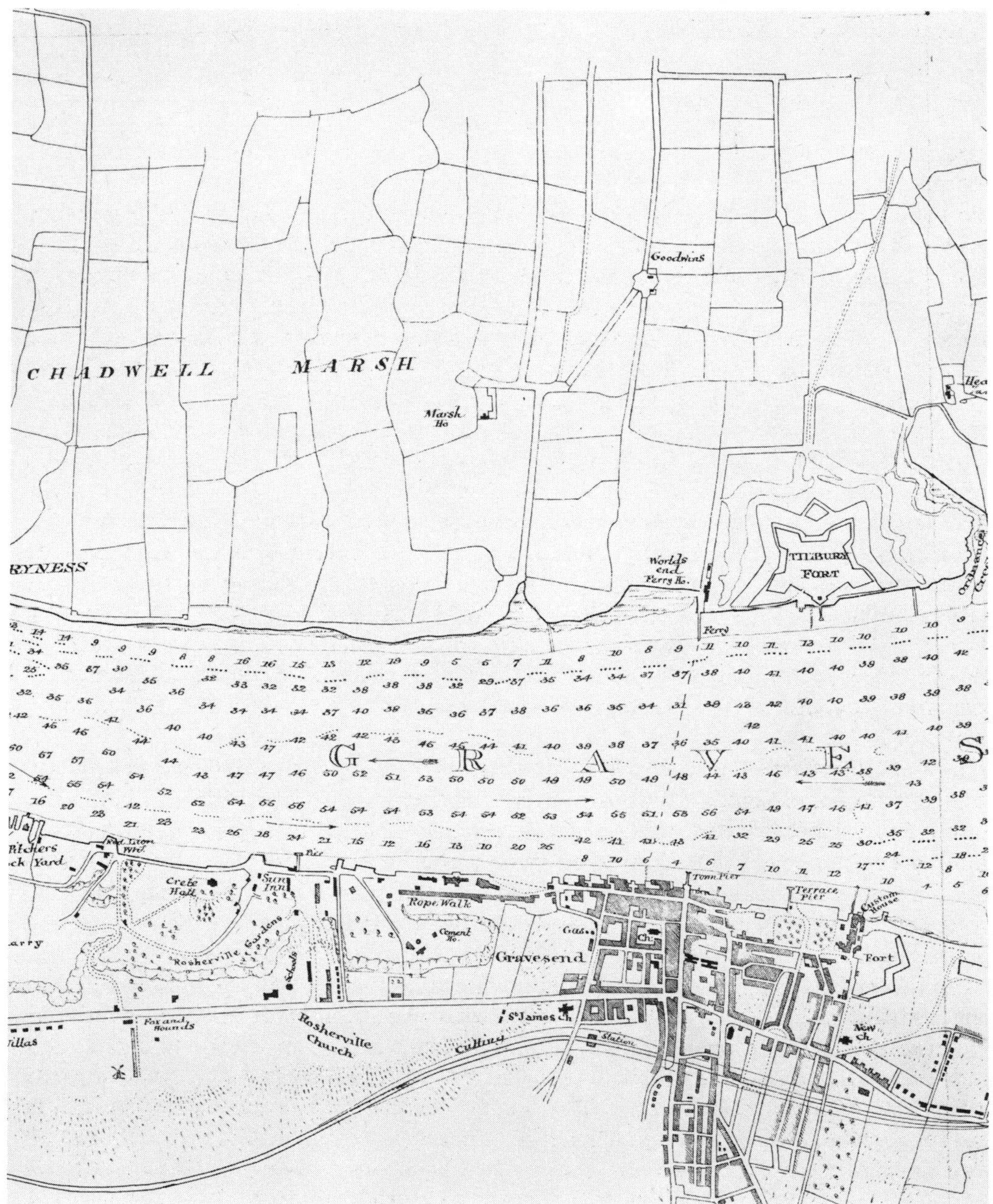

Figure 1.14. FRANK BULLOCK'S SURVEY OF THE RIVER THAMES, PART OF THE 'GRAND SURVEY OF THE BRITISH ISLES'. (Source: as for Figure 1.13)

charts are kept. The official Admiralty *Catalogues of Charts* commenced in 1825, though there are some gaps in what is basically an annual series. They list charts by standard number giving details like scale and notes of corrections and revisions.

Photographs and pictures

> Direct comparisons of landform configuration at different times are most readily made using photographs. (Church, 1980)

Unfortunately most early photographs were taken for purposes other than landscape recording and at best only about 100 years of record can be recovered using their evidence. The availability of landscape photographs also varies greatly from one part of the world to another. There is certainly no equivalent in the United Kingdom, for example, of the great photographic archive produced by the United States and Canadian Geological Surveys in their nineteenth- and early twentieth-century explorations of western North America. Also in the United States of America some of the earliest experiments with the use of air photographs for topographic mapping were made shortly after the First World War, so perhaps fifty years of United States' landscape history might be encompassed by such records. Most of the world, however, has been photographed from the air only since 1950. In the United Kingdom good Royal Air Force 6-inch (1:10 560) coverage is available from the 1940s. Aerial photographs in Britain are held and catalogued by the Department of the Environment. Private survey companies may also be helpful in supplying early prints.

Landscape paintings, engravings and drawings can extend the timescale back into the eighteenth century and in some special circumstances right through to prehistory. Some prehistoric cave paintings pre-date the earliest written languages and have provided physical scientists with information about early environmental conditions. For example, a classic study of the variety of wild life represented on the Tassili Frescoes of the central Saharan massif confirmed that conditions were once very much more humid than today (Butzer, 1965).

In Great Britain the effective origins of landscape painting are to be found in medieval times and there was a peak of output in the nineteenth century (Clark, 1949; Abbey, 1952; Hardie, 1966–1968). In the county of Devon, for example, a great upsurge of landscape painting began about 1780. From then there was a short period when most paintings contained a building as their prime focus but by the 1820s pure 'natural' landscapes arrived, though almost entirely confined to valley scenes; by the 1850s the heights of Dartmoor and Exmoor were very attractive to both amateur and professional artists (Howard, 1979).

The value of art as a source of evidence on past conditions has been recognized for some time, particularly for providing factual statements about the landscape, for identifying what was, or was not, present. Although some paintings may be an attempt to depict a real landscape on the spot, or be compiled from a number of field sketches, others may be partly or completely fictitious (Heathcote, 1972). Some of these difficulties can be overcome, or at least highlighted, by establishing a number of parameters, for example, the extent to which contemporary technical conventions such as the exaggeration of relief by the Picturesque School, the media of publication or the means of reproduction may have distorted the image in specific paintings.

Probably one of the most useful roles for all pictorial sources is to corroborate evidence of facts and events known from present-day field or contemporary written evidence. In the history of climate (see Chapter 4) pictures have been much used for charting the phenomenon of glacier advance and retreat.

Most libraries and archives which contain manuscript maps also house collections of pictures and photographs (Royal Commission on Historical Manuscripts, 1979). For example, the Department of Manuscripts of the British Library possesses a considerable collection of topographical drawings (British Museum, 1844, reprinted 1962). Generally speaking this and similar collections are richest for the period 1760–1860, which coincides with the hundred years of greatest output (Wright, 1957). In England, the southern counties, the Home Counties and East Anglia were especially popular with landscape artists. There are two particularly useful guides to collections in Britain. G. W. A. Nunn's *British Sources of Photographs and Pictures* was published in 1952. Though some of the technical guidance is now dated its museum and library descriptions can provide a useful starting point for a search. For photographs alone, John Wall's *Directory of British Photographic Collections* is more up-to-date (Wall, 1977). Some counties are fortunate in possessing published lists of topographic prints; that for Devon, for example, covers the period 1660–1870 (Somers-Cocks, 1977). Enquiry at a local level is best addressed in the first instance to the respective county record office.

1.2 WRITTEN SOURCES AND ORAL EVIDENCE

This category of source material is extremely heterogeneous in nature and encompasses the whole range from, say, a small marginal note in a manuscript diary to carefully researched, printed scientific reports. Between these poles there is the whole spectrum of manuscript manorial and estate records, legal documents and property deeds, files of documents accompanying official surveys, newspapers and published topographies.

Printed topographies and descriptions

In studies with a firm areal base it is obviously worth while finding out from other than primary sources what material has already been consulted on the district before initiating original research among manuscript sources. For many places a useful first stage might be to consult the appropriate volumes of the *Victoria History of the Counties of England.* The Victoria County History's work is still continuing; it is engaged on the virtually endless task of providing a standard reference on English local history (Stephens, 1973). In recent years the survey has paid a lot of attention to 'topography', but some of the volumes compiled in the early years of the century are now rather out-of-date and have an antiquarian flavour. One of the most useful features of the Victoria County History is the careful way its authors have footnoted sources; recent volumes provide a useful guide to what has been done in an area and what sources are available. They are particularly useful for identifying the names of local landed estates for which records and surveys have survived (Pugh, 1967). Most university libraries have fairly complete sets of the Victoria County History, while the local studies rooms of principal public libraries are the places to find local volumes.

A next step will be to scan the older county histories and books published on an area. Quite invaluable in this respect is J. P. Anderson's *The Book of British Topography* (1881, reprinted 1976), which is a list of such works held in the library of the British Museum at the date of its compilation in the last quarter of the nineteenth century. An earlier work, now also available as a reprint, is William Upcott's *A Bibliographical Account of the Principal Works Relating to English Topography* (1818, reprinted 1978), which describes, county by county, all the major county and local histories and lists their illustrations.

As well as the 'histories', articles in the published proceedings of local history, archaeology and field societies which proliferated in the nineteenth century can also be useful. Harcup (1968 edition) has published a 58-page list of these organizations as *Historical, Archaeological and Kindred Societies in the British Isles.* There were, for example, a series of articles published in the late nineteenth century in the *Proceedings of the Woolhope Naturalists' Field Club*, which include details of flood levels reached on the river Wye at Ross and observations on the great drought of 1887 (Potter, 1978). Nationally circulated periodicals also contain articles on natural environment phenomena but they are scattered among other scientific papers. The British Library's *Catalogue of Printed Books* may help to locate some of these journals and its periodical entries, arranged by place of publication, can help track down local scientific societies and their transactions. The printed *Proceedings of the Royal Society* for the eighteenth and nineteenth centuries contain many potentially useful accounts. The trick is to be able to locate the apposite wheat from among much irrelevant chaff.

In the eighteenth and nineteenth centuries it was quite common for travellers to publish *diaries* of their journeys and *topographical descriptions* of the country they visited. These can provide a lot of detail on local topography which is not available in more 'official' records. A number of diarists are well-known national figures like John Evelyn, Daniel Defoe and William Cobbett. But there were also many works published of more local importance. Many of these are listed by Anderson and there are also a number of other guides available (Fussell, 1935; Cox, 1949; Matthews, 1950). Some counties (Dorset, for instance) are fortunate in possessing modern bibliographies of local topography which cover subject areas such as geology and natural history (Douch, 1954, 1960; Carter, 1974).

Although map sources and statistical series can provide much valuable data on land use it is often useful to obtain information on agricultural methods and practices. Some published accounts of eighteenth- and nineteenth-century farming can help in this respect. At the end of the eighteenth century the Board of Agriculture commissioned a number of writers to describe the agriculture of each county (Darby, 1952; Grigg, 1967). Two reports for each county were published under the title *General View of the Agriculture of*. . . The first reports, written to a uniform plan, were published between 1793 and 1796 and were circulated among local farmers and landowners to allow comments to be made for incorporation in a second edition. The first and last editions of the *General Views* are listed by McGregor (1961). The reports should be used with some caution and only after careful enquiry into the qualifications of the authors. The writers were not always natives of the counties they described and were occasionally woefully unfamiliar with local conditions (Grigg, 1967). A critical review and summary of the whole body of reports was made by William Marshall in his *Review . . . and abstract of the Reports to the Board of Agriculture . . .* (1817). But Marshall was a disappointed and embittered man passed over by the Board in favour of Arthur Young as secretary, and 'his volumes must therefore be approached with the knowledge of the personal relationships and animosities clustering around the Board of Agriculture' (McGregor, 1961). Nevertheless, used carefully these reports have revealed much about farming practices at the time of the Napoleonic wars (Darby, 1973; Horn, 1980).

The body of descriptive literature relating to the mid-nineteenth-century years is very much richer. Soon after its formation the Royal Agricultural Society of England offered prizes for essays which reviewed the state of English farming county by county. The first of the series, the essay on Essex farming, appeared in 1845 and the last, on Middlesex, in 1869. Darby has provided a list of the Prize Essays and their publication dates (Darby, 1954). As he said:

> Taken together this body of information presented the only detailed picture of English agriculture that had been drawn since the county surveys of the Board of Agriculture.

From the mid-nineteenth century onwards the volume of agricultural writing greatly increases, although there was no countrywide survey on the scale of the Board of Agriculture Reports or the Prize Essays until the County Reports of Dudley Stamp's First Land Utilisation survey were published between 1936 and 1946 (Stamp, 1962). There are a number of national, single-author and usually single-volume studies (e.g. Caird, 1852; de Lavergne, 1855; Haggard, 1902) but there is much more in journals such as that of the Royal Agricultural Society and, of course, the fast-multiplying pages of local journals, parliamentary reports and the like. There is no easy way in to this material other than by using the catalogues of major copyright libraries in conjunction with topographic bibliographies (e.g. Anderson, 1881). Some indication of the breadth of nineteenth- and early twentieth-century agricultural writing can be gained from McGregor's survey of the historiography of English farming (McGregor, 1961). Since 1953, the Agricultural History Society has published an annual list of books, pamphlets and articles in the pages of *The Agricultural History Review*. In effect, this amounts to a comprehensive bibliography of post-war work in British agricultural history.

Newspapers

Published topographies and travel diaries can provide useful background information but newspapers, on the other hand, record information about specific, often cataclysmic events in the natural landscape such as floods, landslips and important meteorological events. Local newspapers first appeared early in the eighteenth century (*Norwich Post-boy* 1701, *Exeter Post-man* 1704), but not until the later part of the eighteenth century do they contain useful local news. Before the nineteenth century they usually took the form of local advertisements filled out with reprinted items from London newspapers. In the early nineteenth century the number of local newspapers published increased dramatically. In 1790 there were about 60, by 1808 over 100 and by 1821, 135. More importantly, newspapers like the *Leeds Mercury* and the *Manchester Guardian* began to record local news at some length. The repeal of the stamp duty levied on newspapers in 1855 brought about another 'newspaper revolution'—by 1865 there were about 750 titles. In consequence it is for the second half of the nineteenth century that newspapers are most useful; in fact, their numbers began to decline after about 1900.

An illustration of the beginnings of interest in recording local, natural phenomena is evidenced by the fact that at the end of the eighteenth century newspapers like the *Caledonian Mercury* and the *Edinburgh Advertiser* were recording the abnormally high snowfalls experienced in parts of Scotland from 1782 to 1786 (Pearson, 1973). Although much of this early evidence is qualitative and much that is quantitative is suspect, the accounts do give a clear impression of the severity of winters at this time, particularly when associated with corroborative manuscript evidence. Major flood events were also recorded in the local press. Where they are recorded in daily newspapers, the sequence of events can be followed through several issues. H. R. Potter has demonstrated what can be done in the following summary of events culled from the *Nottingham Daily Guardian* in January 1901:

> A heavy and prolonged downpour on Sunday, 30th December 1900. It was recorded locally as 1.684 inches, and compared with the year's previous heavy fall of 1.286 inches on June 11th 1900. The water-level rose rapidly during the night of Monday 1st January/Tuesday 2nd January 1901, and by the morning of the 2nd, flooding was taking place over both banks of the river. By Tuesday night the flood level had passed the 1869 flood mark and had almost reached the 1852 flood mark.

If newspapers are to be used to provide information on events, it is essential that the searcher knows roughly what dates to look at, otherwise the procedure can be very time-consuming unless good indexes to the particular newspaper have already been compiled: as yet these are rare. Some newspapers printed a summary, either on the last or the first day of the year, of outstanding events of the past year. Often this is the only 'index' available.

Many newspapers and serials are now available on microfilm, and a good place to enquire in the first instance is the local studies room of county libraries. Another local source might be to approach the offices of the local press, which usually maintains files of back issues: occasionally these are indexed. Dixon (1973) has compiled a very useful guide to nineteenth-century local newspapers and periodicals, which is particularly valuable as it covers precisely that period of growing concern for reporting local issues. Her guide lists the papers held by county record offices and local libraries and the dates of runs, etc. Bibliographies of local newspapers for some individual counties like Devon and Wiltshire have also been produced (Smith, 1973; Bluhm, 1975). Earlier local newspapers, though they rarely contain useful material, have been listed by Crane, Kay and Prior(1966).

The principal national repository of newspapers is the Newspaper Library Division of the British Library at Colindale in north-west London. Its titles are catalogued and it contains many serials which are no longer readily available locally. The Bodleian Library also has a good collection of early local newspapers (Milford and Sutherland, 1936; Mellor, 1955). It should also be noted that during the nineteenth century national newspapers contained more and more local news. It is certainly worth consulting the detailed published indexes of *The Times* to check for reporting of natural events and possibly to consult early issues of weekly and monthly magazines like the *Gentleman's Magazine* and the *Illustrated London News*.

Parliamentary Papers

In the nineteenth century Parliament frequently employed the device of setting up select committees or royal commissions to investigate particular problems. A number of those which relate to hydrological questions are listed in Table 1.2. There are others which relate

Table 1.2
EXAMPLES OF ROYAL COMMISSIONS WHICH EXAMINED HYDROLOGICAL TOPICS
(after Potter, 1978)

Royal Commission on Water Supply	1828	Drought of July–August 1827 and Thames flood of 1821
Metropolitan Water Supply Enquiry	1856	1852 drought
Water Supply Report following	1868	
on Commission	1866	River Lea gaugings in 1850–1868 Cirencester well levels 1863–1868
Report of the Select Committee of the House of Lords on River Conservancy	1877	
Duke of Richmond's Commission on Water Supply	1887	
Royal Commission on Coast Erosion	1906	
Royal Commission on Land Drainage	1927	(led to 1930 Land Drainage Act)

to sewers and drainage in towns, agricultural land use, coasts and harbours, etc. From 1808 they are all effectively indexed. Sets of House of Commons papers (sometimes incomplete) are to be found in most university libraries. The papers have been reclassified for each session into bills, reports from committees, reports from commissioners and accounts and papers. The House of Lords volumes are much rarer; probably only five sets were issued to libraries and other institutions in the early nineteenth century. To a considerable extent the House of Lords volumes reproduce the material in House of Commons sessional papers (Bond, 1971).

Geological Survey District, Sheet and Water-supply Memoirs

The Geological Survey of Great Britain was formally instituted as the Geological Ordnance Survey in 1835 and is generally regarded as one of the oldest national geological surveys in the world (Flett, 1937). The various series of published memoirs produced in the course of the Survey's work provide information not only on the facts of geological structure and lithology but also contain a record of some of man's changes to the physical environment.

In 1839 Henry Thomas de la Beche, first Director-General of the Survey, marked the completion of his mapping of the south-west peninsula of England by publishing the first District Memoir of the Geological Survey—his *Report on the Geology of Cornwall, Devon and West Somerset.* He devoted some of its 648 pages to a discussion of agriculture and past mining activity. W. Whitaker's *The Geology of London* (1889), a revision of the author's 1872 district memoir, has 352 pages of records of wells and bores for water.

One of Roderick Murchison's first reforms when he took over the Survey in 1855 was to supplement the district memoirs by introducing Sheet Memoirs or Explanations to accompany individual 1-inch maps; the first to be published was for Cheltenham-by-Hull in 1857. One of the objects of the sheet memoirs was to expedite publication by producing more, but less bulky, reports. Publication of sheet memoirs in regular fashion had to await the twentieth century, when Director-General Jethro Teall decreed that both memoir and map should be published simultaneously. However, more memoirs did mean slower map production (Bailey, 1952).

In 1899 the Survey's County Water-Supply Memoirs were instituted. The first, by Whitaker and Clement Reid, was the *Water Supply of Sussex from Underground Sources.* Concern over water supply was part of the Survey's growing appreciation of its responsibilities in applied geology. Eleven water-supply memoirs were published during Teall's Directorship (1901–1914) and covered most of south-east England, where underground water was one of the main economic geology interests. Figure 1.15, redrawn

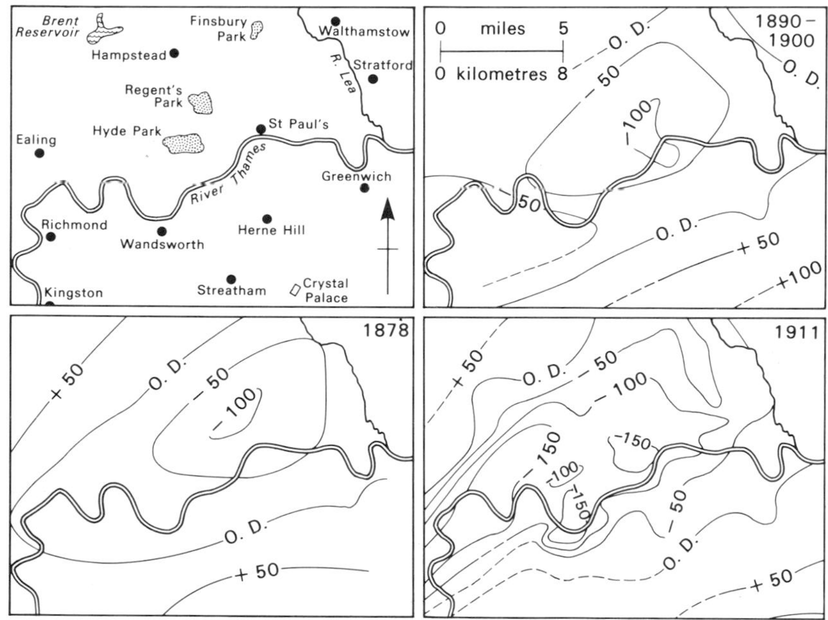

Figure 1.15. THE PROGRESSIVE LOWERING OF THE WATER TABLE IN THE CHALK UNDER LONDON BETWEEN 1878 AND 1911. (Source: Barrow and Wills (1913))

from one of these (G. Barrow's and L. H. Wills' *Records of London Wells* (1913)), illustrates the progressive drop in water level in the chalk below London caused by nineteenth-century pumping. These memoirs also contain rainfall statistics commissioned from the meteorologist Hugh Robert Mill or taken from the publications of his British Rainfall Organisation. Publication of water-supply memoirs flagged during the First World War but resumed afterwards. The memoirs, though, discontinued publishing rainfall data after Mills' organisation was taken over by the Meteorological Office of the Air Ministry. In the 1930s the Geological Survey's work on underground water was supplemented by the Inland Water Survey Committee, which was charged with collecting records of both rivers and underground waters. The Geological Survey and the Committee debated the relevance of the county basis which the Survey used for its water-supply memoirs, the latter preferring realistic catchment units. The boundaries of all officially defined catchment areas were inserted on maps accompanying Geological Survey water publications and the Survey compromised to the extent that all underground data were subsequently tabulated and published with reference to 1-inch maps. During the Second World War pamphlets were produced instead of full memoirs; forty-eight of the ninety-three published during the war dealt with water-supply matters. Twenty-seven authors contributed to this war effort and, besides reports on hydrogeology, the pamphlets include catalogues of more than 12 800 annotated well records.

Manorial surveys and estate papers

Published accounts, maps, and statistics can usually only provide the bare bones of a description of the natural and man-made environment in the historical period. More detailed evidence to supplement maps and published material can often be found among the manuscripts relating to particular landed estates. In the medieval period, manorial rolls contain the administrative details of land management on feudal manors. In areas like the midlands of England, where open-field farming flourished, the manor survived through to the seventeenth century (Kerridge, 1966).

The fullest of these manorial surveys detail the arable and pasture on each of the tenant holdings (Thirsk, 1955). From some it has even proved possible to extract weather information as indicated by its effects on the sowing, growing and harvesting periods of the year. Historians working through the rolls of south-coast manors have established almost complete series of wet and dry seasons for the thirteenth to fifteenth centuries (Brandon, 1971a and 1971b). The rolls of the Bishops of Winchester estates have proved particularly fruitful in this respect (Titow, 1959–1960, 1970). The PRO is the principal repository for this kind of material. The content of many medieval documents has also been published either as *texts* or *calendars*. The former are full transcriptions, often with translations, while the latter are summaries. There are two invaluable guides to this printed medieval material: R. Somerville, *Handlist of Record Publications* (British Records Association, 1951) and E. L. C. Mullins, *Texts and Calendars* (Royal Historical Society, 1958).

From the seventeenth century onwards, material relating to events associated with, but outside the actual sphere of, the landed estate also becomes more common. The *personal notes and diaries* of parish priests and occasionally also of farmers survive in county record offices. Such interested observers concerned with matters like tithe payments and the size of the harvest frequently described the performance of crops and made monthly, weekly or even daily notes on weather patterns (Jones, 1964). The parish priest observer was also joined in this period by a number of, in Ingram and Underhill's terminology, 'proto-scientific' individuals who began to make relatively systematic meteorological observations. They refer particularly to the diary of Christopher Sanderson from County Durham and illustrate the way in which he kept a progressively more detailed account which begins with random notes on the more remarkable droughts and floods in the 1650s and becomes more systematic until, in the period 1682–1689, he produced a detailed meteorological journal (Ingram and Underhill, 1979).

Evidence in manorial rolls and estate manuscripts on meteorological phenomena is nevertheless scattered and in surrogate form. On the other hand, such papers can provide a first-class, detailed record of land use. Landowners' estate records have more frequently survived than the notebooks of farmers and evidence, when it is available, often covers a good-sized tract of country (Thomas, 1966). Moreover, larger estates which maintained a fair degree of continuity over time and space have usually preserved a fuller range of estate source materials. The actual survival, though, of a very good collection of estate papers may indicate superior estate administration, and such sources may not, therefore, always provide a representative picture of conditions over a wider region.

Maps of such properties (the estate maps referred to in the section above) have often been detached and are catalogued and preserved separately. If land use is not recorded on the face of the maps then it is usually available for the time of the survey among the accounts and papers. A selection of such material has been published for Suffolk and indicates the nature of estate manuscript material which can be found to accompany the maps (Thirsk and Imray, 1958). Historical geographers and economic historians have made much use of estate accounts and surveys in conjunction with maps to trace the development of agriculture and land use (for example, see the studies in Darby (1973)). To take just one example, W. J. Smyth has traced the progress of moorland reclamation in County Tipperary, Ireland, using estate records. Evidence from leases shows that enclosure of marginal mountain lands accelerated around the mid-eighteenth century, while the main 'colonial' phase of upland settlement stretches from 1813–1814 to the early 1840s. Subsidies were granted for digging up mountain land, most of which activity was concentrated on the flatter areas between 400 and 800 feet (Smyth, 1976).

For the most part, land-use data for the nineteenth century can be more easily retrieved from 'standard' map sources like tithe surveys and, from 1866, the agricultural statistics. Only if it is desired to reconstruct the detailed vegetation history of a particular place or to work retrospectively to periods earlier than the nineteenth century, should it be necessary to turn to quite probably uncatalogued and often unsorted cartons of estate papers.

Commissioners of Sewers' records

Commissioners of sewers were local bodies instituted by the fifteenth century and charged with the management of low-lying (usually tidal) lands. Their papers are of particular value in relation to fluvial studies and the study of drainage alterations. 'Sewers' were 'Fresh water Trenches, or little Rivers, encompass'd with Banks on both Sides, to carry the Water into the Sea' (Emmisson, 1947). At first there were no fixed procedures except for a general encouragement for commissioners to adopt the custom of Romney Marsh as a standard. All their activities were regularized by the statute of sewers in 1532. A history of Romney Marsh which details the sewer records is provided in Derville (1936).

Commissioners of sewers were established for all the main marshland areas of England and Wales. They were appointed by the Crown and worked through courts of sewers. These were assisted by juries of twenty-four men in each of the ancient hundreds which contained some marshland. They had to inspect the drains and banks and apportion responsibility for their upkeep.

When the improvement of rivers for navigation purposes became more widespread in the seventeenth century, courts of sewers became involved in numerous disputes occasioned by navigators' proposals which affected the livelihood of other proprietors such as mill owners (Willan, 1964). Many of the papers generated by these disputes and the minutes, accounts and assessments of the various levels contain references to flood events and the like, while the surveys often took the form of maps of each 'level' or district. In this respect the records of the commissioners of sewers complement some deposited plans and surveys which relate to navigation and canal works. As a record of flood events they should also be read in conjunction with local newspapers. For example, commissioners of sewers' papers record that some of the worst floods on the low-lying Hoo peninsula at the confluence of the Thames and Medway rivers occurred in 1843, 1854 and 1897 (Kent Archives Office S/NK/SM8). Of these the last is by far the best documented. During Sunday 28 November high winds were first recorded: these gradually increased and reached such a force that the *Chatham Observer* reported it as 'a gale of almost unparalleled severity'. At 3 p.m. on Monday afternoon the first of the sea walls gave way and one breach after another occurred in the earthen embankments. Within a very short space of time, something like half the peninsula was under water and several villages cut off.

The Land Drainage Act of 1930 wound up the courts of sewers, whose powers and documentary collections were transferred to the new catchment boards. Many of these boards have deposited their earlier records with county record offices. Also useful in a search for documents relating to the early period of their operation is William Dugdale's *History of Imbanking and Drayning* (1662), in which he catalogued all the documents relating to each district that he could find in public and private records.

Miscellaneous manuscripts

There is a miscellany of sources which occasionally contain material useful in environmental analyses but, because that which is apposite is usually so scattered, a general search is rarely warranted. With some prior knowledge of its likely location derived perhaps from other sources noted in previous sections it may be possible to abstract specific information. These miscellaneous records are mostly manuscript and include legal, administrative, taxation, ecclesiastical and personal records.

Legal records are usually classified by the court to which they refer. Examples of the use of legal records in environmental analysis are where a boundary dispute has involved a physical feature such as a river or coastline the nature of which has been carefully documented as a result (de Vorsey, 1973). Disputes over property or rights to use of land often contain details of land use, and these have been used in studies of vegetation history (e.g. Peterken, 1969; Nicholls, 1972). De Boer's studies of coastal movements using the records of disputes over the siting of a lighthouse are discussed in Chapter 4.

Administrative records are a particularly heterogeneous group. Taxation accounts are a specific type which has been used in glaciological and climatological studies. Patterns of fluctuations in the taxation record can be related to natural phenomena. Payments which detail crop yields may also give some indication of land use. Otherwise, this group is too diverse to describe generally but this does not mean that they should be discounted. Trade records have been found useful, for example, to indicate the decline of a port consequent on silting of a harbour (Cozens-Hardy, 1924).

Ecclesiastical records are equally varied. Many parish registers and parish records such as diaries kept by clergymen have been deposited in record offices and are catalogued by name rather than by provenance. Evidence for the occurrence of natural disasters may be found in notes of visits by clergy, the saying of special prayers and the like.

Personal diaries have already been discussed above, but other personal letters and papers may occasionally prove useful, perhaps where they contain descriptions of unusual events or areas visited. Again, though, it is really a matter of luck if relevant material turns up: a systematic search is not really recommended.

Oral evidence

In western Europe the earliest 'histories', sagas from about 3000 years ago, set down the oral tradition about the then-distant past: only much more recently were events of the 'present' chronicled in writing. Once documentation was an accepted way of recording, the human memory was demoted in status until the written word became the final authority and the guarantee of transmission of facts and opinions to future times. In other words, in the world's long-literate societies there has been a gradual decline in the prestige of an oral tradition which itself partly explains the radical tag with which modern oral historians are labelled. Today, 'oral history' implies not only the use of a particular type of evidence but also assumes a particular methodological stance with respect to the nature of history itself. This reflects the growing concern amongst social historians for understanding the lives of ordinary individuals in the past rather than with assessing the roles and influences of 'important' persons.

To an earth scientist concerned with past events or processes such methodological issues are probably only of passing interest. Major concern is with the extent to which oral evidence can provide trustworthy information on past environmental conditions witnessed by persons still living. Thompson (1978) has written an excellent guide to the exploitation of this kind of evidence. In physical geography, oral evidence is best viewed as a potential corrective and supplement to other sources. It is unlikely to open up whole new fields of enquiry as it has done in social history. Its most common and likely applications are first, in connection with settlement and development processes in the New World and in areas where large-scale changes in environmental conditions have taken place within living memory. Although 'pioneers' no longer survive, their oral evidence has been used in some earlier physical geography studies, particularly those associated with the development of arroyos (Bryan, 1928b). Second, and probably most importantly, oral evidence can provide eye-witness accounts of events such as earthquakes, floods and landslides. Here allowance has to be made for exaggeration but effects described can often be corroborated by other accounts, photographs and illustrations as well as the physical field evidence. Third, oral evidence may be used where change has been rapid but not sudden. People who have a professional concern with such phenomena are often very reliable sources of evidence amongst whom coastguards, water bailiffs and local surveyors might be numbered. Long-term residents, especially farmers, can also help since they are often closely aware of changes in their environment and property. But how trustworthy are eye-witness accounts in general?

To resolve this problem, two subsidiary questions must be addressed. The first of these concerns the memory process itself and the second, the potential bias and subjectivity of witnesses. On the first point the evidence of repeat questionnaire surveys in the social sciences can be brought to bear to underscore the perhaps commonsense fact that memory does fade. This occurs quickly up to about nine months after an event but quite slowly thereafter. Certainly historical facts derived from interviews need to be treated with care but, in Thompson's (1978) opinion, such scientific evidence as there is does not suggest that memory is so subject to error as to invalidate the use of retrospective interviewing. If the evidence provided in an interview does seem to be misleading or contradictory it is possible to ask for more, an opportunity which written documents do not offer.

To interview effectively does require skill. There is a whole range of methods which can be employed, from the informal, conversational approach to the use of formal, structured questionnaires. In an earth science context it is probable that there will usually be a relatively small number of individuals to question about the nature of a specific environmental event.

In the last analysis, oral evidence should be regarded as just one of many potential sources of evidence on past conditions. It should be assumed in the first instance to be neither more or less intrinsically trustworthy than the written word, map or picture, but be subject to the usual processes of historical evaluation. Each interview should be carefully checked for internal consistency and cross-checks made with other interviews and sources. Differences between the written and oral record should not necessarily mean that the oral tradition is at fault.

1.3 STATISTICAL SOURCES AND SERIES

There are a number of series of systematic measurements and statistics for which records are continuing but whose origins date back into the historical period. Where early measurements are comparable with present data little guidance will be given here, since most relevant sources will be familiar to the earth scientist or do not entail historical knowledge or techniques. However, in the three areas of meteorological measurements, hydrological data and land-use data, quantitative information is extremely important and is fundamental to explanation of physical phenomena. The methods of making these measurements and the bodies responsible for their recording have changed over time, so some summary of the availability of historical statistics in these fields is considered useful. Other series of statistical data which may be used by earth scientists include land-survey and height data from the Ordnance Survey and tide records obtainable from the Institute of Oceanographic Sciences, Bidston Observatory.

Meteorological data

Important developments in meteorological instruments occurred in Europe during the seventeenth and eighteenth centuries. This period also saw the beginning of systematic measurement by amateurs; measurements on an official and organized basis only began in the nineteenth century. Some very useful background on the history of meteorological observations is provided by H. H. Lamb in Volume 2 of his book *Climate Present, Past and Future* (1977).

Raingauges are known from the fifteenth century in Korea, while in India they were probably used even earlier (Folland and Wales-Smith, 1977). The history of the technical development of raingauges has been outlined by Biswas (1970) and some background, particularly on British applications, is contained in Reynolds (1965). In Europe, raingauges were developed from the end of the seventeenth century onwards but to very many different designs. Many of the early ones had a collecting vessel sited on a roof and precipitation was led by a series of pipes to a receptacle inside the building. Measurements were often made in units of the collecting vessels. In order to interpret these readings and adjust for exposure, time of reading and other factors it is worth while investigating documentary material concerning the instruments and method of observation. Many early observations were made by enthusiastic amateurs, sometimes scientists like Dalton, but often by clergymen and doctors. However, most evidence suggests that they were diligent in their tasks. Some correspondence on instruments and measurements survives and papers were published in journals such as the *Transactions of the Royal Society* discussing the design and siting of instruments. For example, Folland and Wales-Smith (1977) have reconstructed the background to measurements of rainfall by Towneley at Towneley Hall, Burnley during the period 1677–1703 using such sources. During the eighteenth century the effects of exposure were realized and during the nineteenth century agreement was reached on standardized designs and sitings. Symons in particular did much to promote the standardization of rainfall measurement and the compilation of records in the last half of the nineteenth century.

Thermometers were developed during the early seventeenth century and were in general use by its end. Many early instruments were very large and there were some problems of calibration and of deterioration over time. At first, many were sited in fireless rooms and later in boxes on north walls and the readings have to be adjusted for these positional influences. Zeros and scales must also be checked by comparison with other records.

Wind vanes have also been in regular use for some considerable time. Sailors were particularly interested in wind characteristics and Oliver and Kington (1970) have shown how valuable *ships' logbooks* can be for reconstructing weather conditions (Figure 1.16). From 1731 ships were instructed to keep a journal or logbook which included observations of position, wind and weather. Oliver and Kington have discussed some of the problems of interpreting these logs and have defined some of the terms used in them. Captains' and masters' logbooks are now kept in the PRO and lieutenants' logbooks at the National Maritime Museum.

The earliest regular daily instrumental meteorological observations were initiated by the Academia del Cimento in Florence in 1654. From the 1660s onwards the Royal Society encouraged observations in England, mainly under the impetus of Robert Hooke. Among the longest records are those from the Kew, Greenwich and Radcliffe observatories but there are also some from other institutions in major towns. Many early records are kept by local libraries and museums, and some individual records may still be amongst estate and personal papers. Lamb (1977) has published a list of the longest records available and Craddock *et al.* (1976, 1977) have compiled homogeneous records of rainfall from early instrumental sources for various parts of Britain. Manley has produced similar records of temperature for Lancashire (1946), central England (1953) and England as a whole (1959).

By the mid-nineteenth century, measurements were being made on a more coordinated

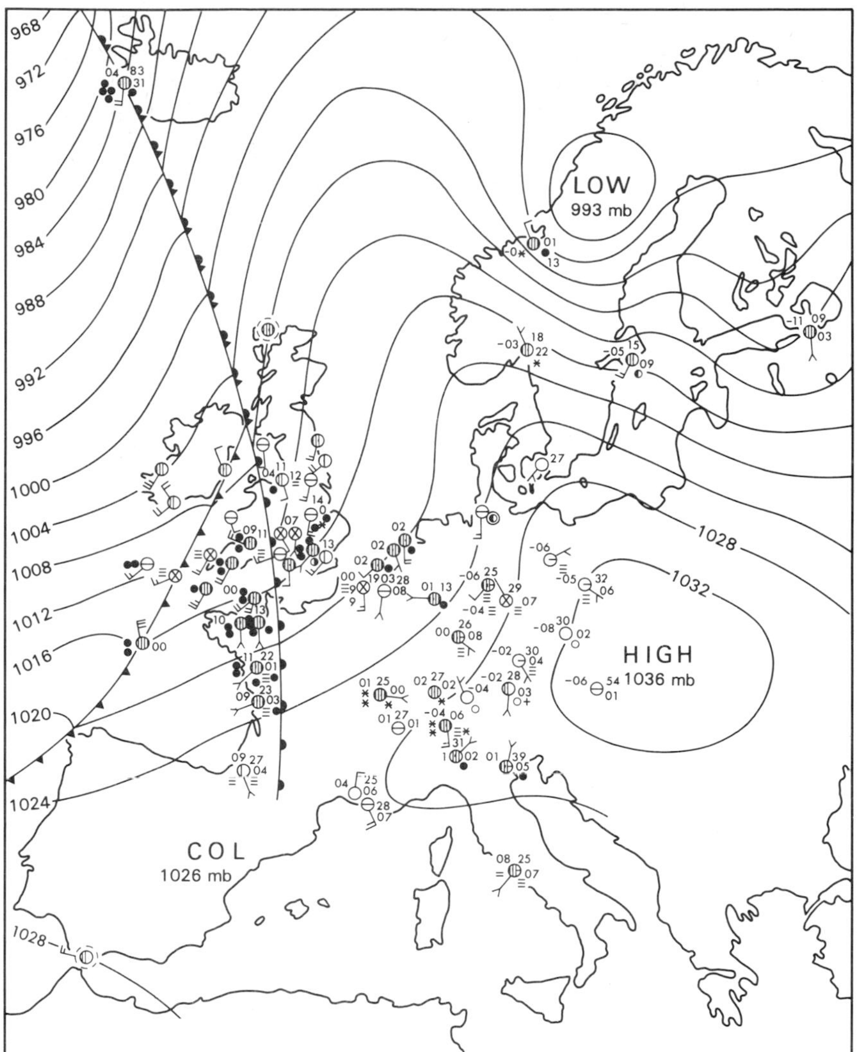

Figure 1.16. SYNOPTIC CHART FOR 1400 HOURS 4 JANUARY 1783.
Reconstructed from land station data and ships' logbooks. (Source: Oliver and Kington (1970))

basis and 1860 saw the first appearance of *Symon's British Rainfall*, renamed in 1899, *British Rainfall.* This annual publication lists stations making rainfall measurements, provides monthly and annual totals and specifies the dates and amounts of maximum daily falls. Each volume also analyses rainfall distributions and highlights some of the main meteorological features of the year. Rainfall records, both old and modern, are also compiled in the ten-year books of the Meteorological Office, which are the official repository for monthly rainfall records. These give details of stations and for early records references to original sources may provide further information (Craddock, 1976).

In the second half of the nineteenth century, specialist organizations such as the

Meteorological Society and the Meteorological Office rose to importance (Ratcliffe, 1978). This period also saw the beginning of many of the specialist British journals. *Symon's Monthly Meteorological Journal* began in 1866, became *Symon's Meteorological Magazine* in 1901 and then the *Meteorological Magazine* in 1920 when publication was taken over by the Meteorological Office. The *Quarterly Journal of the Royal Meteorological Society* was first published in 1872.

Daily Weather Reports were published from 1873 until discontinued in 1980 and contained observations of temperature, pressure, rainfall and precipitation, wind, cloud and sunshine from approved and standardized stations in the British Isles. Manuscript Daily Weather Reports were produced from 1861 to 1872. The Weekly and Monthly Weather Reports have been published since 1878 and 1884 respectively. The Monthly Weather Report in particular is useful for detailed records of districts, and since 1911 has included the *Meteorological Record.*

Hydrological records

In Britain there was no requirement to operate gauging stations on rivers until the river boards were set up in 1948. Since then the total has expanded to over 1200. Some regular daily records do date from much earlier. For example, the Thames at Teddington has been continuously gauged since 1883 and the gauge at Fieldes Weir on the river Lea has been in intermittent operation since 1851. All data from gauges are collected by the Water Data Unit of the Department of the Environment at Reading as well as by the present water authorities. Summary records are also published in the Surface Water Year Book (HMSO). Details of the longest and most reliable records in Britain are given in the Flood Studies Report (NERC, 1975). Catchment details and some historical data, mostly acquired from the water authorities, are also listed in this publication. Regular flow data is rather meagre for Britain pre-1940 but useful hydrological information on a non-systematic basis may be found in the records of early water and drainage undertakings. Much of it, though, concerns floods. These organizations are listed in Table 1.3 (Potter, 1978).

One source of information which has been little used in published studies is evidence from the records of early reservoir and water-supply schemes. A number of reservoirs were constructed in the latter half of the nineteenth century in Britain and their levels have been monitored and controlled. There were also many small impoundments of water for industrial purposes as well as abstractions from streams by industrial concerns whose records may have survived in municipal archives, county record offices or are perhaps in the custody of the firm if it is still trading. Canal-lock records can also provide some information, though there are many difficulties in their interpretation. These records are mostly held by the British Waterways Board and some by the British Transport Historical Records Department. There are many other records of flood events, and these are indicated in the review of applications in Chapter 5.

Regular flow records exist for longer periods in other countries, particularly those with much greater water-supply problems than Britain. For example, stream-gauging was organized on a systematic basis from the late nineteenth century by the Geological Survey in the United States. Biswas (1970) has outlined the development of river-discharge

Table 1.3
RIVER MANAGEMENT ORGANIZATIONS (after Potter, 1978)

Organization and operative dates	Regional coverage and area of responsibility	Reporting policy
Water authorities 1974–	10 to cover England and Wales water supply, river management and river and effluent quality	Very little technical detail in annual report. Flood events are often subject of a separate report
River authorities 1965–1974	29 to cover England and Wales. Water resource planning, river management and river quality	Floods and other hydrological and meteorological reports. Some separate reports for flood events. Internal committee dealing with land drainage and water resources
River boards 1951–1965	32 to cover England and Wales. Primarily river management with some water resources responsibility	Reporting policy similar to river authorities.
Catchment boards 1931–1951	46 to cover about 67% of England and Wales. River management and land drainage	Annual or triennial reports produced spasmodically by some and include mainly flood and flood protection information
Fishery boards reconstituted 1932	Cover major fishing rivers in UK	Annual reports produced and include river level, rainfall and temperature and pollution information
Internal drainage boards 1861–	Currently 314 to cover 12 500 km^2 of low-lying areas adjacent to larger rivers and fen districts	Only the largest authorities produced annual reports, and these concentrate on floods and flood protection measures. Engineer of authority would report internally on notable flood events
River conservancies (1857–1930)	A few catchments including Thames, Lea, Cumbrian, Derwent and Dee. Maintenance of major river channels	Occasional reports on particular floods and annual reports
Sewer commissioners (various acts from 1427)	Responsible for specific river reaches, sponsoring improvement works and involved in cases of dispute	

measurement, including early measurements on the Nile and on central European rivers such as the Elbe and the Rhine. In 1862 Beardmore published a *Manual of Hydrology* which includes tables of discharge for long-gauged rivers and reservoirs in all parts of the world. It gives detailed figures for gauges in the Lake District and Scotland and the river Lea in the 1850s. It also notes the heights of maximum floods on major rivers, flow data from sewers and water supply in London. Details are given on tides and tidal rivers with longitudinal and cross-profiles of several of the major tidal estuaries of the world. There is also a section on rainfall and evaporation.

Groundwater is another major hydrological component for which some records are available. Wells have long been a source of supply of water but concepts of groundwater were not fully elucidated until well into the nineteenth century, and systematic measurement of groundwater has only been undertaken in Britain on a general scale since about 1950. Groundwater study has always been closely associated with geological investigation. The Geological Survey, founded in 1835, published the *Memoirs on Water Supply* the first of which appeared in 1899 (see above). Some measurements were made in the South Downs from 1893 (Thomson, 1938) and after the 1945 Act all groundwater abstractors were required to keep records of abstractions. By 1965 there were about 600 observation wells in Britain and 113 autographic recorders (Ineson, 1966). The *Groundwater Year Book* performs a similar service to the *Surface Water Year Book* and was first published in 1964.

The Water Resources Act of 1963 placed a statutory obligation on river authorities to measure evaporation, but many were slow to implement this. There are few 'historical' records of evaporation or evapotranspiration.

The statistical assessment of land use

It was not until 1866 that the Board of Agriculture made a complete and comprehensive return of crops and land use in England and Wales, though attempts, some partially successful, to attain this aim can be traced back some seventy years. In the 1790s the Home Office made a number of inquiries into the state of harvests (Minchinton, 1953) and in 1801 the much more widely known *Crop Return* was collected. It is generally agreed that although the proportions between the crops recorded are tolerably accurate, the actual acreages were considerably understated. These statistics have been much used in temporal studies of crop-combination changes (Henderson, 1952; Thomas, 1958 and 1963) but their inaccuracy really precludes their use in studies of acreage change.

The first comprehensive source which enables proportions between land uses to be assessed with tolerable accuracy are the *tithe surveys.* It is possible to obtain measures of different land uses by summating the acreages of each field from the schedules of tithe apportionment. The fact that the tithe surveys provide both land-use and information and acreages for individual fields means that the land use of particular tracts of land, for example a drainage basin, can be calculated very precisely.

On 25 June 1866 the first of what were to become *Annual returns of crops and livestock and grassland acreages* were collected by the Board of Inland Revenue on behalf of the Board of Trade (Coppock, 1956; Best and Coppock, 1962). In subsequent years the machinery of collection was changed, the date of return altered (4 June 1877) while the returns have grown in complexity as more categories of information have been surveyed. All the returns were collected on a voluntary basis (except for a short time during the First World War) until 1925, when the occupiers of holdings of more than one acre were required by law to make an annual return.

The accuracy of the returns cannot be assessed as all the early original forms have been destroyed and those which are still extant cannot be consulted since the Agriculture Act of 1947 prohibits the disclosure of details of individual holdings (Best and Coppock, 1962). It is possible, however, to consult returns aggregated by parishes in the PRO. Coppock

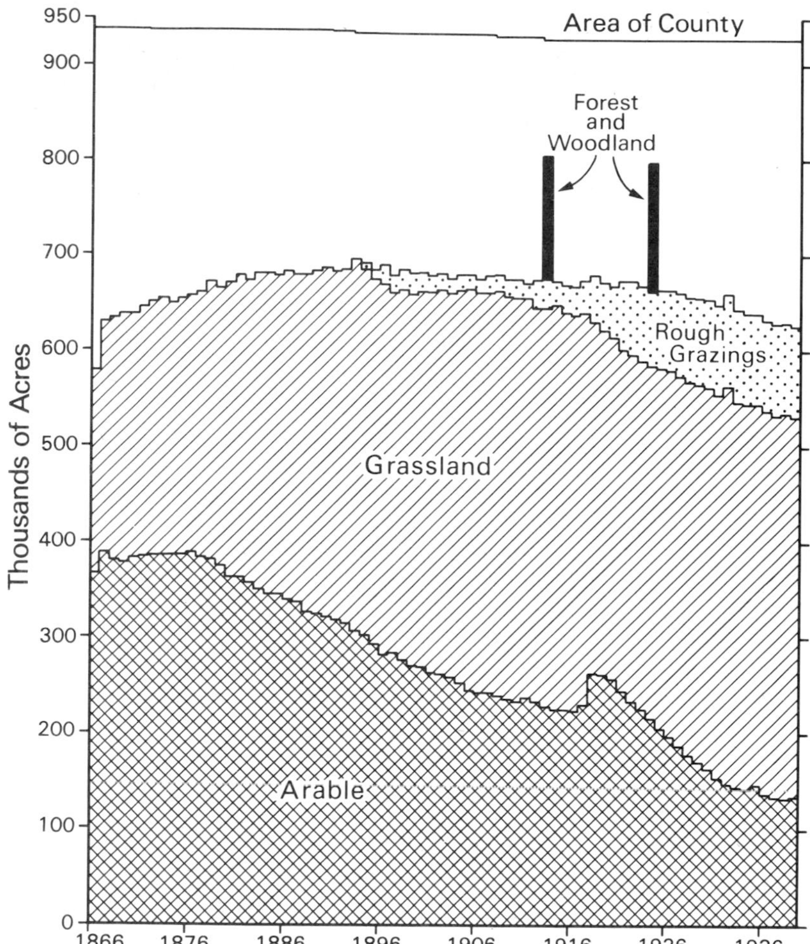

Figure 1.17. LAND-USE CHANGE IN SUSSEX, 1866–1937.
(Source: Briault (1942))

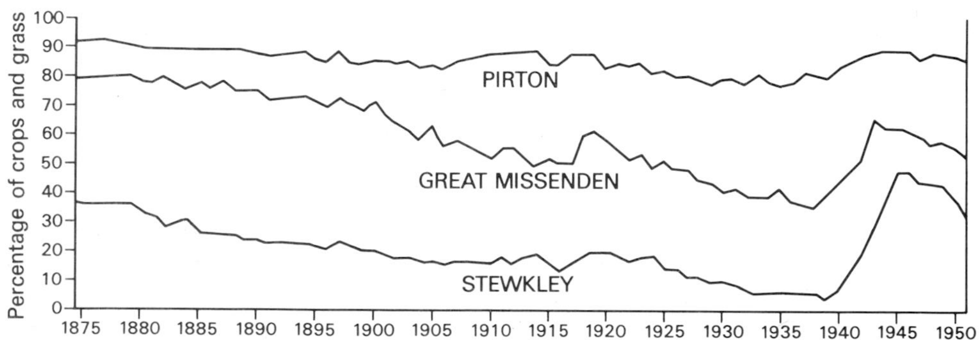

Figure 1.18. CHANGES IN THE AREA UNDER CROPS AND GRASS IN THREE CHILTERN PARISHES, 1875–1950.
(Source: Coppock (1957))

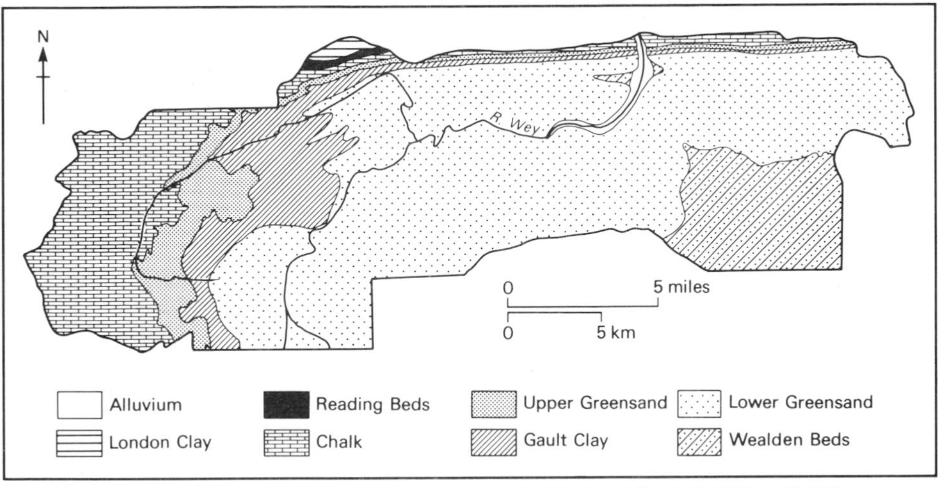

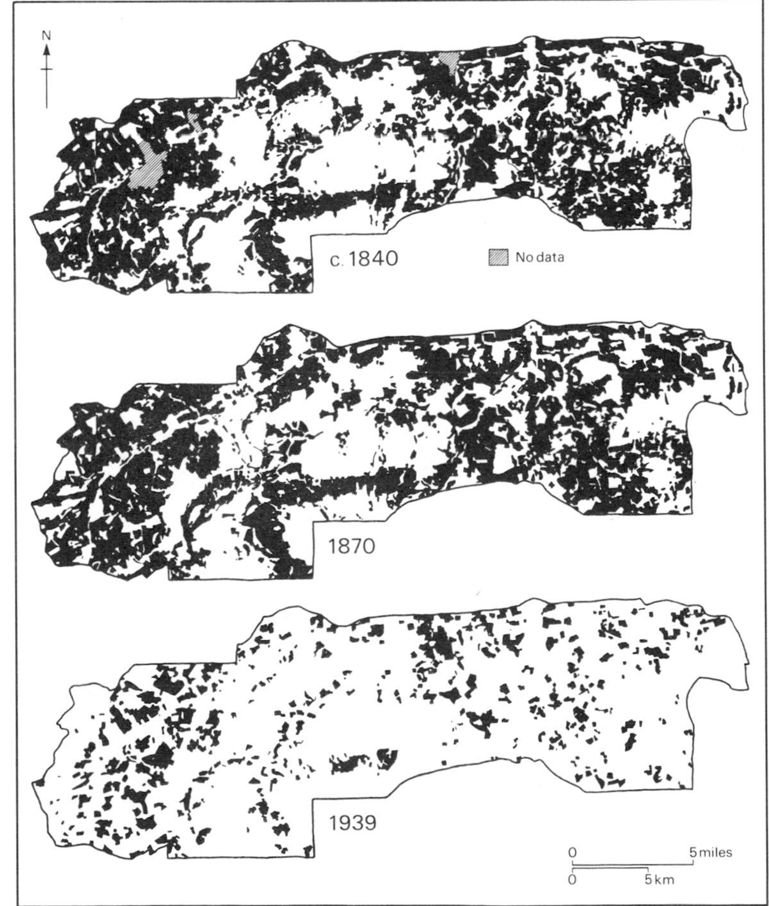

Figure 1.19. ARABLE LAND IN THE RIVER WEY BASIN ABOVE GUILDFORD, SURREY, *C.*1840–1939.
Data for 1840 is from tithe surveys, for 1870 from Ordnance Survey Books of Reference and for 1939 by field survey. (Source: Shave in Stamp and Willatts (1941))

has examined the accuracy of these parish returns and in particular his tests have focused upon the question of the comparability of a run of returns. It seems that most sources of error will result from changes in farm and parish boundaries, from the return of holdings situated in more than one parish in the parish in which the occupier resided and in differences in interpretation by farmers (Coppock, 1956; Robinson, 1981). There are certainly real problems inherent in the accuracy of the returns themselves, but more serious from the point of view of physical geography applications is the fact that the figures are only available as parish aggregations. Where small-scale land forms are not approximately coincident with parish boundaries or in fact run directly counter to the parish pattern, then the use of such aggregate measures may be quite wrong. The larger the land units under study the less this will be a problem, and with regional scale comparisons the aggregate nature of the data may be a positive advantage.

For the years about 1875, the Books of Reference to the first edition of the 25-inch (1:2500) Ordnance Survey plans (see above) can be used in the same way as the tithe surveys to obtain accurate data for small tracts by summing the acreages of individual fields.

Each of the *County Reports* to the First Land Utilisation Survey of Great Britain contains a statistical appendix based on an analysis of the Agricultural Returns. The official statistics are abstracted for each year from 1866 to 1939 inclusive, giving for each county and each year the acreage under crops and grass (i.e. total cultivated land), arable, permanent grass for hay, grass not for hay and rough grazing. In some reports these figures are also shown in the form of a graph. Figure 1.17 is an example. Similar tables have also been constructed for individual parishes from the manuscript returns. Figure 1.18 is an illustration from a study of this type.

Although there are differences from county to county and from parish to parish within counties there is a remarkably consistent trend of land-use change in England and Wales. There was at first a rise from 1866 in total area under crops and grass which coincided with what is sometimes known as the period of 'High Farming', the golden age of arable farming culminating in the 1870s (Stamp, 1962; Orwin and Whetham, 1964; Chambers and Mingay, 1966). From this time on, the well-known decline in the acreage of farm land set in and the decline was most marked with arable. Much heavy land 'tumbled down' to grass, the proportion of rough grazing increased and the moorland edge moved downwards. This trend, once established, has been very persistent with the exception of the special conditions produced by the two world wars (Figure 1.19).

2
Comparative and Complementary Sources

2.1 AN INTERNATIONAL PERSPECTIVE

The main purpose of this section is to provide some comparative context in which to place the material already discussed in Chapter 1. Although the British sources reviewed in the preceding chapter exemplify the types of information available, the nature of the historical record varies from one country to another according to particular political–legal histories and from continent to continent according to the length of settlement.

The Old World

The European countries with broadly similar spans of settlement have broadly similar types of records. Maps become more readily available from the sixteenth century, while throughout Europe the early nineteenth century can be characterized as an 'age of cadastral surveys'. The documents produced by these surveys have for long been important sources of evidence for economic historians and historical geographers. Indeed, the first volume of Marc Bloch's influential periodical *Annales d'Histoire Economique et Sociale* contains a number of articles reviewing the nature of cadastral surveys in France, Germany, England and Denmark (e.g. Bloch, 1929; Vogel, 1929; Aakjaer, 1929). So, at the same time as enclosure and tithe maps were being produced in England, the *communes* (parishes) of France, for example, were being subjected to the detailed ground survey of the *ancien cadastre parcellaire*, while large-scale Land Registration Surveys had begun in Sweden in the seventeenth century. A brief discussion of these continental European cadastral surveys might help to provide some context for contemporary English surveys.

By 1814 about 9000 *communes* of France had been surveyed by elementary trigonometrical methods and by 1850 the whole of the then area of metropolitan France had been surveyed (Herbin and Pebereau, 1953; Clout and Sutton, 1969). Much of this nineteenth-century material has been deposited in *départemental* archives. The *cadastre* of a typical *commune* consists of a map portraying roads, watercourses and settlements accompanied by lists of landholdings and landowners. The maps were drawn at a variety of scales from 1:500 to 1:5000, the usual range in rural areas being between 1:2000 to 1:2500. By relating the maps and landholding lists a map of land use can be compiled. There are, however, a number of difficulties with this. The *cadastre* was compiled for taxation purposes after the French Revolution and the fiscal nature of the classification of land use together with local terminology can cause problems of interpretation. Aimé Perpillou, for instance, found that the three terms *bruyères, chaumes* and *pâturages* were all used to denote waste and uncultivated land in early nineteenth-century Limousin (Perpillou, 1935). On the other hand, as Clout and Sutton (1969) noted, the plans of the *ancien cadastre* are far superior to other contemporary topographical maps and were used in the compilation of the 1:80 000 Carte d'Etat Major. They have been considered sufficiently accurate to be used by geomorphologists in ways similar to English tithe maps. In one such study the maps of the *ancien cadastre* have been used to measure cliff retreat in the Saintonge area of northern France over the past 130 years (Gabert, 1965).

Large-scale maps of arable land in Sweden (the equivalent of English estate maps) date from the 1630s. These early maps cover only land held by peasant proprietors or the state but between the 1690s and the 1720s a new series of maps was made which also cover the extensive estates of noblemen (Bergsten, 1961). From the mid-eighteenth century, the Great Partition or *storskifte* marked the end of the traditional open-field system of farming (cf. the English enclosure movement) by consolidating holdings to improve agricultural efficiency (Jones, 1977). Beginning in the Vaasa area in 1749, a survey and land description was initiated and the work undertaken constitutes the main source of information on topography and land use for eighteenth-century Scandinavia. In Sweden, the Great Partition was replaced by new land reforms in the nineteenth century but it continued in Finland after it was separated from Sweden in 1809. From Ostrobothnia, the Great Partition was gradually extended to most parts of Finland in the nineteenth century, while in Lapland most surveys date from the early twentieth century. In this respect they are quite different from the French *cadastre* or the English tithe surveys which were produced within the span of a few years. They are more akin to enclosure maps whose surveys also covered a century or more. For some more limited areas the *storskifte* surveys are complemented by the surveys of a second major land reallocation known as the New Partition (*nyskifte*), begun in 1848. All these sources, particularly the Great Partition maps held by the National Board of Land Surveys in Helsinki, have been used by Jones (1977) to chart the progress of isostatic readjustment evidenced by emerging land along the Baltic Coast.

In central Europe there is a long scholarly tradition of studying the changing physical environment in historical times. This is founded not only on the intrinsic interest of such enquiries or the availability of uniquely suitable sources but also on the precept that the findings of historical–physical geography are fundamental for understanding human landscape change. A number of these studies are reviewed in later chapters, but in this comparative, international section we might note the review by Glässer (1969) of publications on man's effects on the natural vegetation of central Europe. His work provides a convenient bibliographic entrée to this field. Examinations of climatic change in historical times have been pursued on wide fronts and with a wide range of sources. Again, one convenient summary of sources and methods is an essay by Flohn (1967) published in a Europe-wide review of climatic change. Other work in central Europe has investigated man-induced relief changes, soil-erosion, changes in river courses and coastal change. Much of this work is characterized by its focus on the changing physical basis for man's activities and it uses a wide range of documentary and map sources (Fels, 1967).

In the USSR, historical geography has been viewed as a subdivision of physical geography concerned with changes in physical conditions during historic times (French,

1972). Zhekulin (1968), for example, in drawing up an agenda for historical geography included exclusively features of the physical environment: lakes, swamps, flood-plains, deltas, estuarine marshes, karst, gullies, stream channels, shorelines, soils, vegetation and fauna. Other historical work has been conducted as an essential preliminary to understanding the present. In this category are the soil studies of Gedymin (e.g. 1968) who has used cadastres of the late fifteenth to seventeenth centuries and maps of the eighteenth-century General Survey to analyse changes in forest cover and the spread of arable. Historical sources have also been used by Soviet workers, notably by Kirikov (1960), to examine changing distributions of fauna. His work is based on thorough use of documentary evidence including medieval chronicles, sources relating to hunting, records of taxes and rents paid in furs and the manuscript surveyors' notebooks compiled for the General Survey.

In the oldest of civilizations, for example China, climate reconstructions using pseudo-documentary sources such as inscriptions on oracle bones have been extended back for 3000 and more years through the Yin Dynasty (1400–1100 BC). The documentary period proper begins as early as the eleventh century BC in China when in the Chou Dynasty (1066–256 BC) official documents were inscribed on bronze and then later written in bamboo books. These Books of Annals can be supplemented by odes and epic poems which describe the seasonal rhythm of life and its tasks and record the dates of flowering of plants, particularly fruits such as plums and litchis. Similar evidence on the flowering of cherries comes from Japan where, from the ninth century onwards, the emperors and feudal lords celebrated the blooming of Japanese cherry trees by feasting in Kyoto gardens and recorded the dates of these occasions. Chu Ko-Chen (1973) quotes the following stanzas from the Chinese Ode of Pin (early Chou period) as examples:

In the eighth moon, they knock down the dates;
In the tenth, they reap the rice;
And make the wine for the spring,
To cherish the longevity of the old.
In the days of the twelfth moon they cut the ice with harmonious blows;
And in the first moon they store it in the ice house, making ready for the second moon,
When offering up a sacrifice, presenting a lamb with scallions;
In the ninth moon it is cold with frost.

From the Yang Dynasty (*c.* AD 1400) such 'soft' sources are supplanted by series of gazetteers on regional geography, probably unrivalled in the world in comprehensiveness. They were published by every province or district and contain, as well as historical and geographical descriptions of local interest, information on striking climatological events such as great floods and droughts, severe cold, heavy snow and other phenomena, particularly those which caused damage to agricultural crops. These gazetteers are some of the earliest sources of direct evidence on climate history in the world. Indirect evidence such as the effects of climate on flora, fauna and agricultural practices can only suggest, but not prove, that climate was different.

The above examples illustrate the way in which the nature of the historical record varies from one part of the Old World to another. To obtain detailed information on a particular country reference should be made in the first instance to guides such as the *International Directory of Archives*, published with the aid of UNESCO by the International Council on Archives in *Archivum* volumes XXII–XXIII (Paris, 1975). This invaluable survey briefly summarizes the contents of some 2515 repositories in 132 countries worldwide. Most importantly, it provides for each archive the bibliographic details of published guides. Each enumeration is of necessity very brief but all the necessary information for follow-up enquiries is provided. A complementary instrument is the *Répertoire des bibliothèques, collections . . .*, edited by G. E. Weil in 1978.

The New World

The countries of the New World, the Americas, Australia, New Zealand and South Africa, are regions of comparatively recent settlement. All were relatively sparsely occupied by technologically primitive peoples until European colonization, and then in the late eighteenth and early nineteenth centuries they became tributary to the rapidly expanding and industrializing North Atlantic economy. Although in detail the subsequent development of land and society in each area was unique, regions of recent settlement do form a distinct group on the world canvas (Wynn, 1977). Indeed, the very recency of European colonization and settlement means that the primitive 'natural' landscape or *urlandschaft* lies much closer to the present. Its appearance was described and mapped in advance of permanent settlement and landscape alteration (Prince, 1971). On the other hand, the landscape alterations of aboriginal peoples should not be underestimated. Whatever the actual size of the pre-Columbian population of South America may have been, for example, it was undoubtedly large enough and had inhabited the land long enough to blur the distinction between the natural and the wholly man-made landscape. Fire, which is one of the most potent instruments of ecological change, was known to even Stone-Age Amerindians. Descriptions of the burning of grasslands and savannahs are frequently encountered in the earliest accounts of travel on this continent (Sternberg, 1968).

In some parts of the New World the physical geographer is offered a unique opportunity to reconstruct the natural vegetation cover in its unchanged state. In the United States, for example, mapping of public lands west of the Ohio river preceded the sale of land to permanent settlers. The plats (maps) and notes (descriptions) of this Federal Land Survey provide an invaluable source for reconstructing the nineteenth-century woodland and grass cover over about 70 per cent of the land area of the coterminous United States (Figure 2.1).

The so-called Jefferson–Williamson plan for rectangular survey was embodied in the United States Land Ordinance of 1785, and perpetuated in the Land Act of 1796 (Pattison, 1956, 1957, 1975). In 1804 the rectilinear grid on which the survey was based was extended and a whole army of land surveyors was launched on careers of recording and exploration in advance of settlement (Thrower, 1966; Johnson, 1975).

The public domain was to be divided up into townships 6 miles by 6 miles divided into 1-square-mile lots or sections. All boundaries were to be identified in the field by marking trees the positions of which were to be recorded on a *plat* at 2 miles to the inch. The organizing committee also recommended that watercourses, 'mountains' and other prominent

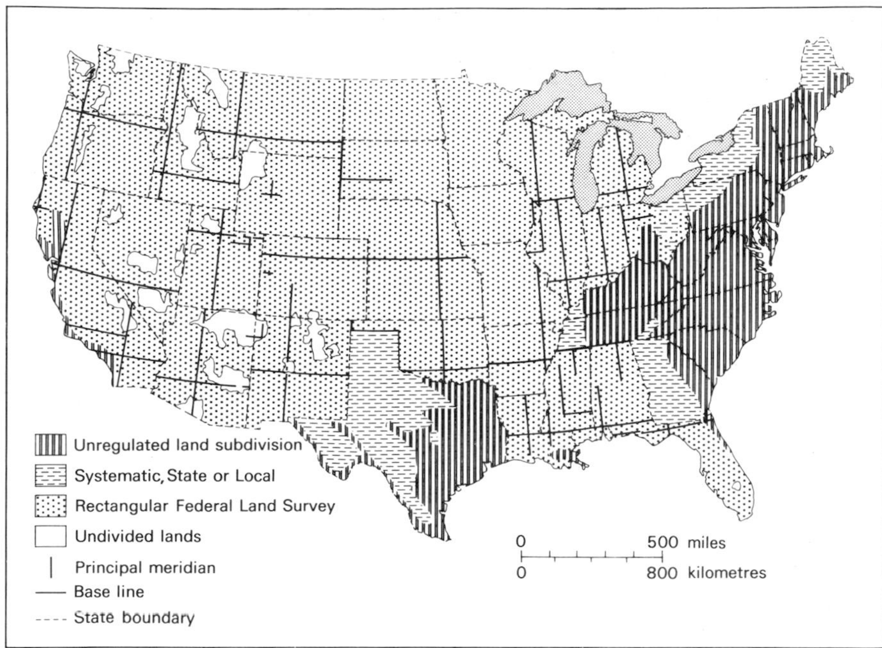

Figure 2.1. LAND SUBDIVISION TYPES IN THE COTERMINOUS UNITED STATES.
(Source: Marschner (1959) and Thrower (1966))

permanent features over or near which boundaries passed should be noted and fixed by compass and chain. The plat was supported by a written description known as the *note*. Surveyors were intended to do much more than just provide a record of the lines of the survey. They were also required to give a picture of the lay of the land, they identified particular forest cover and suggested land-use capabilities. Plats and notes were displayed for inspection at New York land auctions. In their notes the surveyors described, for example, the species and diameters of witness-trees used to mark boundaries, the type of grass in prairie districts and bush on the desert margins (Figure 2.2). The interpretation of these notes in modern studies to reconstruct *urlandschaft* vegetation requires an extensive review of early literature to identify the local common names of plants correctly (Buffington and Herbel, 1965).

The first townships were identified by number and arranged in seven north–south *ranges* with a point of origin at the intersection of the Pennsylvania state line and the Ohio river (Figure 2.3). Survey of these first seven ranges was completed by 1787. The United States acquired the Mississippi Basin from France by the Louisiana Purchase in 1803 and the next fifty years saw further accessions: Florida, the Oregon, the Pacific South-west and Texas. All the time the rectangular survey of the growing public domain went on preceding sale, the only material difference introduced by later legislation being the internal subdivision of the township sections into smaller and smaller lots for sale. Some areas designated as National Parks, Forests and some lands of very low capacity have not been covered by this system of survey. Some studies which use the plats and notes to reconstruct the *urlandscahft* of parts of North America are reviewed in Chapter 4.

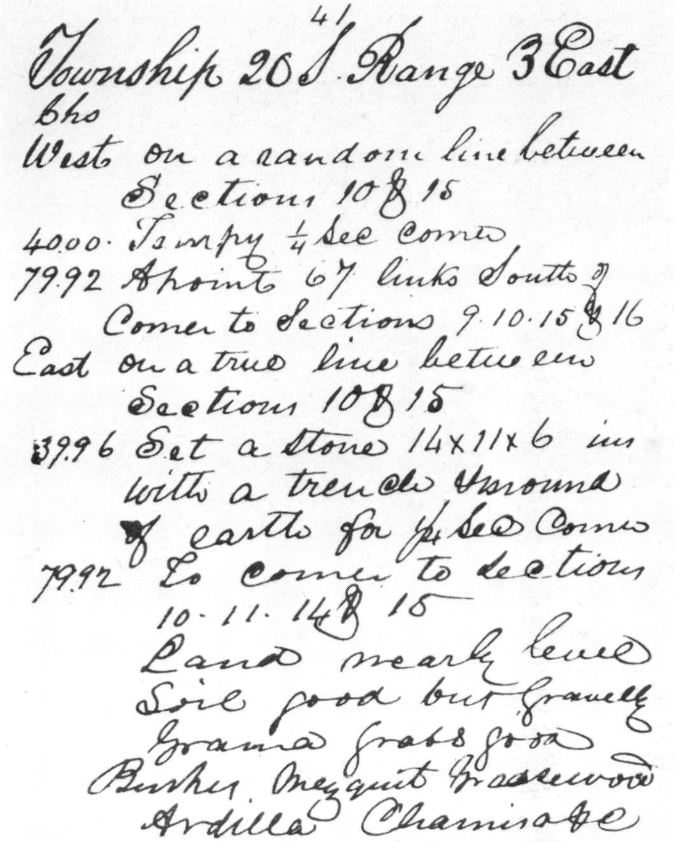

41
Township 20 S. Range 3 East
Chs
West on a random line between
Sections 10 & 15
40.00 Tempy ¼ Sec Corner
79.92 A point 67 links South of
Corner to Sections 9. 10. 15 & 16
East on a true line between
Sections 10 & 15
39.96 Set a stone 14 x 11 x 6 ins
with a trench & mound
of earth for ¼ Sec Corner
79.92 To corner to sections
10. 11. 14 & 15
Land nearly level
Soil good but gravelly
Grama grass good
Bushes Mezquit Greasewood
Ardilla Chamisal

Figure 2.2. A PAGE FROM A FEDERAL LAND SURVEY NOTE.
(Photograph by courtesy of Ecological Society of America)

In other parts of the New World the examination of similar surveys made on the eve of European colonization has scarcely begun. In the Australian states some similar systems of land survey were developed to provide a visible and measurable framework for the expansion of settlement, although the division of land was not everywhere quite as arbitrary as in the United States. Water and road frontages were regarded as essential bases and blocks were usually oriented to these lines with no continuous grid system (Powell, 1970). In comparison with the United States, 'This was no triumph of geometry' (Jeans, 1966). The plans do, though, offer the modern researcher an invaluable guide to official opinions regarding land quality, though some are misleading as a few surveyors were undoubtedly in the pay of local squatters (Jeans, 1966–1967; Powell, 1970). They have been used to construct maps of vegetation at a large scale to complement and refine reconstruction at a small scale of the distribution of pre-European vegetation based on present vegetation remnants, soil indicators, historical descriptions and early general surveys (Jeans, 1978).

On a vast and largely flat terrain like that of the United States or Australia, the

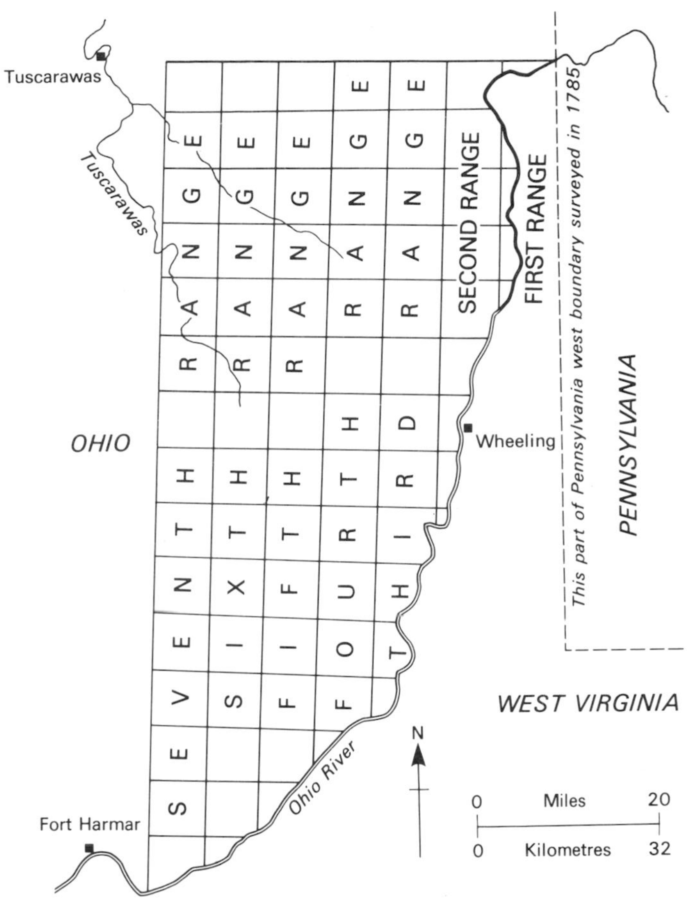

Figure 2.3. THE SEVEN RANGES OF TOWNSHIPS.
(Source: Mathew Carey's plat of the seven ranges, 1796)

rectangular survey worked reasonably well. New Zealand, which in the main is a rugged and partly mountainous country, was rather less suited to the application of a regular grid (de Vries, 1966). However, the tradition died hard, and in the first fifteen years of organized European settlement in New Zealand from 1840, several areas of rectangular grid were superimposed upon the landscape. New Zealand was colonized from a number of isolated regions separated from each other by exceptionally difficult country or by the sea. Each colonization association developed its own pattern of survey and associated settlement. R. K. Dawson, organizer of the English tithe surveys and consultant to the directors of the Wakefield settlements in Australia and New Zealand, advocated a running rather than a triangulation survey. Changes were introduced from 1856 when John Turnbull Thomson was appointed to the post of Chief Surveyor for Otago province. He replaced the Dawson method based on motives of speed and low cost with a slower, more expensive system based on triangulation, which was more suited to the difficult New Zealand terrain. This had been used with success in India for survey before settlement and

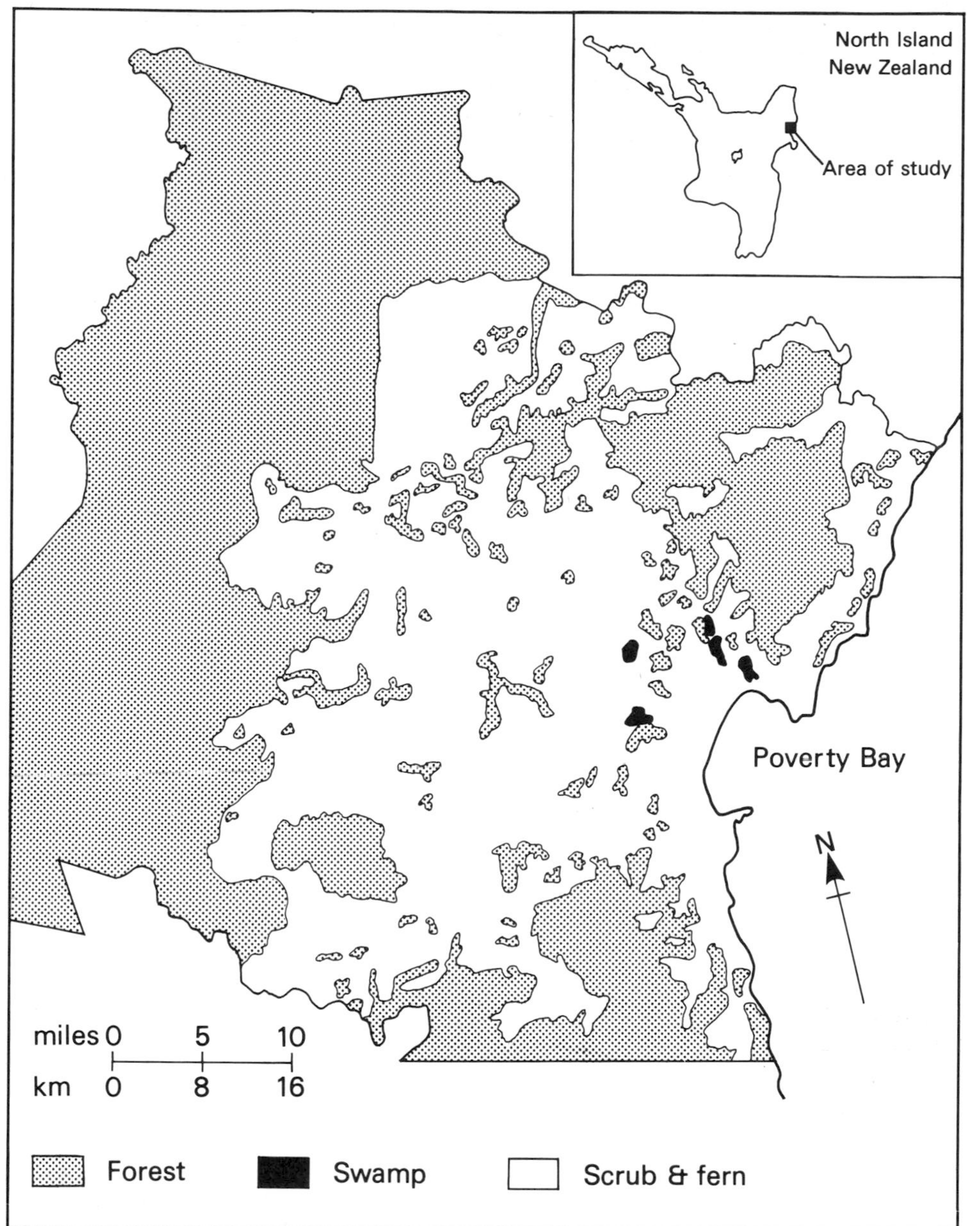

Figure 2.4. THE IMMEDIATE PRE-EUROPEAN VEGETATION ON THE EAST COAST OF THE NORTH ISLAND OF NEW ZEALAND. (Source: Murton (1968))

became the basis for the whole of New Zealand after 1876, when Thomson was appointed Surveyor-General. Rectangularity was not condemned but rather, 'all sections shall, as far as the features of the country will admit, be of rectangular form' (Kelly, 1947, quoted in de Vries, 1966).

The potential for biogeographical–historical studies in New Zealand using one or other of the early land surveys is reviewed briefly by Perry (1969). B. J. Murton, for example, has used 850 Maori block plans kept in the offices of the Lands and Survey Department (Murton, 1968; Powell, 1972) (Figure 2.4). Surveyors were required to map vegetation as well as buildings and boundaries. The amount of vegetation data on each plan varies considerably and they are not uniform in scale, date or detail, but most of them show the boundaries between forest of different types and fern and scrub.

Maps like Figure 2.4 constructed from New World original land surveys have the advantage that they are independent of those based on other sources and they establish formally and on a large scale a datum point for the reconstruction of early vegetation patterns. They also provide a fairly accurate picture of one aspect of the environment encountered by European settlers. Most importantly, they can be used to provide some 'ground truth' in which to set both explorers' accounts and pioneers' and other impressions of the uncharted wilderness sent back from frontiers in the New World. This material is very diverse both in nature and content and discussion will concentrate on just one element by way of example: the reports and commentaries of early explorations of the American West.

Accounts of exploration in the New World

The first of the great government-sponsored expeditions to the American West was authorized by President Jefferson in 1803 to chart a route across the newly acquired territory of the Louisiana Purchase. Lewis and Clark, leaders of this expedition, were the first to make known the vast extent and principal physiographic regions of the north-western quadrant of the United States. In 1805–1806 Lieutenant Zebulon Pike's expedition set out from St Louis, Missouri, up the Mississippi river to search for its headwaters. He did not find them but the map and journal of his route are landmarks in the history of exploration (Friis, 1958, 1965, 1975). Most of the official correspondence, reports, maps, journals, diaries and sketches of these activities and of later explorations in the West like those of Stephen Long and Lieutenant John Frémont are preserved in the United States National Archives in Washington DC. The National Archives' collections are exceptionally well catalogued and it also publishes a periodical: *Prologue, Journal of the National Archives.* Specialist guides include that compiled by Ashby of the records of the Forest Service (Ashby, 1967; see also Pinkett, 1970). The cartographic holdings have also been catalogued. Muntz (1969) has written a general review of the holdings and their cartobibliography in *Prologue.* In addition, there is an excellent review of sources for American historical geography by Ehrenberg (1975). Evidence culled from the National Archives has been extensively used in studies of the formation of images of the settlement potential of the West and particularly of the promulgation of a myth about the existence of what came to be known as the 'Great American Desert' west of the Mississippi (e.g. Lewis, 1966; Watson, 1967). Pike's assertion that this whole area might in time be as celebrated as the African deserts, or Lewis and Clark's general description of much of the land between the Missouri and Columbia as 'desert and barren', or Stephen Long's assertion that the 500–600-mile strip of country east of the Rockies 'bears a manifest resemblance to the deserts of Siberia' (which he had never seen!) should caution against the *uncritical* use of explorers' reports and pioneers' notebooks. Observers first on the scene are not always the most reliable witnesses.

Nevertheless, as long as such cautions are kept in mind it is possible to use both the reports of government explorers and commissioners and the impressions of colonists to reconstruct past climates, soils, vegetation and landforms in New World countries. Some workers, like Moon (1969) in her study of environmental perception and appraisal in South Australia during the early phase of colonization in the 1830s and 1840s, have looked only at the diaries and papers of colonists, considering that those who actually used the land would know more about its precise ecology. On the other hand, others, like Johnston (1963) in his study of past and present grassland in Texas and Mexico, noted much imprecision in laymen's use of terms like *monte, chaparral, mesquite*, prairie and *llano. Llano*, for example, had either vegetational or physiographic meaning, or both. Much more reliable in his opinion are the day-to-day notes of military men and surveyors. But even their observations on vegetation can be highly selective; many failed to comment on shrubs. Travellers ordinarily used horses and were, therefore, dependent on grasses while sheep and cattle on a range will utilize shrubs as well as grass. Most commentators noted the easily recognized gamma grass but other important forage grasses are not mentioned in early accounts. This was probably because their value was not recognized rather than that they were absent or only present in small quantities (Leopold, 1951b). Despite all this potential bias, selectivity and sometimes downright inaccuracy, such reports can provide a useful picture of vegetation in the West and have been used in studies of landscape change, particularly of the development of gullies and alteration of stream channels (Bryan, 1925, 1928b; Burkham, 1972).

Photography was not introduced into the West until after the Civil War and did not come into general use until the 1880s, so that many of even the earliest photographs are too recent to show conditions before heavy use by stock. However, some attempts have been made to examine change with photographs. One of the most thorough photographic 'retake' exercises, that undertaken by Homer L. Shantz, compares pictures of the same scenes in the northern Great Plains states in the 1950s with original photographs, the earliest of which date from 1908 with most from around the First World War (Phillips, 1963).

2.2 NON-DOCUMENTARY SOURCES

Historical sources are frequently used in conjunction with other types of evidence such as that deduced from sediment or biological remains. Such evidence is generally familiar to the earth scientist and provides some of the accepted tools of dating and reconstruction. For this reason the physical techniques themselves are not discussed in detail here since this information can be found in specialist reviews such as those of Cullingford, Davidson and Lewin (1980) and Goudie (1981).

There are three major ways in which physical evidence is used in combination with historical sources. Historical sources can be used to provide primary information and field evidence used in corroboration. For example, a river movement may be ascertained from maps and supported by the morphological and sedimentological evidence of old channels. The second situation is the converse, where historical sources are used to confirm field evidence. Recession of a glacier might be established from moraines and then the movement substantiated by written descriptions of glacier positions. A third area of application is where historical sources are used to establish dates for natural or man-made features which are then used in the interpretation of physical changes. Buildings of known age may show subsidence damage and thus provide a maximum date for slope movement.

Field evidence can provide information on a number of timescales, some much longer than that for which historical documents are extant. Thus they may be used to extend a document-based chronology retrospectively. Sometimes both field and historical evidence overlap in time and one type may corroborate or complement the other, although their relative importance depends partly on the richness of the record and length of the historical period. Thus for some Mediterranean and Middle Eastern areas, such as Greece and Egypt, documentary records extend to several centuries BC, whereas in North America very little exists for periods earlier than the last four centuries.

The main types of non-documentary evidence which are used in conjunction with historical sources are listed in Table 2.1 together with a note of the timescales to which they apply, the uses to which they can be put and their connections with historical sources. Surface structures, lichenometry and caesium- and lead-dating cover comparable or even shorter periods to historical evidence and the first two in particular usually require historical evidence for their effective use. Living botanical evidence such as age of trees may also refer to a few centuries or so but dendrochronology can now be used to at least 7000 BP. Pollen, morphological and sedimentological evidence can be applied to time periods

Table 2.1
NON-DOCUMENTARY SOURCES OF EVIDENCE

Evidence	Useful time-scale (years)	Use	Relationship to documents
Caesium-137 and lead-210 radiometry	10–50	Absolute dates; sources of sediments	As corroboration
Lichenometry	10–500	Date of exposure of surface, e.g. frequency of flooding	Calibrated by documentary dating of objects on which they grow
Man-made structures	10–1000	Positional evidence	As a bench mark for measurements of movement
Tree-rings	10–7000	Date, climatic conditions and events	Provides absolute date where documents unavailable, date and occurrence of events
Archaeological features	50–5000	Mark and date horizons; indicate climatic conditions and human activities	As direct evidence and as corroboration for documents
Morphology	5+	Location and nature of processes; relative dates of features	As corroboration
Sedimentary stratigraphy	5+	Evidence of processes; climatic and erosional conditions; marker horizons	As corroboration
Pollen analysis	100–100 000	Vegetation and climatic conditions, relative dates	To extend land-use documents
Carbon-14	300–40 000	Absolute dates	To extend beyond historical period

varying from a few years to several thousand at least, though to be of most value they require some absolute datum. Carbon-14 provides absolute dates to about 40 000 years BP but because of the margins of error involved is not always sufficiently accurate for very recent times. In the immediate past, historical evidence can be complemented by process

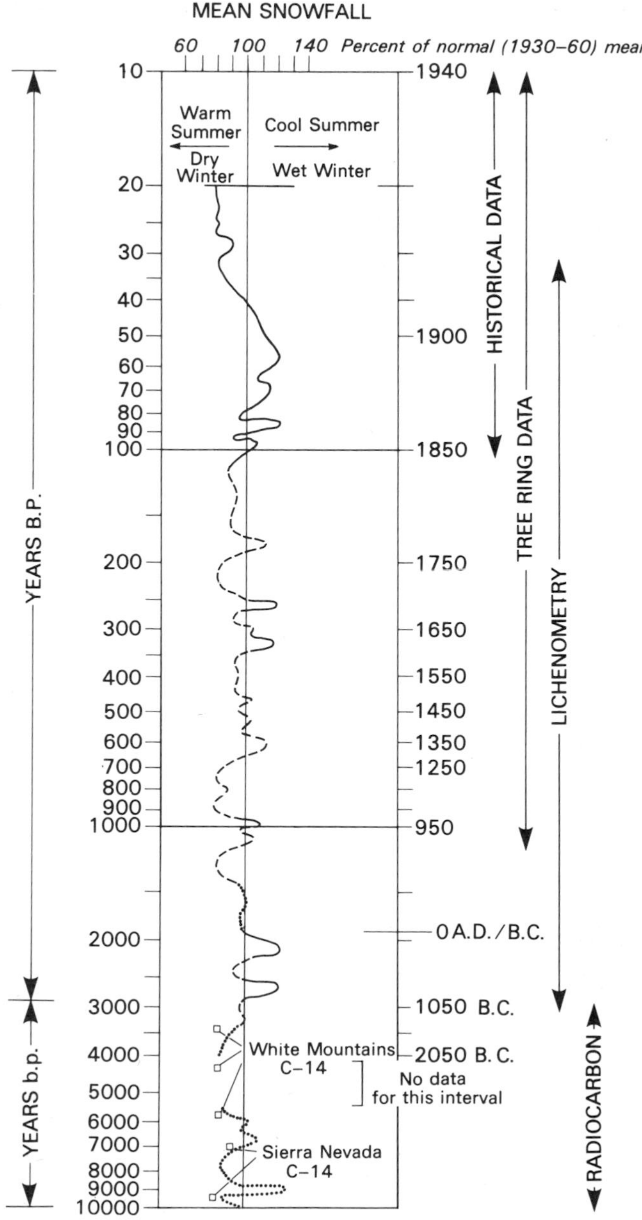

Figure 2.5. VARIATIONS IN SNOWFALL IN THE SIERRA NEVADA OVER THE LAST 10000 YEARS. (Source: Curry (1969))

studies. These may confirm, illuminate or explain apparent historical trends. Measurements of currents in coastal studies, for example, can be used to confirm the movement of bars detected by historical investigations. Equally, if not more important, is the use of historical evidence to extend process studies to provide further evidence of the nature, rates and consistency of processes and change. The value of historical evidence for these purposes is increasingly appreciated by earth scientists.

The various techniques noted in this introductory section are described briefly below under headings of archaeological artefacts, morphology, sedimentary and depositional evidence, dendrochronology, pollen and other biological evidence, lichenometry and radiometric dating. The complementary nature of various types of evidence is indicated in Figure 2.5, which illustrates the data used by Curry (1969) to study snowfall in the Sierra Nevada of the United States. The timescales indicated are specific to his evidence, so the precise dates quoted are not necessarily generally applicable today.

Archaeological artefacts

This heading includes a variety of forms of evidence, ranging from buried ruins of buildings and pottery to existing fences and lighthouses, all of which have been used to demonstrate the position of features and the nature of human activity at a particular time. Dates may be obtained from the internal evidence of particular characteristics, for example, the style of building or pottery, by relation to other datable material, such as sedimentary or biological evidence, or from historical documents. Archaeological evidence has been most widely used in studies of soil erosion, sedimentation and coastal change. The links between archaeological evidence and environmental conditions and techniques of interpretation have been discussed by Butzer (1965) and Vita-Finzi (1978).

The position of man-made structures in relation to present landscape features is commonly used to infer the occurrence of environmental change. In North American arroyo studies the presence of gullies cutting through Indian ruins has been used to confirm that the arroyos were not present at the time of Indian settlement (Duce, 1918; Bryan, 1925). Similarly, abandoned or destroyed settlements and buildings have been used to confirm historical evidence of river movements in India (Login, 1872; Wood, 1924; Wilhelmy, 1966; Gole and Chitale, 1966). On coasts, erosion can sometimes be substantiated by evidence of the destruction of buildings and even whole settlements (Sheppard, 1912; Williams and Fryer, 1953).

Evidence of siltation and sedimentation is much more abundant than that of erosion. It takes the form of buried artefacts which can be excavated or have been subsequently exposed naturally. Akeroyd (1972) in her assessment of archaeological evidence of changes in sea level around Britain warns:

> Mistakes easily arise if it is assumed that an artefact is necessarily contemporaneous with the deposits with which it is associated (unless there is very clear evidence that this must be so), especially when the deposits have formed in peat or marshland. Artefacts may have fallen into creeks cut below the contemporary land surface, they may have been intentionally or unintentionally buried in artificial or natural hollows; they may have been moved from their original positions, for example by downward movement through the peat bed, by lateral erosion along coasts or water channels, or by the reworking of deposits.

The presence of sediments above datable structures such as Indian remains in North America or Roman remains in Europe has frequently been used to infer the period, amount and rate of sedimentation and thus of erosion on the slopes above. Such buried structures may also indicate past environmental conditions in an area. For instance, remains of dams, irrigation and drainage works have been used as evidence for past land use and agricultural practices, low rates of soil-erosion or greater water flow at earlier times, as Thornthwaite *et al.* (1942), Jacobsen and Adams (1958) and Vita-Finzi (1969) have demonstrated. The type of buried structure may indicate particular activities. Trimble (1970) used a fish trap now buried under 5 metres of sediment to indicate the level and gradient of a stream in Georgia early this century. On coasts and estuaries there are examples of former ports and marine activity at locations that are now some distance from navigable water (Sheppard 1912; Vita-Finzi, 1969). Gottschalk (1945) has cited physical evidence such as wharves, posts and a boat exposed in mud deposits to confirm the sedimentation which has taken place in Chesapeake Bay since European settlement began. Historical sources complement such physical evidence and confirm the decline of port activities. Gottschalk encountered references to problems caused by siltation throughout the historical record. Kidson (1950) and Brookfield (1952) have both used the evidence of bridge-remains to confirm the former position of water where there is now dry land. On coasts, lighthouses (Redman, 1852; de Boer, 1964; Shepard and Wanless, 1971), lifeboat houses (Athearn and Ronne, 1963; Hardy, 1966), Martello towers (Steers, 1951; Lovegrove, 1953; Carr, 1969) and sea walls (Martin, 1876; Homan, 1938) have been used to verify changes in coastal position. By analagous reasoning, Field (1932) dated glacier limits in Alaska by their relationship to a railway line and Duce (1918) has dated some gullies in the south-west United States by the way that they cut across the lines of old trails. The fact that buildings can provide a datum against which to measure sediment-accumulation or removal is illustrated in Figure 2.6, which shows blown sand on the Norfolk coast at Eccles in relation to a church tower.

Smaller artefacts and objects such as pottery and tools are commonly used to date horizons or indicate the nature of past human activities, as in Van Zuidam's (1975) study of the Zaragoza region of Spain and Leopold and Snyder's (1951) and Miller and Wendorf's (1958) work in New Mexico, in which they used pottery to date river terraces. Geomorphologists now find that modern objects such as tin cans (Beaumont and Oberlander, 1973), wire fences (Alexander and Prior, 1971) and car-licence plates (Costa, 1975) can help to date the sediments in which they are buried.

The relationship between certain datable structures and the features which they cross or intersect may enable the date of movements or changes to be confirmed. For example, the presence of ridge-and-furrow has been used by Chandler (1970a, 1971) to assess slope stability (Table 2.2) and by Roberts *et al.* (1973) and Rackham (1975) in analyses of land use. Likewise the presence of a sea wall crossing beach ridges in Sussex confirms that they pre-date the wall (Lovegrove, 1953). In the United States, Alexander and Nunnally (1972) used Indian mounds to confirm stability of the Ohio river, while Ambraseys (1979) substantiated his suggestions of the seismic stability of part of Iran by the presence of very old buildings. Buildings have also been used to

Figure 2.6. TOWER OF THE BURIED CHURCH AT ECCLES, NORFOLK IN 1839 (TOP) AND 1862 (BOTTOM) SHOWING REMOVAL OF BLOWN SAND OVER THE PERIOD.
(Source: Lyell (1867 edition))

date terraces by providing an earliest date for incision, as Womack and Schumm (1977) showed in their work on terrace sequences in Douglas Creek, Colorado. The destruction of parts of buildings (Hutchinson, 1965b; Hutchinson and Gostelow, 1976) and walls (Skempton, 1964) can show the effects of slope instability and, if the structures are datable, allows a maximum age to be obtained. The effects of gradual processes on datable structures may enable estimates of their rates of operation to be made, as with weathering on buildings (Emery, 1960) or siltation behind a barrier (Bird, 1979).

'Psuedo-documentary' evidence can also be noted in this section; markings on objects have been used to indicate flood levels (Jones, 1975) and to mark the position

Table 2.2
CHRONOLOGY OF MASS MOVEMENTS ON THE ESCARPMENT AT ROCKINGHAM, NORTHAMPTONSHIRE (after Chandler, 1971)

Phase of landsliding	Date	Event
1	Late-glacial	Active degradation of upper slope; formation of contour ridges; movement of lower slope apron
2	Iron age	Deforestation of escarpment by burning; hill-wash resulting in accumulation of charcoal soil; degradation of upper slope; movement of lower slope apron
	Medieval	Development of ridge-and-furrow systems
3	Post-1615	Local minor degradation of upper slope; minor movements of lower slope apron on pre-existing shear surfaces

of glacier snouts (Grove, 1966). Piest *et al.* (1977) used the records of a farmer who inscribed dates and channel depths on a cellar door in Missouri to monitor stream entrenchment. Plaques and inscriptions may also provide useful evidence; Cottam and Stewart (1940) discovered a plaque commemorating a battle in a 'meadow' in south-west Utah through which an arroyo now runs, and Sheppard (1912) described a stone set in a wall of an east-Yorkshire village on which the distance of the building from the sea at the time of construction is stated. More tenuous is the use by Martin (1893) of inn names like 'Ship' and 'Anchor' to confirm the presence of a harbour at Starcross, Devon, indicated by documentary and morphological evidence.

Morphology

Morphological evidence is one of the physical geographer's main tools and has been extensively used for identifying and interpreting landscape features over a wide timescale from a few to many thousand years. Morphological evidence is frequently used to confirm inferred changes. For example, in studies of fluvial morphology the presence and position of old river channels in the field has been matched with map and aerial photograph evidence (Eardley, 1938; Sundborg, 1956; Lewin and Hughes, 1976; Hooke, 1977a) as a method of checking the accuracy of historical documents. A succession of datable channels can confirm the direction and characteristics of movement and Klimek and Trafas (1972) have dated ox-bows from their relative state of siltation. Although the possibility of reworking can make such inferences difficult, generally freshness of form can be considered at least as a crude indicator of age, as Price (1980) has noted in relation to moraines.

In river studies the size of old channels can help to corroborate a change in flow characteristics (Login, 1872), while the level of an old channel may indicate whether there has been incision since formation. Similarly beach ridges are indicative of coastal changes. On spits and bars these ridges commonly represent successive positions of a shoreline and may be dated by cartographic evidence or used to support historical information (Redman, 1852; Lewis, 1932; Lovegrove, 1953).

The morphology of slopes on which mass movements or other events have occurred can help confirm the type and nature of movement. Such evidence has been widely used in landslide studies where the position, size and tilting of slipped blocks described in historical sources have been corroborated by field evidence (e.g. Arber, 1941; Pitts, 1973). Historical sources can occasionally provide a means for dating such forms and subsequent modification can be studied (Hutchinson, 1965b). Brunsden and Kesel (1973) have dated the cessation of active undercutting of Mississippi bluffs from maps and from this were able to demonstrate the sequence of degradation over time.

Sedimentary and depositional evidence

The characteristics of sediments associated with particular processes can be used to identify the probable environments of sediment deposition, particularly topographic and climatic conditions. Interpretation of sedimentary sequences is usually founded on the principle of superimposition, but, as Thornes and Brunsden (1977) warned, there are dangers in inferring relative age and chronology from height alone, so care must be taken with, for example, the interpretation of cut-and-fill sequences. Ideally a stratigraphy should be confirmed both relatively from pollen, biological or artefactual evidence, and be absolutely dated by radiometric techniques, dendrochronology or historical documents. In general, the characteristics of deposits may indicate the supply of sediment and properties of the transporting agent, the environment of deposition, the relative distance of transport, the weathering conditions and the direction of movement. The stratigraphic relations of different units can indicate the position in time and space of discontinuities and the sequence of events and phases.

The presence of datable horizons enables the depth of deposition in a particular time period and rate of sedimentation to be calculated (Butzer, 1965; Vita-Finzi, 1969). Akeroyd (1972) warned, however, that included material such as artefacts may not necessarily represent the horizon in which it is found, though it can be used to provide a minimum date. Layers of volcanic ash (tephra) and other distinctive and widespread deposits are commonly used as synchronous marker horizons in comparisons of sediment-sequences from various sites. They have been used in the historical period (Wolman and Leopold, 1957) and the tephra layers can sometimes be dated by historical records of eruption. For more distant periods, buried soils and organic layers can form important reference horizons while erosion can also be distinguished by truncated soil profiles (Bennett, 1931) and exposed structures, roots and underlying materials. In studies of the soil-erosion and sedimentation which occurred in nineteenth- and early twentieth-century North America, the evidence of location, depth and nature of deposits has been widely used in conjunction with historical sources. For example, Nelson (1966) used sedimentary evidence to assess the effects of human activities on rates of deposition in the Chemung Valley, New York State and Pennsylvania.

Flood deposits in river valleys have been dated from historical evidence (Anderson and Calver, 1977) and deposits have been used to confirm the occurrence of floods and variations in discharge (e.g. Shvets and Zaika, 1976). Bluck (1971) has used map evidence to date meander-deposits and to interpret sedimentary features in Scotland, while sedimentation patterns and amounts have been used by Fisk *et al.* (1954) to confirm their dating of the development of the Mississippi delta.

On coasts, discontinuities in deposits can indicate former shorelines or limits to processes. Barnes and King (1957) and Kestner (1962) have related sedimentary evidence of the phases of accretion of salt marshes at Gilbraltar Point, Lincolnshire, and in the East Anglian Wash respectively to map evidence of the shore positions through time. Kaye (1976) has shown how the presence of organic deposits in combination with historical records can be used to draw the limits of Boston Harbour at the time of European settlement. The presence of deposits around structures can confirm siltation of harbours and estuaries. Such evidence has been used by Gottschalk (1945), Kidson (1950) and Brookfield (1952). Similarly, the effects of structures on beach morphology and sediment transport have been confirmed by depositional evidence (e.g. Shepard and Wanless, 1971; Rawls, 1972; Bird, 1979). Coastal deposits have been used to reconstruct and interpret former sea levels, though such evidence is rarely applicable in the historical time period.

Morainic deposits, like morphological evidence, can indicate past positions of a glacier and may be datable with documentary evidence. Likewise mass movement deposits on slopes are usually readily distinguishable and can suggest the type and nature of movement and may supplement historical information on the properties of the material involved and the conditions of the movement. Such methods have been mostly applied to landslides, mudflows and other rapid movements, but Imeson *et al.* (1980) have used deposits in combination with ^{14}C, pollen and historical evidence to date colluvium in Luxembourg and to interpret the conditions of its formation.

Dendrochronology

The width of tree-rings varies with growth conditions, so their measurement can provide a useful surrogate for climatic records. The sequence of variations in growth patterns and, therefore, of environmental conditions has been established to about 7000 years BP by the use of a number of tree-remains which overlap in time. Dates of other tree-trunks can now be established by reference to these standard patterns of variations and used to date the sediments in which they are found. Dendrochronology can be a major tool in establishing the sequence of climatic changes as Fritts *et al.* (1979) have shown in their study of climate in North America since 1602. Fritz (1976) provides a comprehensive review of the relationships between tree-rings and climatic conditions. Tree-ring evidence can be compared with historical sources in more recent times (e.g. Arakawa, 1957) or be itself corroborated by historical sources (Schove, 1961; Lamb, 1965). The conditions which variations in tree-rings indicate do require very careful interpretation. As Matthews (1951) noted, it may be water or temperature which is the limiting factor to growth, depending on the particular environment.

Asymmetrical growth or the presence of damage to tree-rings may indicate the occurrence of slope movements and dramatic events such as landslips, avalanches and floods. These last can be dated from the position of scars in the tree-rings. Alestalo (1971) has reviewed the use of dendrochronology in interpretation of geomorphological processes and Potter (1969) and Carrara (1979) have used such evidence in their analyses of snow avalanches. Moore and Matthews (1978) have used tree-damage to confirm the date of a landslide in British Columbia which was also reported in historical documents.

The age of living trees can be calculated by counting tree-rings, and this can be used to date deposits and features on which the trees are growing. It only provides a minimum age, and allowance must be made for some lag in colonization since formation of the feature (Lawrence, 1950). Tree-ages have been used to date moraines and glacier recessions (Table 2.3) and river bars, flood deposits and gullies have all been dated by the age of the trees growing on them (Bailey, 1935; Dietz, 1952; Helley and LaMarche, 1973; Womack and Schumm, 1977). The dates of landslide and mass movement deposits have been similarly determined (Skempton and Petley, 1967; Skempton and Hutchinson, 1969) as have colluvial and soil deposits (Trimble, 1970), and lake levels (Lawrence and Lawrence, 1961; Thomas, 1962).

Table 2.3

LOCATION OF THE NISQUALLY GLACIER FRONT IN RELATION TO THE PRESENT NISQUALLY BRIDGE, MOUNT RAINIER, WASHINGTON STATE (after Harrison, 1956)

Year	Distance from bridge (in metres)	Source of data
1750	−500	Estimated age of trees from incomplete ring-count
1825	−107	Ring-count of trees in area not covered by 1855 ice
1855	−107	Ring-count inside trimline; A. V. Kautz description
1870	− 46	Descriptive report by S. F. Emmons
1883	− 46	Moraine 3 metres high with three closely spaced loops
1884	− 30	A. C. Mason photograph
1885	0	James Longmire report
1893	+350	E. T. Allen photograph
1903	+335	O. A. Piper photograph
1905	+290	Joseph N. LeConte measurement
1907	+259	Single moraine 1 metre high
1908	+267	Single moraine less than 1 metre high
1910	+317	F. E. Matthes measurement
1915	+380	Asahel Curtis photograph

Pollen and other biological evidence

The type of pollen in a deposit can be used to infer vegetation and, therefore, climatic conditions at the time of inclusion of the pollen. Indeed, much of the post-glacial sequence of climatic changes has been based on pollen evidence. However, the amount and type of pollen present in a particular sample is influenced by many factors, including local site conditions and the pollen-producing characteristics of vegetation. It is only by detailed analysis of many samples and comparison of different layers that a reliable stratigraphy can be worked out. Pollens are resistant to degradation and may be subject to much reworking and transport. With such cautions in mind, pollen evidence can be and has been used in a number of ways to interpret past environmental conditions. It is used directly to indicate the vegetation assemblage or land use of an area at a particular time, to establish a sequence of changes or to indicate climatic conditions at a particular time. By comparison with known sequences, pollen evidence can be used to correlate sediments, or, if dated absolutely, it can define conditions at a fairly precise time.

Palynological evidence has been widely used for the prehistoric period but much less so in the historic period. However, Oldfield (1969) has demonstrated the value of using pollen in conjunction with historical evidence. Documentary sources have been used by Oldfield (1970) and Roberts *et al.* (1973), Tinsley (1976) and Walker and Taylor (1976) to explain variations in pollen type and frequency through sequences of sediment by providing information on land use and agricultural practices.

Various other flora and fauna are used as indicators of conditions and for relative dating, especially in the prehistoric period. These include fauna such as molluscs, gastropods and coleoptera. Most of these are of limited value for establishing historical chronologies, though they may confirm conditions where there have been distinct changes. Bones and other animal remains have also been used in the prehistoric period. Remains of plants and their seeds can provide direct evidence of past vegetation assemblages.

One of the more unusual ways in which flora have been preserved is in the adobe bricks of Indian buildings in North America. The plant remains in these bricks have been used by Hendry (1931), Burcham (1956) and Clark (1956) to confirm vegetation reconstructions. Living vegetation in the form of species remnants can also assist deductions about past vegetation or land-use conditions, as the work of Sears (1925), Peterken (1969) and Rackham (1975) demonstrates. Colonization sequences may also confirm relative ages of exposed surfaces such as river bars, coastal deposits and moraines (e.g. Everitt, 1968). Thornes and Brunsden (1977) have warned of the dangers of circular arguments in using biological evidence such as pollen to date sediments. Suppose, for example, that a sample is deduced as glacial in age because of its characteristic 'cold' fauna and flora. It might then be quite wrong to infer that a sample from another site is of glacial origin simply because its stratigraphic position suggests an age the same as the first sample.

Lichenometry

This technique depends on relating the age of lichens to their size and thus dating the exposure of the substrate on which they are growing. Useful reviews of this technique and its application have been made by Mottershead (1980) and Worsley (1981a). Locke *et al.* (1979) have produced a technical manual on lichenometry, though not all practitioners agree with the advice it contains (Innes, 1981). One of the main problems is to establish the correct age–size relationship of lichens. The main role of historical documents in relation to lichenometry is to date surfaces from which the calibration curve can be defined. Gregory (1976) has used gravestones as datable surfaces and datable buildings and walls can also be used. Denton and Karlén (1973) and Karlén (1973) have used mining deposits, structures associated with a railroad, glacial drift and a building to establish the calibration curve by dating the surfaces from historical sources (Figures 2.7 and 2.8). Lichen sizes were then measured and, by comparison with the calibration curve, a date was obtained. The

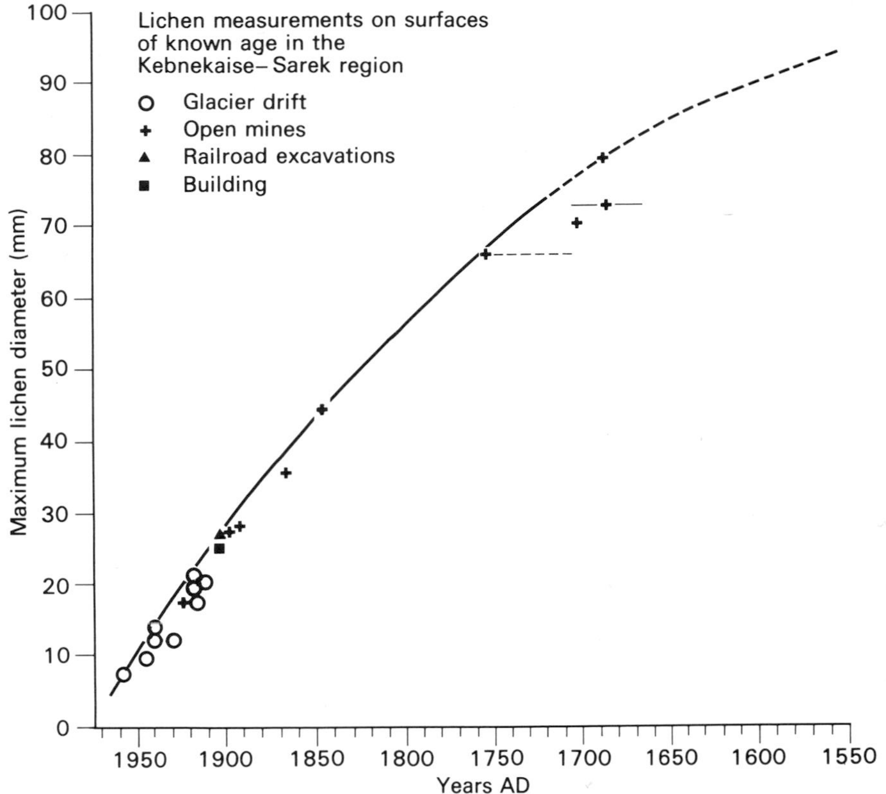

Figure 2.7. LICHEN GROWTH-CURVE BASED ON MAXIMUM THALLI DIAMETERS OF *RHIZOCARPON* SPECIES FOUND ON SURFACES OF KNOWN AGE IN THE KEBNEKAISE MOUNTAINS, SWEDEN.
(Source: Karlén (1973))

technique has been widely used to date moraines and also river-flow levels (Gregory, 1976). Mottershead (1980) has also noted its potential in coastal studies. Lawrence (1950) has used the absence of lichens to corroborate the recent date of glacier recession which he derived by tree-ring counting.

The reliability of dating based on lichen measurements can be checked against further historical evidence. Some of the problems of using lichens concern decisions about the species to use, the number to count, the sampling framework to employ, what to do about coalescent and decayed examples and how to measure the surface areas. Notwithstanding these difficulties, the technique appears to have potential for periods up to 500 or more years.

Radiometric dating

Radiometric dating techniques are based on known rates of decay of radioactive isotopes and measurement of remaining amounts in samples of material. Various isotopes are suitable for different time periods depending on their half-life. These techniques provide

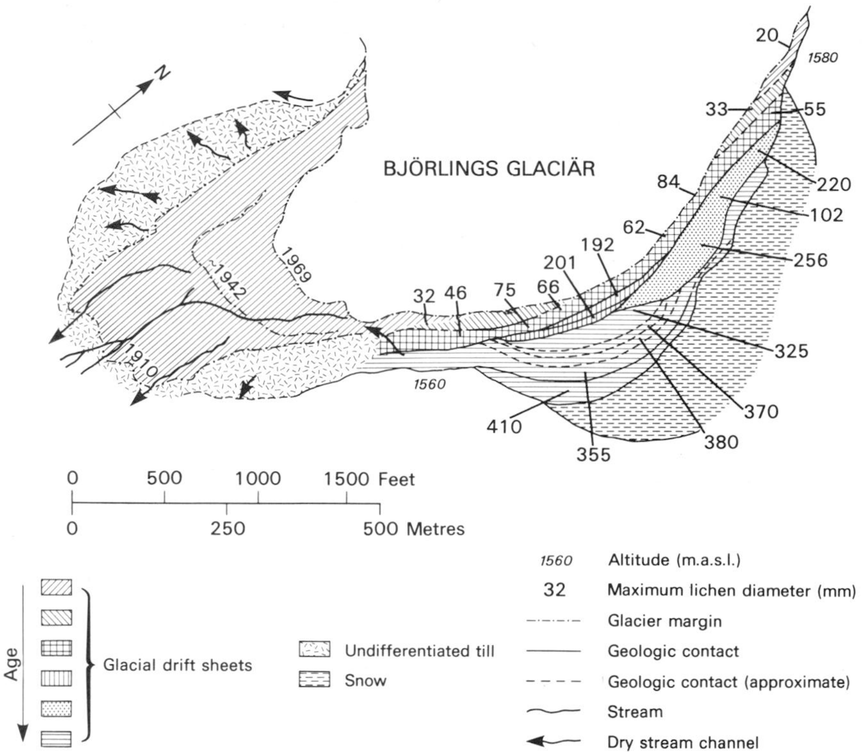

Figure 2.8. SUCCESSIVE HOLOCENE DRIFT SHEETS AT THE MARGINS OF THE BJÖRLINGS GLACIÄR SWEDEN, DETERMINED BY LICHEN DIAMETER.
(Source: Karlén (1973))

absolute dating (albeit with some margin of error) and have proved extremely important in fixing environmental chronologies. The technique particularly appropriate to the historic timescale and at present most widely developed is ^{14}C (Andrews and Miller, 1980; Worsley, 1981b). It requires organic material for dating and covers the period to about 40 000 BP. Its half-life is 5730±40 years, and this margin of error means that it is often not sufficiently accurate for the last few centuries. For the recent past, Ladurie (1972) considered that ^{14}C may best be used to establish a broad time period and then historical sources can pinpoint precise dates.

There are a number of problems in using this technique, notably contamination of the sample and variations in the atmospheric concentration of ^{14}C. The dates of many deposits or features established by ^{14}C dating have been corroborated by documentary evidence: for example, Alexander and Nunnally (1972) and Knox (1972) have corroborated dates and amounts of alluviation in this way. Fisk *et al.* (1954) have used ^{14}C and historical sources to trace the growth of the Mississippi delta, and Grove (1966) and Bray (1971) have used ^{14}C to corroborate dates of glacier variations. Absolute dating techniques such as this also allow variations in different areas to be correlated, and thus historical evidence in one area may be correlated with ^{14}C and other evidence in another area. In some cases ^{14}C has been used to date sediments and has shown them to be of an age very different from that suspected, and often, particularly with fluvial, coastal and slope deposits, to be much younger than originally thought. For example, Legget and Lasalle (1978) cited the case of a peat layer in the St Lawrence valley overlain by clay where the peat was thought to be of interglacial age. It was found from ^{14}C dating, however, to be only about 400 years old. The clay layer was formed as a result of a landslide and the occurrence of an earthquake and slope movements in the year 1663 is confirmed in historical descriptions.

Other radiometric dating methods are now being developed which cover much shorter periods than ^{14}C. Probably the most important of these are caesium-137 and lead-210, the use of which is reviewed by Wise (1980). They have half-lives of thirty and twenty-two years respectively. Caesium-137 is produced as a result of fusion from nuclear explosions, which first occurred in 1954. It is washed out from the atmosphere by rainfall and into the soil where it is absorbed in surface layers by clay and organic colloids. Eventually, as a result of erosion and transport, particles containing caesium-137 may be deposited in lakes. Its especial value is in indicating rates of soil erosion since 1954. Less is known about the behaviour of lead-210 but, used in conjunction with caesium-137, it can provide information on source areas of sediment and sedimentation rates. It might be possible to relate these to documented changes in land use. Church (1980) has provided a table of other radioactive isotopes with short half-lives which provide information on the 10–100-year timescale.

Also useful for distinguishing the history of sediments, though not radiometric techniques, are measurements of the magnetic properties of mineral iron oxides present in sediments (Oldfield *et al.,* 1979; Oldfield, 1981). Materials which have undergone different geomorphic and anthropogenic processes exhibit different magnetic properties, and thus sediment from different sources can be distinguished.

3
Accuracy and Analysis

The evaluation of accuracy is an essential stage in the use of historical sources, yet scrutiny of the ways in which sources have been employed reveals that some physical geographers have neglected or evaded this question despite their use of increasingly detailed and sophisticated quantitative analyses. This may be partly due to unfamiliarity with methods of testing and also with the type and scale of potential error.

The level of accuracy which is acceptable and the appropriate means of its assessment depend on the particular questions posed, on the source itself and also on the availability of other evidence. In this chapter, the stages at which error can arise are outlined and some common types of error are identified. Methods of evaluating accuracy by measurement are discussed with respect to maps and techniques of corroboration by comparison with external evidence are detailed.

Error may arise at any or all of three stages in the compilation and storage of documents:

(1) During *collection* of information; examples might be error in original survey measurements or in the recording of observations. These all depend upon the compilers of the documents and will be affected by their attitudes, perceptions and methods.
(2) During the *transfer* of information both in producing the original and in copying and editing.
(3) During subsequent *use and storage* of documents.

Some of the more common types of error are discussed in the section below.

3.1 CAUSES OF ERROR

Data collection

Errors can occur through lack of understanding or awareness of *date conventions.* This applies particularly to early historical periods and can lead to misdating or duplication of dates. In Europe use of the Julian calendar prior to 1752 and subsequently the Gregorian calendar can cause confusion, as Pfister (1979) has indicated. Monasteries in medieval times often had their own style of dating (Bell and Ogilvie, 1978) and in the early days of settlement in the western United States time was not standardized, and this has caused difficulty in analysis of events like earthquakes (Townley and Allen, 1939). This problem of dating is particularly acute in the study of climatic phenomena (Brooks and Glasspoole,

1928; Ingram *et al.,* 1978). Some details of dating conventions are given by Potter (1978) and Lamb (1977).

The *frequency* and *timing* of observations may cause error in the assessment of event frequencies or calculations of rates of change. An apparently increasing incidence of events may often be due to increased frequency of reporting as, say, population and the extent of settlement increases and communications, such as the number of newspapers, improve. Chu (1973) has identified these problems in relation to the reporting of floods and droughts in China, and Lamb (1970) encountered similar difficulties when studying world volcanic eruptions. Bias in reporting hurricanes has been identified by Prentiss (1952) and Galway (1977). Woolley (1946) has neatly summarized these sorts of problems:

> Any discussion of historical cloud-burst floods must necessarily be based on man's experience with such floods and the available records of that experience. This involves such factors as dates of settlement, population growth and its encroachment upon nature's flood plains and watershed areas, the efficiency of recording mediums, the most common of which is the newspaper, and the ability of the recorder or reporter to make factual records rather than verbose, sensational and highly colored ones.

Bias may also occur as a result of the time or date of observation in relation to inherent changes in phenomena. For example, both Leopold (1951b) and Thornthwaite *et al.* (1942) cited errors in the interpretation of past vegetation conditions because historical descriptions on which the studies were based were written at one particular time without regard to seasonal and yearly variations in vegetation characteristics.

Awareness of the *perception and attitudes* of observers is also critically important. Attitudes can be closely related to interest or motivation; Trimble (1970) reported that some nineteenth-century surveyors in the United States were under a $10 000 bond to ensure accuracy. Bryan (1928b) and Sundborg (1956) have both discussed the backgrounds and likely bias of surveyors, and Sundborg (1956) concluded that highly valued land was probably surveyed more accurately than that of lesser value. Humphrey (1958) demonstrated that 'different individuals can pass through the same country and receive in part at least, different impressions'. These will depend to some extent on where they originated, where they travelled and on their previous experiences: this is exemplified in Brooks' (1926) study of climate and Leopold's (1951b) study of vegetation. Chavannes (1941) considered that immigrants' descriptions of nineteenth-century North American vegetation were the best source for its study, and Leopold (1951b) suggested that observations made on the first trip to an area were most satisfactory. Leopold discovered bias in the description of the types of vegetation, with more attention paid to grass because of its importance as grazing for horses. Similarly, Randhawa (1952) demonstrated bias in vegetation descriptions due to the careful noting of trees of particular religious significance.

The filter of previous experience pertains particularly to records of events, as Louderback (1947) has shown in relation to earthquakes. Such events can appear much more dramatic and intense to those who have not experienced them before. Another difficulty which may be encountered in the use of sources relating to areas of new settlement and development is that the image presented in reports may be deliberately biased to make an area appear favourable. Malin (1946) considered that this may have been a factor contributing to the general lack of reports of dust storms in mid-nineteenth-century Kansas.

The *style and conventions* of writing and drawing can influence the communication of content. The style of language has changed over time, as Lowe (1870) clearly realized: 'Formerly the descriptions given were very different from those we are now accustomed to . . . Errors will creep in between "old" and "new" styles as, for some years after the change, there is frequent ambiguity as to which has been used.' Brooks (1926) noted how literary records range from terse statements to flowery verse, and both he and Ingram *et al.* (1978) have indicated how language is closely related to perception and attitudes. Reid (1914), in reviewing evidence of the 1755 Lisbon earthquake, also demonstrated how the nature of description has changed:

> The imagination was drawn upon very much more freely in 1755 than at the present time, to give vividness to narratives, and many remarkable phenomena were described which could not have occurred; moreover reports were not examined very critically, and sufficient care was not taken to avoid referring to the Lisbon earthquake events which took place at other times and which had no connection with it. As a consequence, many false conceptions of the effects of this earthquake have become general.

Not only written records, but maps, pictures and related sources are very much subject to fashions of style and convention. For example, the conventional representation of features on maps may cause problems. An example is hachuring on early Ordnance Survey maps, which caused Steers (1926) and Brunsden and Jones (1976) difficulty when interpreting steep slopes (Figure 3.1).

Figure 3.1. HACHURING OF STEEP SLOPES ON THE ORDNANCE SURVEY OLD SERIES 1-INCH MAPS CAN MAKE THE INTERPRETATION OF SLOPES DIFFICULT.
This extract from sheet XVII (published in 1811) portrays landslips on the Dorset coast east of Charmouth

Written sources may be prone to exaggeration. Bryan (1923) reckoned that a seventeenth-century topographic description of a valley near Tucson was 'coloured by missionary zeal' and Brooks and Glasspoole (1928) remarked on the 'sensational reports of early chroniclers'. Sheppard (1912) encountered some overestimation by contemporary reporters of Yorkshire coastal movements when compared with other more reliable evidence. He quoted the following passage from an unnamed 'popular' magazine dated 7 April 1906:

> On the coast of Yorkshire there are two noses of hard rock that the sea can eat but slowly. They are Flamborough Head and Spurn Point [!], and between them lie 33 miles of coast, which the North Sea is swallowing at the rate of 3 yards in every 12 months. At Withernsea, just to the north of Spurn Point, houses go over the cliff almost daily [!].

Such exaggeration is not confined to modern times; Vita-Finzi (1969) has suggested that the highly lyrical style of some classical writings might distort the truth.

Very closely related to style and also to perception and attitudes are the *semantics of descriptions.* Lewis (1962) has analysed the changing emphases in description of the American Great Plains, and Malin (1953) also discussed the meaning of the term 'desert'. Likewise Brookfield (1952) has questioned the interpretation of 'sea' in certain documents. Many authors stress the importance of understanding the classifications which were used and the precise meaning ascribed to words. For example, in relation to climatic analyses Ingram *et al.* (1978) stated that the significance of categories such as 'frost', 'drought', 'flood' and 'sea-ice' must be established. Similarly, the consistency and importance attached to terms of degree such as 'large', 'severe' and 'greatest in living memory' must be analysed. The use of technical terms, particularly in vegetation surveys, can pose problems, as Buffington and Herbel (1965), Mason (1963) and others found when analysing descriptions of nineteenth-century vegetation in North America. Their work involved identification of plants to which certain common names had been applied in explorers' and travellers' notes. York and Dick-Peddie (1969) have listed the common names which are found in historical accounts with their present names, thus demonstrating the confusion which can arise (Table 3.1). Watts (1960), Johnston (1963) and others have also complained of the lack of conformity and precision of some terms.

Table 3.1
PLANT NAMES USED BY EARLY SURVEYORS IN NEW MEXICO
(after York and Dick-Peddie, 1969)

Surveyors' plant names	Present common name	Scientific name
Cedar	Juniper	Juniperus sp.
Chamisal	Chamisa	Atriplex canescens
Edeondilla	Creosotebush	Larrea tridentata (DC.) Coville
Gatuno	Catclaw	Acacia greggii Gray
Greasewood	Creosotebush	Larrea tridentata (DC.) Coville
Hediondilla	Creosotebush	Larrea tridentata (DC.) Coville
Hediondo	Creosotebush	Larrea tridentata (DC.) Coville
Largoncillo	Whitehorn	Acacia constrica Benth
Mimbres	Desert willow	Chilopsis linearis (Cav.) Sweet
Palmilla	Soaptree yucca	Yucca elata Engelm.
Palmira	Soaptree yucca	Yucca elata Engelm.
Ponil	Apache plume	Falugia paradoxa (D. Don) Endl.

Lastly, a major factor influencing accuracy at the observation and measurement stage concerns the *methods and instruments* used. This is of prime concern where detailed measurements are involved, as in the production of maps, plans and sections, and the measurement of meteorological phenomena. Available instruments set limits to the precision with which measurements could be made. The way the instruments were used could also affect accuracy. In land survey, the technical limitations of a particular model of theodolite may be known, but the accuracy of a survey will also depend on whether, for example, triangulation was used. Likewise in measuring temperatures, the characteristics of early thermometers may be known but the placement of the thermometer and timing and frequency of readings can also affect the values obtained.

Compilation and transfer

Several types of error could have been entered when documents were compiled from the original observations. Error can be incorporated in maps during construction from original field-survey measurements due to misreading of figures, misplacing of points or the incorrect calculation of scale conversions. There may also be errors of omission. The types of drawing instruments used also pose a limitation on achievable accuracy. Any copying of information may also incorporate error as Monmonier (1980) has shown in the printing of maps. Bell and Ogilvie (1978) have pointed out that numbers are more frequently miscopied than words, which has important implications for quantitative observations. Craddock and Craddock (1977) recognized this problem and Peterssen (1914) produced evidence of incorrect copying of climate data. Other errors can be produced when original sources are paraphrased or edited:

> . . . many editions, especially those produced before the mid-nineteenth century, but also some more recent ones, include serious errors or distortions caused by misreading of the original text; the use of imperfect copies, the unwarranted conflation of texts, and unsignalled editorial interventions such as the rearrangement or abridgement of the original. (Ingram and Underhill, 1979)

Error may also occur in language translation, as Mikesell (1969) has shown in his analysis of deforestation in the Lebanon in which he casts doubt on whether the ancient Egyptian term 'ash' should be translated as cedar or fir or more generally as conifer.

Treatment and storage

Lastly, the accuracy of historical documents can be affected by their treatment and mode of storage. This is a special problem with maps, since shrinkage and stretching of paper or other materials is a very common source of error and affects planimetric accuracy. Sundborg (1956), Carr (1962), Kishimoto (1968) and Bird (1974) have all recognized this form of error, and Carr (1962) also noted that modern forms of copying such as photostat or xerox may distort maps. Factors affecting the likelihood of distortion at this stage are type of material, age, method of storage, and conservation or preservation treatments. Braund (1980) has analysed the effects of temperature and humidity on a variety of drawing materials.

3.2 INTERNAL CHECKS OF ACCURACY

There are two main methods by which accuracy can be assessed: by checking certain characteristics, such as the date, of the source itself and/or by comparing the basic information with other evidence. In this section internal checks of accuracy are reviewed while external corroboration is treated under a separate heading below.

Table 3.2 is a checklist of the characteristics of historical sources which should be considered when undertaking internal assessments of accuracy. These are explained in the following paragraphs.

Table 3.2
INTERNAL CHECKS OF ACCURACY

Characteristics	Checks
(1) Originality	Whether 'original'—check author, date, source
(2) Contemporaneity	Closeness in time of document to reality portrayed
(3) Propinquity	Closeness in space of document to situation it describes
(4) Generalization	Level of detail and style
(5) Transmission of information	Reliability of secondary reporting
(6) Purpose	Reason for compiling the record
(7) Observer	Nature of compiler of the record
(8) Intellectual and social milieux	Social, intellectual and technical attitudes and knowledge
(9) Methods and instruments	Type and design of instruments, methods of use
(10) State of document	Whether damaged, distorted or treated

(1) The *originality* of a document must be established, that is, whether it is the original document or whether it is perhaps a contemporary copy or one produced some time later. If the document is not the original, then the source may be subject to the kinds of errors of copying, editing and misrepresentation indicated earlier in this chapter. Many secondary works fail to acknowledge or to date precisely the original sources from which they are compiled. Published sources such as maps and books may be reprinted or revised, and care must be taken, as Carr (1962) has stressed, to ascertain the dates relating to the various categories of map content.

Particular types of error can arise from editing: Ingram *et al.* (1978) have identified four main categories.

(a) Inaccurate or uncertain dating of particular events.
(b) Spurious multiplication of events through failure to recognize that accounts recorded under different dates in various sources in fact relate to a single event.
(c) Acceptance of accounts which are distortions or amplifications of original observations.
(d) Inclusion of events for which there is no reliable evidence whatsoever. If edited material has to be used at all, then Ingram and Underhill (1979) have stressed the importance of checking the credentials of the editor.

(2) The *contemporaneity* of a document is an important characteristic to check. This is the date of the document in relation to the events or conditions with which it is concerned. A description of an area or a painting of a scene made perhaps several years after a visit are much more likely to incorporate error and distortion than one produced at the time. Leopold (1951b) commented on this in relation to explorers' and travellers' notes from nineteenth-century North America, and Messerli *et al.* (1978) have demonstrated its significance in relation to drawings and engravings.

Establishing contemporaneity involves verifying whether a document is the original and investigating the background to its compilation. Often a document will carry a date, though, as pointed out above, with editions and secondary sources this must be verified. Once dated it is then a matter of judgement as to whether the amount of time which elapsed between the event or situation and the completion of the document is sufficiently short to render the source probably reliable. Historians employ various techniques to date documents; for example the use of watermarks, palaeography, calligraphy and the analysis of style. Some details of these methods and further reading are given in Emmison (1966) and Stephens (1973).

(3) Similar to contemporaneity is *propinquity*, or closeness in space to the subject-matter. If a description is written or a drawing is made from the memory of a scene at which the author is no longer present, then the contents may well be distorted. In some cases an author may not have actually been present at the scene but has written or drawn from hearsay; such sources should normally be considered of low *prima facie* reliability, though obviously other characteristics must be checked. Where the precise location to which a document refers is not stated it may be possible to infer the general area, but detailed locations are often required, particularly for analysis of changes in features. Verification of propinquity again involves investigation of the background of a document and the circumstances of its compilation.

(4) *Generalization* is a characteristic of a source which greatly influences the accuracy of information which can be obtained from it. Concern with detail can vary even within a document. Steers (1926) remarked that a map of the Orford Ness area is 'unreliable as far as the upper course of the Ore is concerned, but the coast appears to be carefully drawn'. The level of generalization may depend on the purpose for which the document was constructed, the style employed and, with maps, on the scale at which they are drawn. However, even at the same scale or at a particular level of detail the degree of generalization may vary. On a map, lines may be carefully drawn in or merely sketched with a blunt instrument. Here generalization is closely related to style and convention.

(5) Ingram *et al.* (1978) have identified another major principle on which assessment of reliability should be based and term it the *faithful transmission of information*. This mainly concerns situations where observations of others are being reported, as in the use of eye-witness reports of an event. 'Faithful transmission' depends very much on the circumstances of time and place and on the perception and attitudes of the author. Evidence based on hearsay or information acquired indirectly is not likely to be wholly reliable. As Grove (1966) noted in relation to tales and traditions, they 'cannot be accepted at face value, especially as in the course of time they lost nothing in the telling . . . but they are symptomatic'.

(6) It is already evident from discussion of points 1–5 above that the *purpose* for which a

record was compiled exerts a profound influence on what is recorded and the nature of the recording. The more rigorous the requirements of the purpose of the document and the closer that purpose is to the research questions being asked of it, then the more reliable will content tend to be. Observations incidental to the main theme of a document may not have as high a degree of reliability as those it directly concerns. Harley (1968) discussed the importance of elucidating purpose in evaluating maps and said that 'we should not expect more of maps than was intended'. He indicated that it is important 'to uncover reasons for biases in detail', but added that there may be difficulties in identifying purpose because 'motives may be camouflaged and painstaking research needed to reveal unsuspected purposes'. Purposes also include consideration of the nature of the record's intended audience. This applies particularly to written, descriptive sources and may affect style and level of detail.

(7) Closely related to purpose is the *identity* of authors and investigation of their motives and interest in the subject-matter. The character of observers is particularly relevant where subjective judgements of, for example, the magnitude of an event are involved. Chavannes (1941) has spoken of the 'competent sober men' who compiled some of the early land surveys in the United States, and some early meteorological observations were made according to Wahl (1968) by 'educated and dedicated people, some with instructions'. Similarly Craddock and Wales-Smith (1977) remarked that 'early observers were mostly scientists, doctors, parsons and country gentlemen who took their scientific activities seriously'.

(8) Other background information is also helpful for the assessment of accuracy. Ingram *et al.* (1978) have indicated how knowledge of *intellectual and social milieux* can help in understanding terms and style and thus 'personal and cultural bias'. Some internal checks of consistency with which terms and symbols are applied within a source can be made but external comparison will probably also be necessary. The technique of content analysis outlined below is also relevant to this matter. Some awareness of the general state of technical and scientific knowledge at the time a record was compiled may also provide useful background for assessment of reliability.

(9) Knowledge of the kinds of *methods and instruments* both available for use and actually used is a vital key to the assessment of accuracy. Technical manuals and texts such as Richeson's (1966) *English Land Measurement to 1800* can indicate which instruments were available and their proper methods of use, but it is also of critical importance to identify the instruments and methods that were actually used to construct a particular document. This aspect of accuracy is of most concern in areas where detailed instrumental measurements were made, namely land survey and climatic observation. On maps, the presence of survey lines, pencil marks or even pinpricks may provide clues as to the method of draughting and survey. This is an advantage of using the originals of documents, since such markings are usually not present on copies. Documents accompanying original measurements or maps can provide very useful information. Surveyor's notebooks can prove invaluable for testing accuracy. Likewise any accompanying details of how meteorological measurements were made can prove immensely helpful, as Folland and Wales-Smith (1977) found in interpreting Towneley's early measurements of rainfall. They used descriptions and measurements of the instruments made in a letter from Towneley to the Royal Society to determine the size, configuration and siting of the instruments.

(10) The effects of *subsequent treatment* on a document can be assessed by investigating its history. This will involve finding where and how it has been stored, whether it has been treated to preserve it and what effects these may have produced. Many of these questions can be answered directly by its archivist.

3.3 EXTERNAL CORROBORATION

The main principle of external corroboration is that if there is agreement on a particular point between two or more independent sources then this probably verifies the information and establishes its reliability. The originality of apparently independent documents must, however, be carefully checked. If they are derived from an earlier common source and incorporate the same errors they will be of little corroborative value. As an alternative to agreement between sources, evidence may be compared for contradiction and consistency. Le Roy Ladurie (1972), for example, argued that an increasing number of cold winters should be accompanied by a corresponding decrease in evidence of mild winters. If mild winters increased also then this implies simply a greater overall frequency of evidence and not of the particular conditions.

Corroboration may depend not only on coincidence of evidence in time but also in space. Spatial comparison may be made between two regions or sites, but this is fraught with difficulty because of natural spatial variability. For example, climatic change can only be corroborated by comparison with another area if that area could reasonably be expected to show the same pattern of variation.

The particular method of comparison and interpretation which is appropriate will depend on the type of data. Sometimes vague statements have to be compared or data may be available only for widely separated times and locations. Much corroboration is inevitably subjective, involving weighing up the relative merits of sources and the likelihood of occurrence of the situation inferred. Historians consider this as part of their acquired 'skill' in using historical sources for which it is difficult to lay down definite guidelines. If data are in, or can be converted to, a quantitative form then more rigorous comparisons can be made.

Corroboration, then, is based on comparison of various types of evidence: each of these is now discussed in turn.

(1) *Comparison of like historical sources* such as two maps, two pictures, or two descriptions requires that they be as nearly contemporaneous as possible and certainly close enough in time so that no significant changes could have taken place. But rarely is such an abundance of sources available and, when extant, are often not sufficiently comparable in level of detail, scale and style. The precise method of comparison used will vary from source to source, but will probably include some visual inspection for similarity of contents. Corroboration may only be possible at a lower level of content than actually contained in the documents because of generalization. For example, it may only be possible to corroborate the presence of a feature rather than its exact size and shape. Nossin (1965), in his study of the geomorphic history of the Pahang Delta in Malaysia, said of the evidence about the morphology of Tingoram Bay that 'the reliability of the maps is certainly questionable, but since all the maps examined for the period 1640–1825 give this

information, and because the area discussed is known to have been frequently navigated in ancient times, it can be concluded that this bay . . . did indeed exist'. He added, however, that 'its actual size and shape are depicted somewhat differently on earlier maps'. Sedgwick (1914) compared weather diaries covering the last quarter of the seventeenth century in Britain and Chu (1973) and Kemp (1976) have used climatic data from other areas to check the reliability of their sources and confirm their findings about climatic conditions in China and Dunfermline, Scotland, respectively.

Sources compiled separately for adjacent areas such as parishes or counties can be tested against each other especially if common boundaries can be compared directly. Watts (1960), compared descriptions of vegetation in neighbouring townships of southern Canada and Brandon (1971b), remarked of medieval English Compotus Rolls that 'their testimony can be regarded as trustworthy when several manors are in agreement'. On the other hand, Carr (1962) and Bell and Ogilvie (1978) have warned that if two supposedly independent sources are too much alike then they may have a common origin.

An example of the difficulty of comparing maps and reconciling apparently conflicting evidence is illustrated in the following quotation from Steers' (1926) study of Orford Ness, Suffolk.

> Old maps of the district are somewhat contradictory. The earliest one the writer has examined is of the time of Henry VIII. It bears no date, but is probably about 1530. . . . On this map the entrance to Orford Haven is certainly shown near to Stonyditch Point. On the other hand, however, this ditch is not shown. Havergate Island is not detached from the mainland. The next map is a chart of the time of Elizabeth, and again no particular date is given. Here the North Weir Point is shown to be about midway between Boyton Hall and Cauldwell Hall. . . . A. Appleton's map (1588), which is very similar, and may have been copied from the former, shows the same general outline. Both of these maps are unreliable as far as the upper course of the Ore is concerned, but the coast appears to be carefully drawn . . . in 1601 Norden surveyed the area for Sir Michael Stanhope. . . . This survey is the most reliable old map of the district extant, and is in keeping with Norden's high reputation as a cartographer. The North Weir Point is here shown rather more than half a mile south of the present Dove Point—in fact, in practically the same position as on Appleton's and the other Elizabethan map. Saxton's map of Suffolk, 1575, is again quite different, and would seem to represent the coast as it appeared at a much earlier date. The Stone Eye is shown as a separate stream, and in this and other respects is not in agreement with the map of Henry VIII's time, though both of them show the mouth of the Haven in much the same position relative to the town of Orford.

(2) A frequently employed form of corroboration is comparison of *evidence from different types of historical source.* This may, for example, involve corroboration of map evidence by written sources or *vice versa*, and/or comparison between written documents of different types. Messerli *et al.* (1978) compared drawings and engravings with ground photographs taken from approximately the same spot to substantiate their reconstruction of glacier positions, and Coleman (1969) has compared maps of the Brahmaputra river with aerial photographs. Carr (1969) and many others have used written evidence to corroborate maps. Both de Boer (1964) and Brookfield (1952) in their studies of Spurn Point, Humberside, and the river Adur estuary, Sussex, have effectively illustrated the importance of literary criticism—of sifting evidence carefully—and show how sources may be misinterpreted.

Comparison of several sources of evidence may enable the use of what appears to be fragmentary or even 'valueless' information. Steers and Jensen (1953) in their study of the east coast of England quoted from a 1616 Report to the Commissioners for Sea Breaches, from a history of Norfolk and from a parish survey and concluded: 'These three notes, slight though they may be, are consistent in suggesting that the cliffs were being washed by the sea, at least by storm tides, up to the beginning of the eighteenth century.'

(3) Many of the more rigorous tests of accuracy, especially of maps, depend on *comparison with a modern equivalent.* Obviously this cannot be based on a comparison of features which may have changed in the intervening period, and it depends, therefore, on the identification of relatively stable features which can act as reference points. These tests also assume that the modern equivalent is of sufficient or known accuracy to act as a standard. However, this is not always the case, as Kondrat'yev and Popov (1967) in a study of Russian rivers have indicated: 'If one compares the old pilotage charts for the beginning of this century with the present day charts one sees that the older ones are superior with respect to the completeness of representation of morphometric characteristics.' Stearns (1949) has corroborated the accuracy of mid-nineteenth-century land surveys by finding close agreement between them and present-day maps and air photographs with respect to stable features such as lakes. Thorarinsson (1943) indicated how knowledge of present situations may help in interpretation of the accuracy of a particular source. Discussing travellers' narratives, he said,

> such information naturally varies in value and should be critically examined, which in the first place requires a thorough knowledge of present day topographical and glaciological conditions. In the light of such knowledge apparently meaningless information may often prove of great value.

(4) *Field survey* can provide important corroborative information in physical geography. These techniques have been reviewed briefly in Chapter 2.

(5) The results of *present-day process studies* can also be used to verify historical information. These will often indicate whether the type and magnitude of changes shown by the historical documents were at all likely or even possible. This method has been used in coastal studies by Barnes and King (1957), Hardy (1966) and Robinson (1966) to confirm directions and amounts of movement of sediment in relation to changes in spits and bars and in fluvial research by Kondrat'yev and Popov (1967), amongst others, to confirm when, where and how erosion and channel changes took place.

(6) *Circumstantial evidence and proxy data* are loose groupings of indirect evidence which have been used to corroborate historical conditions. Much of the information is derived from activities which are affected by the phenomena under study: for example, records of farming activities have been widely used to corroborate evidence of climate at a particular time. The most widely used data are those concerning dates of harvesting and ripening crops, dates of flowering of trees, freezing of lakes and other such phenological data (Arakawa, 1957). These data must be subject to the same critical analysis as other historical information.

3.4 TESTS OF ACCURACY

For those historical sources such as maps which are amenable to quantitative analysis it is possible to make rigorous comparisons and to measure some aspects of 'accuracy' precisely. As indicated at the beginning of this chapter, errors can arise in a number of ways and at various stages; tests of accuracy do not necessarily distinguish the sources of error though they may show if they are systematic or not. Most tests are based on comparison of the source to be tested with a like source of known or assumed accuracy; for example, a more recent map edition (Murphy, 1978). Methods used firstly for maps and secondly for climatic data are outlined below.

(1) *Maps.* Blakemore and Harley (1980) distinguish four elements of map accuracy, i.e. 'chronometric', geodetic, planimetric and topographical accuracy. Here we consider only tests of planimetric accuracy. In order to compare measurements, maps must be brought to a common scale and stable reference points, which can be closely defined, must be used for the measurements. Some errors may arise from the projection, scale and graticule used and

Table 3.3

RESULTS OF ACCURACY MEASUREMENTS ON SOME DEVON TITHE MAPS

(A) *Percentage error in scale bar length*

Scale (in chains)	Copy	Map	Error	Scale (in chains)	Copy	Map	Error
6	DRO	Axminster	−1.00	4	DRO	Silverton	−1.24
		Broad Clyst	0.00			Stoke Canon	−2.56
		Colyton	+0.97		PRO	Newton St Cyres	−0.44
		Gittisham	−3.05	3	DRO	Colaton Raleigh	+0.50
		Kilmington	−0.72			Harpford	−1.60
		Rewe	−1.70			Honiton	−0.55
		Salcombe Regis	−2.48			Luppitt	+0.75
		Shute	+0.34			Monkton	−0.48
		Stockland	−0.44			Northleigh	+1.60
	PRO	Kilmington	−0.06		1st Class	Ashprington	−1.36
		Rewe	−0.32		PRO	Brampford Speke	−0.87
		Shute	0.00				

(B) *Group means and means of standard deviations of percentage errors in areas*

Scale (in chains)	No. of maps	Test 1a $\lvert\bar{X}\rvert$	Test 1a $\lvert\bar{s}\rvert$	Test 1a $\bar{X}$	Test 1a $\bar{s}$	Test 1b $\lvert\bar{X}\rvert$	Test 1b $\lvert\bar{s}\rvert$	Test 1b $\bar{X}$	Test 1b $\bar{s}$	Test 1c $\lvert\bar{X}\rvert$	Test 1c $\lvert\bar{s}\rvert$	Test 1c $\bar{X}$	Test 1c $\bar{s}$
6	6	2.49	2.57	−1.78	3.08	2.73	2.87	−1.33	3.63	3.55	3.52	−2.43	4.15
4	3	2.26	2.22	+0.50	2.88	3.16	1.97	−0.90	3.88	3.88	2.33	−1.38	4.15
3	5	3.74	2.80	−0.84	4.54	3.06	3.27	−0.52	4.64	4.69	3.29	−1.26	5.26
All	14	2.99	2.62	−0.94	3.88	2.95	2.80	−1.07	4.14	3.94	3.05	−1.90	4.70

(C) *Group means and means of standard deviations of percentage error in length*

Scale of group (in chains)	No. of maps	$\lvert\bar{X}\rvert$	$\lvert\bar{s}\rvert$	$\bar{X}$	$\bar{s}$
6	9	2.22	2.12	−0.20	2.87
4	3	4.06	3.33	−3.87	3.54
3	7	3.01	1.86	−1.31	3.42

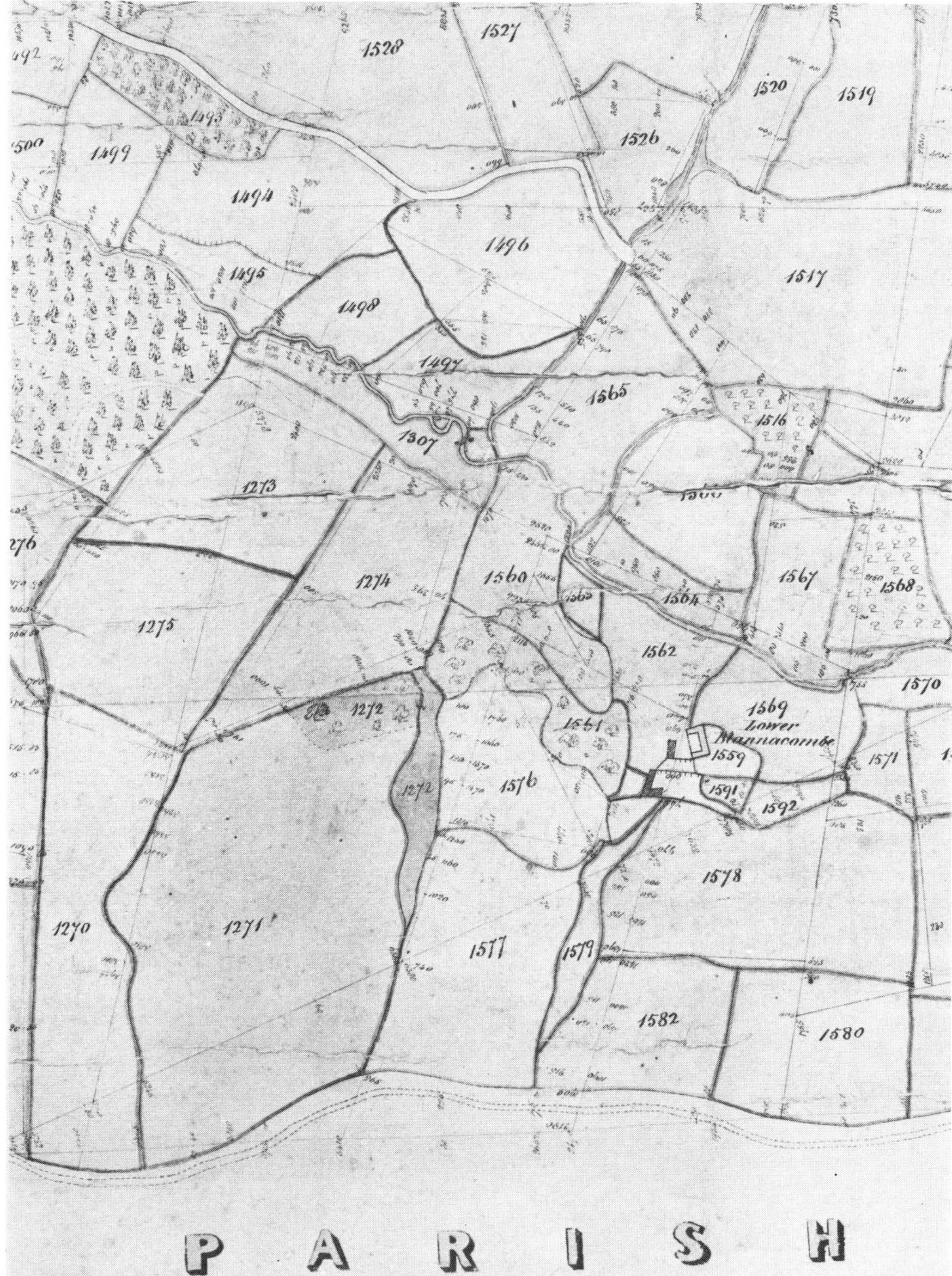

Figure 3.2. PART OF THE TITHE MAP OF HONITON PARISH, DEVON.
Showing construction lines and chainage distances. (Source: Devon Record Office)

these have been tested by plotting vectors of differences in projection (Stone, 1972; Ravenhill and Gilg, 1974). Kishimoto (1968), in an important discussion of cartometry, has identified four types of measurements necessary for testing maps. These are point locations, distances, areas and counts of items. Measurements should be made in random directions and on different parts of maps to reveal any systematic error such as that due to stretching of the material in one direction.

Hooke and Perry (1976) have tested the planimetric accuracy of English tithe maps by comparing them with 1:2500 Ordnance Survey County Series maps. Comparisons were made between areas listed in the tithe apportionment and those marked on the Ordnance Survey map; between the planimetred tithe area and the same area on the Ordnance Survey map; and between the planimetered tithe map area and the acreage listed in the tithe apportionment. (Tests 1a, b and c respectively in Table 3.3). Straight-line distances on the two maps were also compared and, for all of these measurements, the tithe survey errors were calculated as percentages of the same measurements from Ordnance Survey maps. In general, errors averaging 3–4 per cent were found, though they varied between maps and even within a map (Table 3.3).

It is difficult from these tests to isolate the sources of error, whether in survey, plotting or due to subsequent distortion. Measurements of the scale bars on tithe maps and discrepancies from the stated lengths provide some indication of distortion, though these will also incorporate any cartographic error in the original plotting. On certain tithe maps, survey lines of a standard 30 chains are marked and the lengths of these can be measured to test for distortion. A majority of the Devon maps tested exhibited shrinkage but only of the order of about 1 per cent.

If triangulation or other survey lines are present (Figure 3.2) the likely errors arising from the survey control network can be calculated. Additional information available in survey notebooks can provide a further check and may help to distinguish between errors made at the survey stage and those made in plotting or occurring subsequently.

The methods and instruments used for testing can influence the results obtained and operator error must also be taken into account. Kishimoto (1968) and Maling (1968) have discussed some of these problems and the accuracy of different instruments and methods. Accuracy values have been expressed in a number of different ways and Murphy (1978) has shown that results vary according to the analytical methods used. The propriety of using some statistical parameters has been questioned by Blakemore and Harley (1980) who review a number of theoretical and logistical problems of answering the question 'How accurate is that map?' and argue the case for a scientifically defined model of map error.

Few users of historical maps for physical geography appear to have made direct measurements to assess accuracy of maps, though amongst the exceptions are Carr (1969) in relation to coastal changes, Sundborg (1956) and Mosley (1975) in river studies, and Rackham (1975) in his analysis of vegetation change in Cambridgeshire.

(2) *Climatic data.* Accuracy limits can be established for almost any quantitative data derived from historical sources providing that suitable reference data to act as a standard are available. Such data, which must be either comparable in time and place or refer to stable features, tend to be rare, but in climatological work early instrumental records have been checked against each other and referred to modern standards. Because these early

instruments varied in design it is often necessary to determine their units of measurement: early rainfall measures are often expressed in units of the container used. Similarly with thermometers, it is necessary to check that the zeros and scales were correct. The effects of siting and exposure of the instruments must also be calculated. Some information on the instruments may be obtained from extant writings and correspondence of the observers (Folland and Wales-Smith, 1977) but other techniques for assessing accuracy and determining whether there was any change over time can be applied. These techniques are associated with standardization and adjustment procedures and Craddock (1977) has discussed a suggested procedure for rainfall values:

> the ratio of the total rainfall amounts caught at two sites in the same year must have an expected value, which is approximately the ratio of the climatic values at these sites. . . . If the fluctuations are not at random about the ratio of climatic values but tend always to be positive or negative, or to change with time, then either one of the climatic values has been wrongly estimated, or one or both of the rain-gauges has been wrongly installed, or the quality of one or both of the records is changing with time.

Manley (1961) tested and adjusted the values of W. Derham's early eighteenth-century record of temperatures in the following way:

> it was first necessary to deduce the probable characteristics of Derham's exposure by comparing, for each of the 12 months, the mean rise and fall of his shade temperature between the hours of observation, with the averages published for each hour in the same months at Kew and Greenwich observatories. Second, the probable value of Derham's degree can be reasonably checked against the average and absolute extremes of temperature recorded. Third, the over-all mean temperature of the six winter months can be checked, approximately, against the frequency of snowfall during the period. Fourth, there is a very tenuous overlap of a single year (1706) with Labrijn's tables for Holland. Scrutiny of Derham's record leads to the view that his zero did not significantly change.

Lamb (1963) recounted how he estimated the errors of early pressure maps by comparing maps reconstructed with only the network of barometric stations existing at the time of the early maps with those constructed using the full modern network of stations.

3.5 LEVELS AND LIMITATIONS OF ACCURACY

Having attempted to assess the accuracy of a particular source, a decision has to be made about whether it is sufficiently accurate and reliable for use. This decision will depend on the purpose of the inquiry and possibly on the availability of other data. If a particular document is the only one available and its information is vital in elucidating a situation or sequence of events, then it may become a choice between using this source of doubtful reliability or obtaining no data at all. On the other hand, if an unreliable source is used, it may produce spurious results. This was certainly the conclusion of Brice (1974), who stated in his paper on the analysis of meander development that 'most old maps have sufficiently great planimetric errors so that they are not used for this purpose'. Edgar and Melhorn (1974) rejected five early maps as inadequate for examining changes in channel networks. Brookfield (1952) considered maps of the Adur estuary, Sussex, 'far too inaccurate for any firm conclusions to be drawn from them'. Dolan and Bosserman (1972) and Karlén (1973) have also acknowledged the limitations of scale and generalization on maps and have discussed these matters in their studies.

On the other hand, in a number of studies the inaccuracies and inadequacies of sources have been acknowledged but the general situations that the documents indicate have been accepted. The following short extracts from various fields of inquiry illustrate this point. In a study of changes of the river Mississipi Stevens *et al.* (1975) said '1821 maps are not exact but the map errors are believed to be less than the changes in areas', and in relation to the Brahmaputra river Coleman (1969) said 'although the accuracy of the older maps is probably not as reliable as the later ones, the general trends indicated by the comparisons are probably correct'. Peterssen (1914) stated in his discussion of climatic change that 'although most of the evidence is qualitative and that which is quantitative is suspect, a clear impression of the severity of those winters is gained'.

3.6 ANALYSIS AND INTERPRETATION

In this section some general techniques for treating historical sources are discussed. These include methods of *transcribing* data from historical sources and data *processing*. Some guidance to appropriate techniques of data *analysis* and the *interpretation* of results is also provided. More specific applications are reviewed in subsequent chapters. The main types of analysis of historical data in physical geography range from the *reconstruction* of a situation at a particular time and the *dating* of particular features to analyses of *causes* of change. Analyses may be made in the context of a single case study or of a more comprehensive and perhaps comparative survey of extensive areas. At the outset it is perhaps worth noting Rackham's (1975) warning of the dangers of making general inferences from one site:

> . . . it is seldom possible to produce a satisfactory account of just one site. Many historical materials are sporadic in their occurrence. The historical ecologist, concerned with the way in which present-day vegetation is influenced by the circumstances of its origin and by subsequent changes, needs to study a wider range of variation than a single site is likely to provide. It is best to make a comparative study of all available sites within a small area which is as uniform as possible in climate and soil.

He added that use of historical sources involves 'more or less circumstantial interpretations which can be carried out with confidence only within a comparative study . . .'. Where the phenomenon under study covers only a small area an attempt should always be made to collect as much information as possible so that evidence from one source can be corroborated with that from another. Judgement on whether to use all available sources will obviously have to be made within the practical constraints of time and money and in relation to the likely value of the data. More usually, though, earth scientists will find their problems stemming not from an embarrassment of documentary riches but rather from a fragmentary historical record. Furthermore, this fragmentary nature often conditions which techniques of analysis are appropriate.

Data transcription

The precise information to be abstracted from historical sources will depend on the phenomenon being studied and the nature of the source, but basic information on dates and locations is likely to be required in all studies. In relation to geomorphological features, information on size, shape, extent, pattern and distribution, surface and topographic characteristics, composition and associations between features may be required. Analysis of climatological data usually involves chronological ordering of occurrences and information on conditions and individual events may well be required so that fluctuations, trends, means and extremes can be identified. Information on intensity, effects, type of influence and frequency are relevant for the analysis of landscape events and processes.

The appropriate method of transcription will also depend on the nature of the source, whether written or graphical, verbal or numerical, as well as on the type of information and level of detail required. There are several basic techniques by which data may be abstracted: by direct verbatim quotes, copies or tracings of the material, by paraphrasing or summarizing written information, or by measurement from maps and similar sources. It may be necessary to make a pilot study to construct a framework for transcription. All transcribed data must be carefully referenced with repository press marks, date, author, title of source, subject-matter and area of coverage and other relevant characteristics, such as the scale of a map. In the following sections some notes on transcriptions from graphical and written sources are presented.

Graphical sources. Cartographic information may be transferred directly by either tracing or digitizing, and aerial photographs are usually treated in a similar way after a 'map' has been made from the photographs. Parts of maps can also be reproduced directly by photographic means or information transferred onto base maps at the same or different scales. Pictures, drawings, ground photographs and other illustrative sources can be used directly as evidence of a situation, or the conditions depicted may be described in words.

Tracing and digitizing have the advantage that information can be abstracted selectively but still in correct planimetric form, and these methods provide readily accessible data stores. Care must be taken in tracing so that lines are drawn clearly and carefully and on a stable material. Digitizing involves the use of a machine which transforms map information into numerical data in the form of coordinates of selected points or lines. Data can then be used directly in calculations or to produce computer plots. Counts of objects or measurement of areas or attributes can be made. Information on a map can also be summarized as a written description, though this may underutilize its potential.

Written sources. The method of transcription used will again depend on the type of source and the amount or proportion of material to be abstracted as well as the purpose and type of study being undertaken. Whole passages or parts of documents may be copied verbatim, providing that care is taken not to select phrases and sentences out of context so that they bias the meaning. It can be tempting to paraphrase or summarize the contents of a document at the transcription stage, but this is not usually advisable since interpretation can be affected by further source material and later study. Information can be tabulated directly where it is found in numerical or non-prose form, as in accounts or lists of items, or where facts are clearly reported and there is no possible ambiguity. It may be possible and appropriate to plot or map information directly from written sources where sufficient details of location and characteristics are provided. Where information takes the form of incidental references amongst reams of irrelevant text, then use of a coding technique such as content analysis, discussed in the section on data processing below, might be appropriate.

Compilation and measurement

In many applications information must be compiled from a number of different sources, and to do this it must be put on a common basis, classified and ordered. This may involve a variety of techniques, such as listing information chronologically, transferring information to a base map or plotting values on a map. An essential prerequisite is that the information for different dates should be comparable, so as a first step the bases of measurement should be standardized. This is not usually difficult if comparison is being made between different versions or editions of the same source, though the exact basis of their construction should be checked. There was, for example, a major change in projection between the Ordnance Survey County Series maps and the later National Grid Series maps in Britain. To compare pictures and ground photographs accurately it is important to ascertain whether they were taken from the same position and at the same angle. When analysing dynamic phenomena like vegetation which is subject to seasonal or short-term variations, comparisons must be made at similar times of the year and allowance made for inherent annual and seasonal variations. Comparisons can, of course, be made without direct measurement or quantifying change, but such comparisons are likely to rely heavily on subjective judgement and interpretation. A useful method of comparing maps is to bring them to a common scale and superimpose the outlines. Identical reference points, which must be stable features such as prominent buildings, bridges or road and hedge junctions, can be matched. If maps do not fit exactly, then they can be adjusted to minimize the errors. This is best done for small areas at a time. Changes can then be measured from the features superimposed on the base map.

Data processing

Much historical information is qualitative and fragmentary, but often in the earth sciences we require or desire quantitative data. *Content analysis* is a useful technique for abstracting the essential 'content' or meaning from qualitative statements made by contemporary commentators on past environments and processes (Osborne and Reimer, 1973). It can yield numerical data from written and other descriptive sources such as topographical accounts, diaries, journals and written surveys. By converting elements of the historical record into numbers, comparison is facilitated both within past contexts and with the results of present-day surveys. Content analysis has found many and varied applications in the social sciences and the humanities, though rather fewer in the environmental sciences. Yet the need for such a technique in those physical–geographical studies which have an historical dimension is implicit in much discussion of sources and their possible treatment. Much of the basis for climate reconstruction, for example, is contained in descriptions of weather phenomena in unpublished texts and documents residing in manorial, ecclesiastical, legal and administrative archives. The need is for a

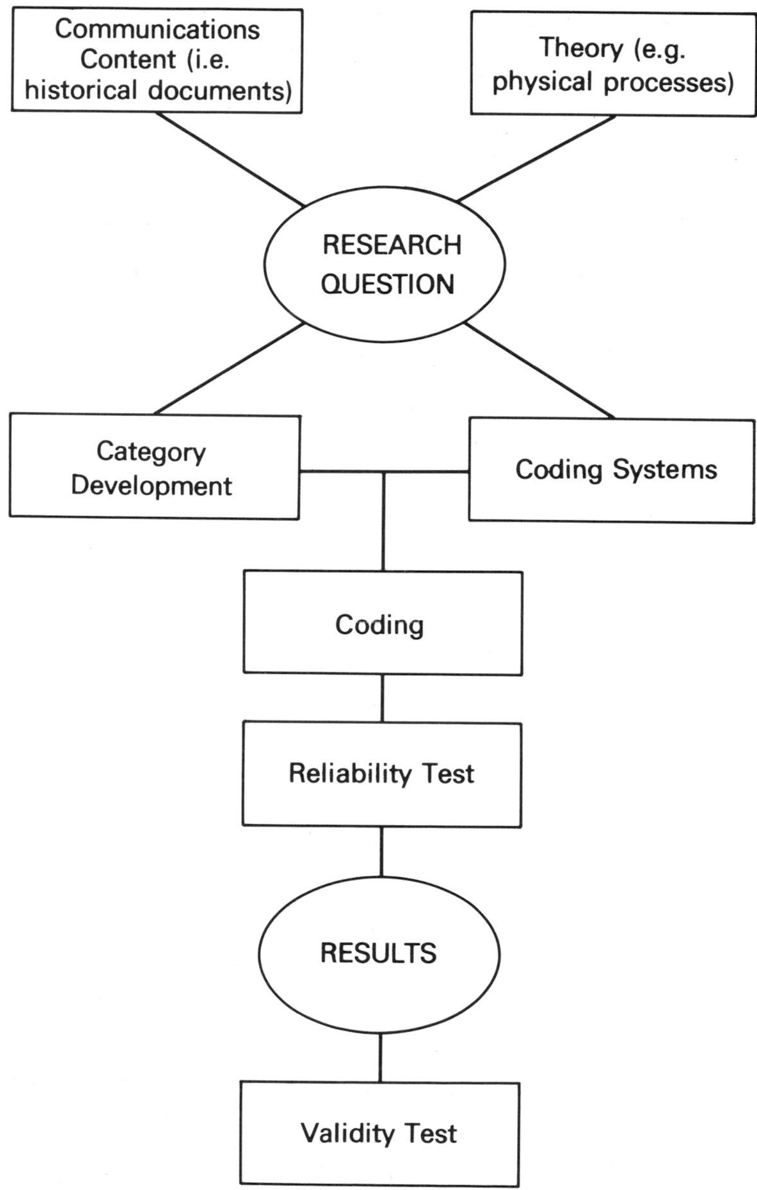

Figure 3.3. STAGES IN A CONTENT ANALYSIS

systematic method of analysis to extract this information, and this is a task which content analysis can usefully perform.

It is not possible here to provide a detailed account of the operational procedures of content analysis; these can be obtained from a number of comprehensive textbooks on the subject (Pool, 1959; Stone *et al.*, 1966; Holsti, 1969; Carney, 1972). Instead, the technique is described briefly and its potential is demonstrated by reference to the results of one seminal study in the environmental science field.

In content analysis the usual way of transforming information in an historical document into numerical form is by counting the frequency of occurrence of particular symbols, i.e. words, phrases or specified themes. Greater power can be obtained by combining frequency counts with contingency measures to assess relationships between symbols by counting how often particular words or themes appear in conjunction. For example, the symbol 'woodland' may take on added significance if it is found to occur in association with the symbol 'birch and pine'.

The main components of a typical content analysis in an environmental science context are set out in schematic form in Figure 3.3. This shows that the research question must be formulated in relation to both present-day theory and the nature of the source material. For content analysis it is essential that the question be posed in ways that will provide systematic, numerical answers. This requires that the question be broken down into a number of categories in which the symbols for counting and correlating are defined. A set of coding procedures has then to be established to control the way in which raw information in the documents is systematically converted into numbers. In many content analyses the categories are simply equated with single-word symbols which are coded on a presence/absence basis. Themes are more difficult to enumerate consistently than single words because their identification requires interpretation on the part of the coder. The objectivity and reliability of the coding procedure can be tested by repeating the coding of a sample of the same documents by several coders. If the frequency counts, i.e. the results of the analysis, are to have any scientific status they have to be tested for validity to establish whether the analysis has, in fact, measured what it set out to measure. This is usually done by comparing the results with some independent or external criteria (Moodie, 1971; Moodie and Catchpole, 1975).

D. W. Moodie, A. J. W. Catchpole and collaborators have applied the technique of content analysis in their investigations of the dates of freeze-up and break-up of ice in the estuaries of Hudson's Bay in the eighteenth and nineteenth centuries (Catchpole, Moodie and Kaye, 1970; Catchpole, Moodie and Milton, 1976). The sources used in their study are the daily journals which the London-based Hudson's Bay Company required each of its overseas settlements to keep (Moodie, 1977). They analysed the content of journals for four settlements by firstly transcribing verbatim all references to ice and water in spring and autumn. Categories were then established and coded to obtain measures of the frequency of references to partial and complete ice cover in autumn and to broken and completely ice-free surfaces in spring. Reliability and validity were assessed, the latter by testing for homogeneity of the results based on the assumption that, within a small geographical area, climatic data will respond in unison to large-scale weather processes; i.e. one set of measures should mirror trends in others. Their results are in the form of an almost continuous sequence of the date each year when water was first partly frozen, first completely frozen and first broken for the four estuaries. A sample of these data for two decades at one site is reproduced in Table 3.4. They are data which are ideally suited to quantitative analysis by time-series techniques and demonstrate the way in which information on past conditions from descriptive and often fragmentary accounts can be transformed into a format suitable for comparison with instrumental data from more recent times and the present day.

Table 3.4
SAMPLE FREEZE-UP AND BREAK-UP DATES AT SITES AROUND FORT ALBANY ON THE ALBANY RIVER, HUDSON'S BAY
The numbers given are days after the beginning of the year; those in brackets have been inferred and were not yielded directly by the content analysis (after Moodie and Catchpole, 1975)

Year	First date partly frozen	First date completely frozen	First date broken
1750	302	305	131
1751	299	(311)	136
1752	298	319	128
1753	301	330	127
1754	303	313	127
1755	282	303	127
1756	299	319	125
1757	299	311	(131)
1758	302	307	132
1759	303	309	134
1760	295	313	130
1761	309	310	126
1762	(298)	303	126
1763	302	303	126
1764	299	309	109
1765	295	(308)	113
1766	297	318	132
1767	309	(319)	144
1768	310	(313)	132
1769	290	298	124
1770	287	299	130

Another way of obtaining quantitative data is by scaling and the construction of indices. This has been most developed in relation to meteorological phenomena but has also been used to assess the intensity of events such as earthquakes. Lamb (1963, 1965) was a pioneer in this field with his construction of a winter-severity index and a summer-wetness index. The winter-severity index is defined as the excess number of unmistakably mild or cold months (December, January, February) over months of unmistakably opposite character per decade. The summer-wetness index was calculated by examining descriptions of July and August where drought scored 0, unremarkable months scored $\frac{1}{2}$ and months with substantial evidence of frequent rains and wetness scored 1. The information was derived from qualitative statements in weather diaries. Pfister (1979, 1980) has used similar types of indices, a thermal index and a wetness index, but has weighted them according to qualitative observations and proxy phenological data.

Indexing and scaling are also used to assess the intensity of events. For example, Townley and Allen (1939) used the Rossi–Forel scale to analyse earthquakes on the Pacific coast of the United States between 1769 and 1928. This scales severity according to amount of movement and damage caused.

Data which are already in quantitative form should be *standardized*, either by conversion to standard units of measurement or by calculating the value of the original units. This particularly applies to early meteorological measurements from times when thermometers, raingauges and barometers were not constructed to standardized designs (see 1.3 above).

Other types of measurements must be similarly standardized: for example, measurements of distances and heights must be made from a common datum. Records of the Nile floods have been investigated to check their datum (Bell, 1970) and discussion of the character and magnitude of subsidence and tilting in Britain centres around whether the levelling base can be considered constant (Kelsey, 1972).

Once quantitative data have been obtained and standardized they are then amenable to analysis by *statistical techniques* quite familiar to most earth scientists. The basic types of data—nominal, ordinal, interval and ratio—and their limitations should be borne in mind. Descriptive statistics are commonly used, though attention is often focused on the deviations, extremes and anomalies rather than the average tendencies. Statistical techniques for analysis of time trends and for comparison of data from different periods are particularly appropriate in historical analysis. Computer techniques can be as valuable for analysing historical data as they are for processing modern records. Their particular strength is for data storage, sorting, cataloguing, and statistical analysis. Pfister (1979, 1980), in reconstructions of past climate in Switzerland, has used a variety of sources of data containing different types of information and covering different periods, and has demonstrated how computer techniques can be used to sort and process data.

Data analysis and interpretation

The final stages in the use of historical sources, after transcription and processing, are analysis and interpretation in the light of present knowledge and theory. The basic technique involved in almost every historical study is the *reconstruction* of a situation from extant evidence. This is achieved mainly by careful comparison and corroboration and may involve not only documentary sources but also the weighing of morphological, biological and archaeological evidence. Both direct and indirect evidence may be used. Peterken (1969) neatly illustrated this in a biogeographical context in his study of a Suffolk estate from medieval times: 'Documentary sources rarely contain descriptions of the vegetation itself, but may include information on the use and value of a site or a land-holding containing the site, from which the state of the vegetation may be inferred.' Lambert *et al.* (1960) have used historical evidence of turbary sales and distribution to calculate the rate of peat digging and, therefore, whether such activity could have produced the Norfolk Broads. In climatology such 'proxy' or surrogate data is often used to indicate conditions in the pre-instrumental period. These include dates of harvesting, flowering and freezing and have been applied to the elucidation of past climates in Japan by Arakawa (1956, 1957) and in Europe by Le Roy Ladurie (1972) and many other researchers.

The completeness of reconstruction will depend on the level of information available; for example, it may be possible to establish only the presence or absence of particular features. With physical features the ideal is usually to be able to reconstruct their location, position and orientation at a certain time, and to identify their morphological characteristics including dimensions, shape and symmetry. Sheppard (1957) has done this for the medieval meres of Holderness (Figure 3.4). It may also be possible and appropriate to reconstruct details of the composition, structure, surface landscape characteristics and associations between features. To reconstruct major events and processes, information is required on intensity and magnitude, the type and distribution of effects and the conditions associated with the event.

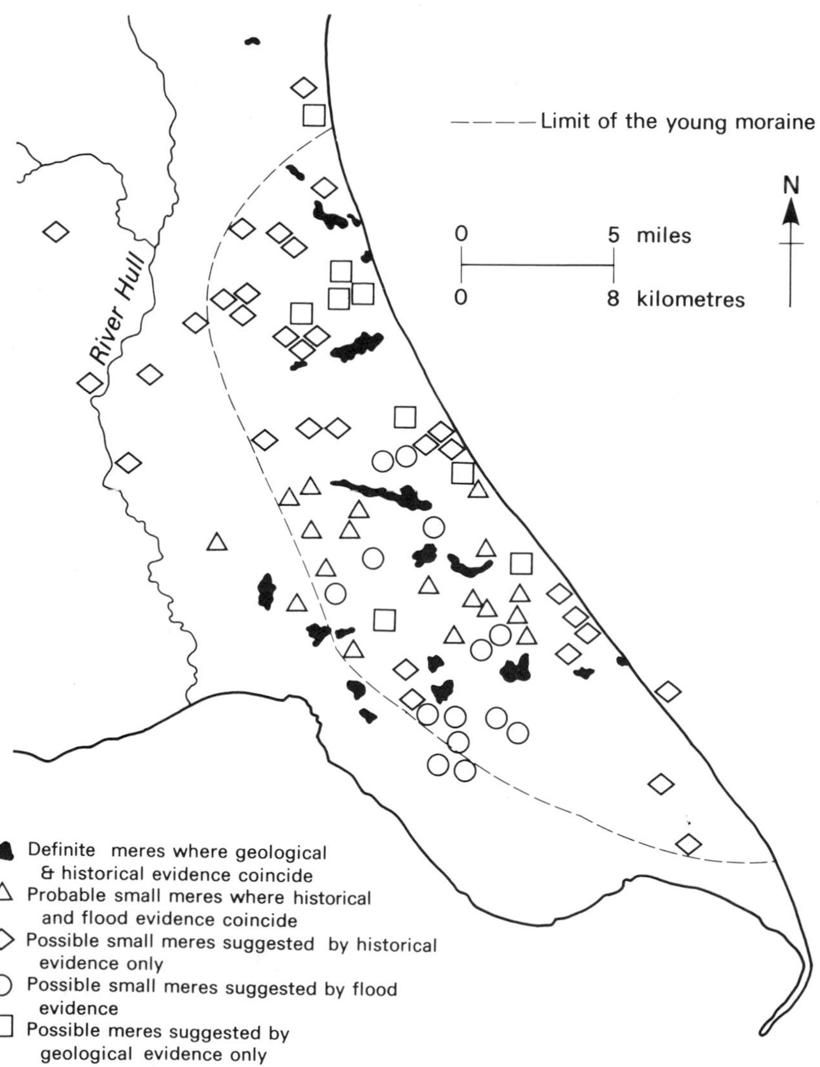

Figure 3.4. THE MEDIEVAL MERES OF HOLDERNESS.
(Source: Sheppard (1957))

Change can be deduced by comparing reconstructions for different dates. The main characteristics usually analysed are date and chronology of change, direction of change, amount of change, pattern of change, rate of change and the processes involved.

One of the main purposes of analysis of conditions at different dates is to elucidate the *chronology* of changes which have taken place. This may vary from a classification and ordering in time of periods with different characteristics described, listed or tabulated to a detailed yearly, seasonal or monthly chronology. Associated with chronological analysis it is usual to examine the type, direction and magnitude of variations. Again this may extend from the use of vague, qualitative statements such as 'colder' or 'wetter' to detailed data on changes in characteristics or conditions. If regular quantitative data over a relatively long period are available then analysis of fluctuations and trends may be made using techniques such as moving averages, regression and time series analysis (Figure 3.5). It is often easier to distinguish direction of change than to measure the amount of change since the latter requires a higher order of accuracy. Spatial characteristics are frequently expressed in the form of maps which may show the situation at different dates as a sequence, or superimposed or with the changes themselves mapped (Figure 3.6).

Once the conditions at various dates have been established accurately, and quantified if possible, then *rates* of change can be calculated by dividing the amount of change by the length of time period. If data are available for several different time periods this may show whether there has been any alteration and variation in intensity and operation of processes, as in Kestner's (1962) study of the advance of saltings at the edge of the Wash over the past 150 years, in which he found that the rate of advance has varied considerably, showing a particularly large increase after 1884 when a river outfall was completed (Figure 3.6). There are, however, a number of problems with interpreting rates of change from sources widely separated in time. It assumes, for example, that change is unidirectional within a period, which is not necessarily the case especially where rapidly changing features are involved. It also assumes that the rate is constant throughout each particular period, whereas much of the total change may take place in one short part of the period. In the opinion of Robinson (1964) the infrequency of the dates of hydrographic charts 'conceals many of the short-term fluctuations which undoubtedly occur'. In relation to changes in Orford Ness, Carr (1969) also discussed this factor: 'Maximum growth or recession may well occur between two surveys and thus not be recorded.' He demonstrated clearly how growth of this spit in the nineteenth century was in fact very irregular (Figure 3.6).

The limitations of evidence encapsulating conditions at a single date also apply to analysis of the *type and nature* of changes and the processes by which they are accomplished. Not only may the amount of evidence through time limit interpretation of the processes and patterns of change but the level of content of the sources themselves may be similarly limiting. For example, Hardy (1966) in his study of Norfolk coastal changes said that 'Soundings on older charts are too widely spaced to be able to establish whether southward movement of the promontory was related to changes in the sea bed'. Where there is sufficient evidence, however, types and processes of change may be described or explained with sequential or superimposed maps, sections or graphs. For example, the nature of changes in river meanders has been examined by Hooke (1977b), who found a general pattern of extension and migration of bends on streams in Devon.

As well as assessing the nature of change there is usually a need to establish whether there

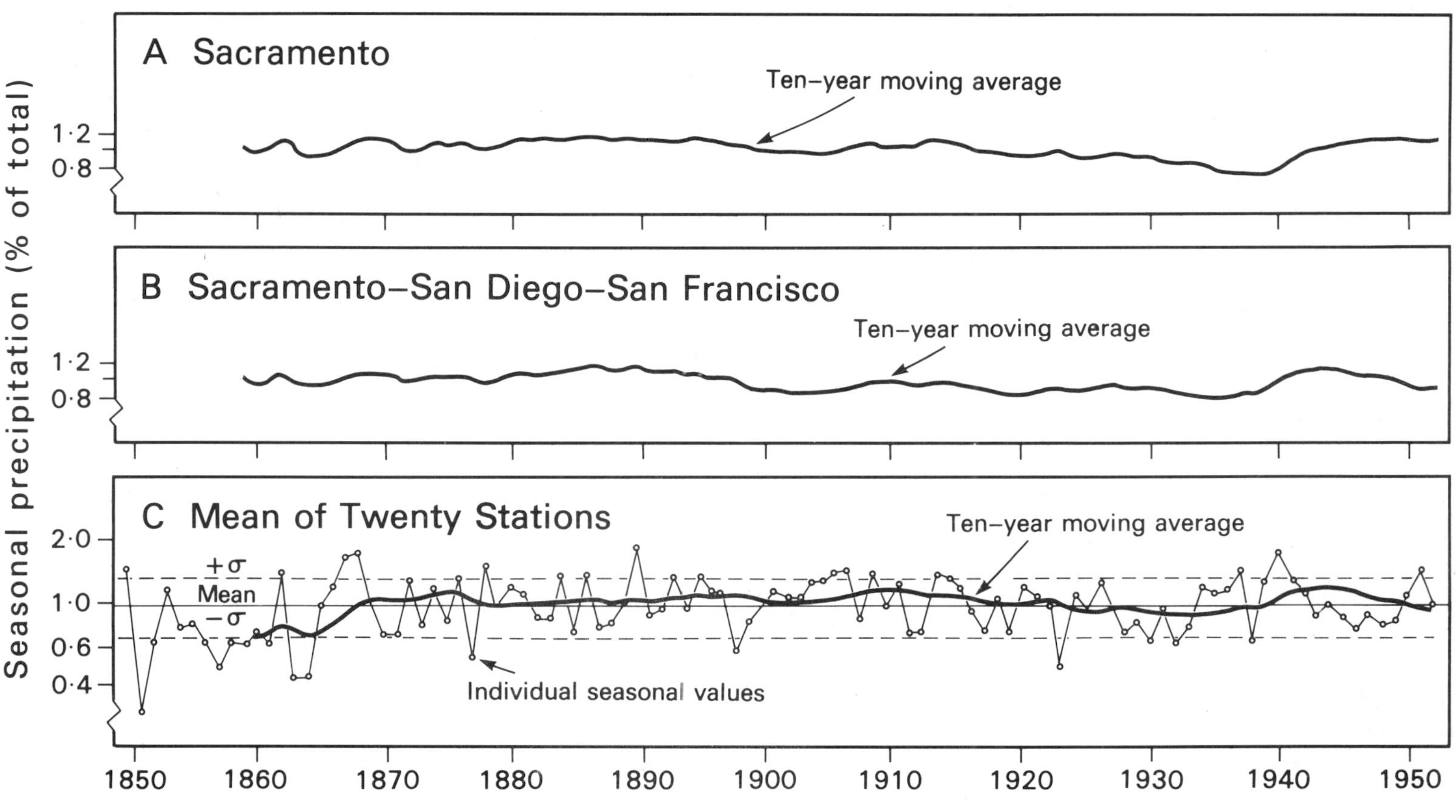

Figure 3.5. TRENDS IN SEASONAL PRECIPITATION ON CALIFORNIA RANGE LANDS FROM *C.*1850 TO *C.*1952. (Source: Burcham (1956))

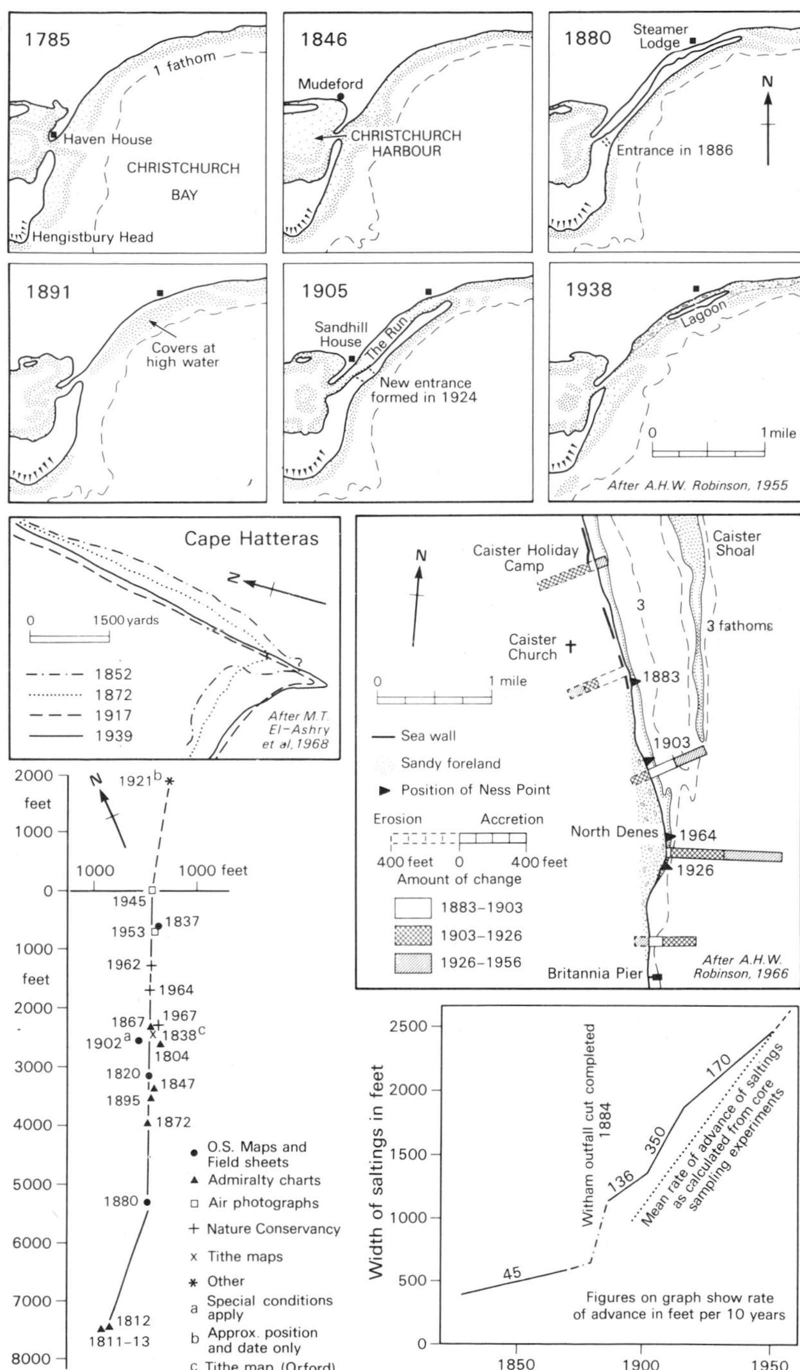

Figure 3.6. METHODS OF PORTRAYING COASTAL CHANGES.
As a sequence — Christchurch Harbour, Hampshire; by superimposition — Cape Hatteras, North Carolina; by graphical methods — Caister coast of Norfolk, Orford Ness, Suffolk, between 1811 and 1967, and the Wash, from *c.* 1820 to *c.* 1950. (Sources: Robinson (1955, 1966), El-Ashry and Wanless (1968), Carr (1969) and Kestner (1962))

was any *association* between particular conditions or features and to establish *cause and effect.* Inference of cause and effect is often based on evidence of temporal or spatial association of activities or phenomena. In some instances it may be possible to employ powerful time-correlation techniques, but more commonly only an approximate comparison, verbally or graphically, is possible. After analysis and interpretation of the historical data, the results may be used for prediction or to derive general models of aspects of environmental development.

4
The Study of Morphology and Changes in Form

In the preceding chapters some of the more important historical sources have been identified and techniques for their use and analysis described in general. In the remaining chapters applications in particular areas of earth science are examined and exemplified more fully. Chapters 4–6 are concerned with the three main applications of historical sources in studies of the physical environment. These are: first, as evidence of conditions at particular times and of change between dates, thus providing the basis for general environmental reconstruction (this chapter); second, as the basis for the reconstruction of processes and events (Chapter 5); and third, in explanation of changes in the historical period (Chapter 6). The main sources used in each of these applications are rather different, with an emphasis, for example, on maps for reconstructing morphology and on written sources for providing data on events.

The applications discussed in this chapter are taken from climate, glacier, river, coastal and vegetation studies. These are a selection of fields of application in which historical sources have been most extensively used to reconstruct situations and to analyse changes. Geomorphological subjects predominate in our review, but the importance of historical applications in biogeography and climatology is also demonstrated. The reviews are not exhaustive or comprehensive but aim to provide bibliographic guidance by selected examples. In each subject area the scope of studies, the aims and methods, the major sources used and the types of results obtained are discussed.

4.1 CLIMATE STUDIES

In 1914 Peterssen said, 'till quite recently the opinion prevailed among meteorologists and geographers that the old records are unreliable and exaggerated and that no real variation in the climate has occurred in historical time'. Historical climatology has since been a developing theme of study throughout this century, as exemplified by Brook's (1926) publication *Climate through the Ages* and Manley's (1946, 1953, 1959) work on establishing temperature trends. A more recent impetus has been provided by the work of the Climatic Research Unit at the University of East Anglia. Lamb (1964) adduced three main reasons to justify study of past climates:

(1) To understand better the development of the environment, including the flora and fauna, and to assess how agriculture and other aspects of the economy have been affected.

(2) To obtain numerical estimates of past variations in climate which might recur under analogous circumstances in the future.
(3) To identify some of the physical processes involved in climatic changes and to assess the magnitude of effects attributable to various influences.

Much work in historical climatology has tried to establish the general pattern and extent of fluctuations in the past, and current research is now providing detailed, often quantitative, data on spatial and temporal variations, though discussion on the basic patterns still continues (e.g. Burroughs, 1981).

Sources of information on weather and climate can be divided into three main types. First, there are instrumental records which mostly date from the seventeenth century but with standardized observations not common until the nineteenth century. Second, there are non-instrumental weather observations, mostly recorded in diaries. Third, there is indirect or proxy evidence, mostly from activities (such as agriculture) which are directly affected by the weather. The main elements recorded in climate studies which include historical changes are precipitation, temperature, pressure and wind direction. As Watson (1962) has remarked, data availability on these varies seasonally, with, for example, more data available on temperature in winter and on rainfall in summer.

A brief history of meteorological measurements and an indication of the major documentary sources containing climate statistics has been provided in Chapter 1. A variety of non-instrumental and non-quantitative sources also contain information on weather and climate, particularly for the sixteenth to eighteenth centuries in Europe. Probably the most valuable are diaries kept by individuals, and some British examples are listed in Table 4.1. Many diaries have been analysed independently: Oliver's (1958) study of Bulkeley's diary from Anglesey falls into this category. This diary is particularly valuable because of its early date (1734–1760) and its reliability. Oliver calculated frequencies and variations in wind direction and numbers of rain-days from Bulkeley's comments. Increasing numbers of diaries are coming to light as their value as sources is realized. All types of diary are worth examining, if only for corroboration. For example, Green (1970) described the type of information recorded in an Aberdeenshire deerstalker's diary of 1783–1792, and demonstrated its practical value in climate reconstruction. Observations were also made by organizations and institutions and are recorded in annals and chronicles. Baker (1932) has used observations made by ambassadors to Britain in the seventeenth century to indicate climatic conditions at that time. However, such observations often record only the most remarkable weather and events and may thus give an incomplete and somewhat biased picture.

There are other types of record which contain direct weather observations, although not necessarily in any regular or consistent way. These include estate and agricultural records, travel journals, explorers' and settlers' papers, military sources and ship records. These often contain only descriptions of abnormal and adverse conditions, but may help to corroborate other information. Other literary works are occasionally referred to in climate studies: for example, the description of a hard winter in R. D. Blackmore's *Lorna Doone* was cited by Lamb (1963) to support other evidence of cold winters. Outside Britain very much older writings have been used. Buchinsky (1963) has employed the writings of Homer in an analysis of climatic fluctuations in the Ukraine, and Seth (1963) analysed the content of hymns and early poetry to establish climatic conditions in India. Reference has already

Table 4.1
IMPORTANT BRITISH DIARISTS WHO PROVIDE WEATHER DATA
(after Potter, 1978, and other sources)

Diarist	Region	Date	Comments
Rev. W. Merle	Oxford	1337–1344	Weekly and monthly weather summaries
Polydore Vergil *c.* 1470–1555	London area	}	All Tudor chronicles and give accounts of various flood, frost and weather events, e.g. tidal surge of September 1555, frost and floods of 1565
John Stow 1575–1605	General	}	
Raphael Holinshed d. 1580	General	}	
William Camden 1551–1623	General	}	
John Dee 1527–1608	London area		Diary of weather records kept for magical purposes
Anthony a Wood	Oxford	Lifetime	Almost complete weather and flood calendar for much of 17th century
Samuel Pepys 1633–1703	London	1660–1669	Tidal surge and Thames breach of March 1660
John Evelyn 1620–1706	London	Lifetime	1683–1684 frost and floods after the thaw
Elias Ashmole	Lambeth	1677–1685	Daily observations of wind and weather
William Sampson	Clayworth (north Nottinghamshire)	1681–1701	Weather account has been compared with Evelyn's diary
Abraham de la Pryme	South Yorkshire	1671–1703	Corresponded with Royal Society and some publictions in their *Proceedings*. Manuscript record in British Museum includes weather and flood reports
William Stukely 1687–1765	Leeds and England		Antiquarian diary
William Bulkeley	Anglesey	1734–1743, 1747–1760	Daily wind direction and weather description

been made in Chapter 2 to the review of Chinese sources by Chu Ko-Chen (1973). Pictures may provide some indication of weather conditions, and these have been used particularly in relation to freezing of the river Thames (Andrews, 1887) and in examinations of the advances and recessions of glacier margins (Le Roy Ladurie, 1972; Messerli *et al.*, 1978).

The third group of data—indirect or proxy data—comprises observations of phenomena which are closely influenced by weather and, therefore, can provide a useful index of conditions. Some of these records are long and consist of regular observations. Included among these are records of Nile flood levels which date from the twelfth century BC (Bell, 1970) and records of times of cherry blossom, lake-freezing and first winter snow dating from the fifteenth century in Japan (Arakawa, 1957). In Europe, records of the dates of wine harvests over several centuries have been used to provide an indication of temperature (e.g. Wright, 1968; Le Roy Ladurie, 1972). Care has to be taken in the interpretation of these records with regard to the time of the year and conditions (e.g. temperature or rainfall) which influence the phenomenon. Other data such as yields of crops, crop prices

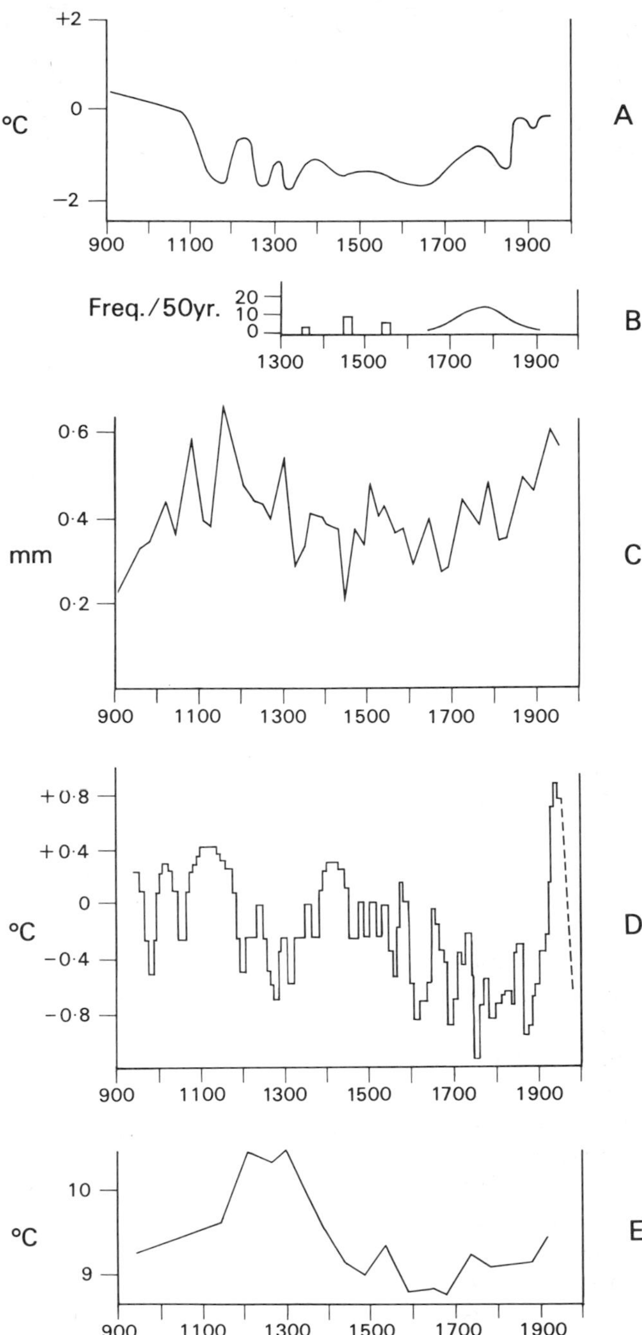

Figure 4.1. CLIMATE CHANGES IN THE NORTHERN HEMISPHERE OVER THE LAST MILLENNIUM.
A — departure of temperature from the present day in China; B — prolonged rains in Japan; C — ring-width of bristlecone pine in California; D — departures from average of mean annual temperatures in Iceland; E — mean annual temperatures in central England. (Source: a compilation of the work of several authors by Barry (1978))

and tithe and tax payments have also been used as indicators of conditions (Beveridge, 1921; Clough, 1933) but the relationships are less direct since they are also affected by economic, social and technical factors. Catchpole and Moodie (1978) have reviewed the major types of sources on climate and have discussed appropriate methods of analysis.

Work on *past climates* earlier this century was principally concerned with ascertaining the general sequence and pattern of variations. Brooks (1926) compiled much of the immediately available evidence and compared variations on a worldwide basis. This established the existence of the medieval climatic optimum and identified the seventeenth- to nineteenth-century 'Little Ice Age' in northern Europe. Since then, work has concentrated on reconstructing conditions in more detail, on providing quantitative estimates and on examining the extent of fluctuations and the processes and conditions under which they occurred. A summary of the pattern of climatic fluctuations in the northern hemisphere as indicated by various types of evidence is given in Figure 4.1.

There are some very early *climate chronologies* but these mostly comprise lists, year by year, of quotations referring to particular events or conditions. Many of these, especially the early ones such as those of Short (1749) and Lowe (1870), contain much unsubstantiated information, but later ones, such as that of Britton (1937), are more carefully referenced. Such compilations are sometimes used as secondary sources, but Ingram *et al.* (1978) and Bell and Ogilvie (1978) have warned of the dangers of duplication of dates and exaggeration and misrepresentation in the use of possibly unreliable chronologies. Lamb (1977) has published a list of these climate compilations.

There are now a considerable number of studies of *particular periods* using historical sources. These vary from analyses of a whole climatic period such as Lamb's (1965) study of the medieval period, to studies of parts of centuries (e.g. Sedgwick, 1914), to periods of a few years (Wahl, 1968) or to single years (e.g. Emery and Smith, 1976) and even particular seasons (Hoyt, 1958; Manley, 1975) and particular months (Kington, 1976a). These often focus on extreme conditions; hence the concentration on the late seventeenth century in Britain. Manley (1975) considered 1684 to be the coldest year for which records can be obtained.

Where conditions have been quantified or indexed then fluctuations can be analysed graphically (Lamb, 1965), as illustrated in Figure 4.2. Weather-type frequencies can also be analysed. These reconstructions of conditions at a particular time are useful for their implications for human activity at that time and in suggesting how such conditions occurred and how they related to conditions elsewhere.

Individual phenomena or parameters of climate have also been examined. For example, the incidence of snowfall in Scotland for much of the eighteenth and nineteenth centuries has been investigated by Pearson (1973, 1976, 1978). Much work has now been done on compiling homogeneous records of rainfall (Glasspoole, 1933; Wales-Smith, 1971; Craddock, 1976, 1977; Craddock and Craddock, 1977; Craddock and Wales-Smith, 1977) and temperature (Manley, 1946, 1953, 1959) for various parts of Britain. Similar types of work have been undertaken in North America. For example, Bradley (1976) has collated records of precipitation in Rocky Mountain states from the mid-nineteenth century onwards, Wahl (1968) compared precipitation in the 1830s in the eastern United States with the present day and Landsberg (1967) has reviewed climatic records for the period 1750 to 1960 in New England. An early study of this type is that by Manley (1946) who compiled

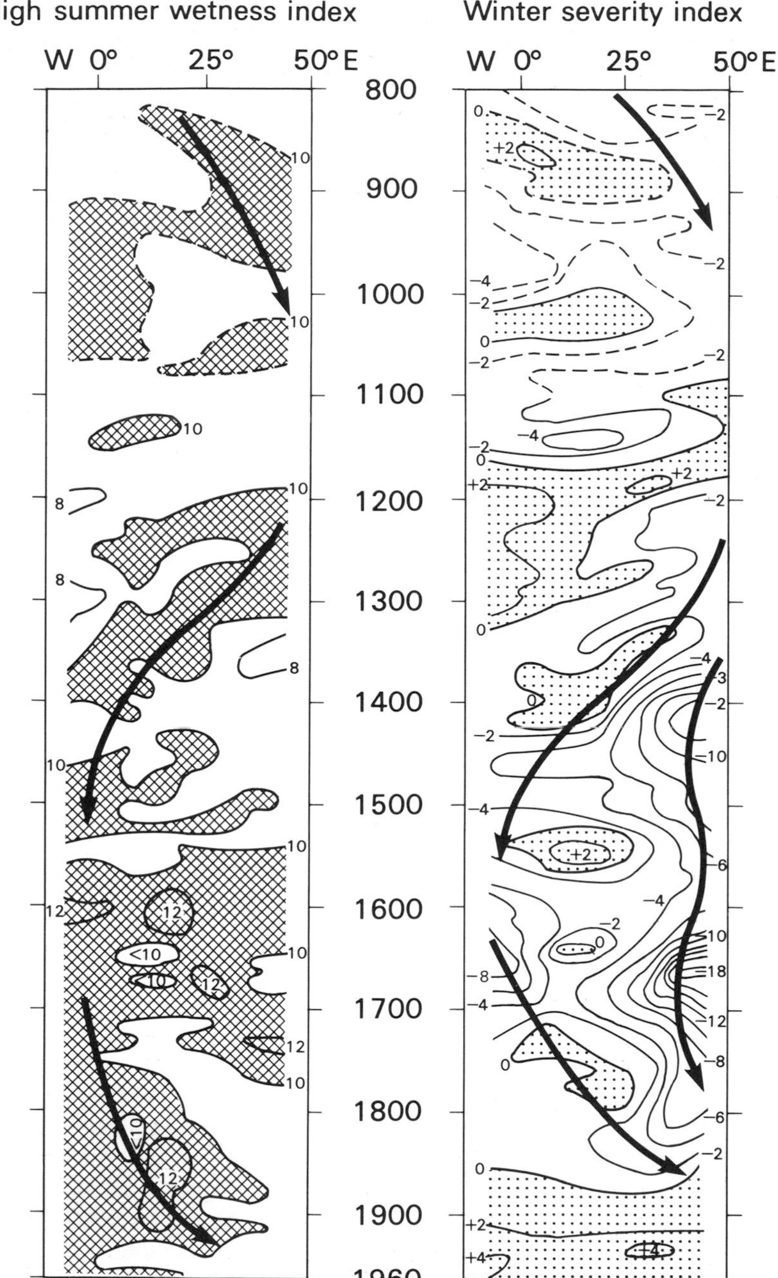

Figure 4.2. SUMMER-WETNESS AND WINTER-SEVERITY INDICES IN DIFFERENT EUROPEAN LONGITUDES NEAR 50° N FROM AD 800 TO 1959.
Excess of wet summers is cross-hatched and excess of mild winters is indicated by dots. (Source: Lamb (1963))

records of temperature for north-west England to produce an homogeneous record. There are a number of early instrumental records of temperature for north-west England but they are not of long duration or complete for a single place. Many comprise the careful measurements made by keen observers typical of this period. He found that from the year 1784 onwards two or more complementary records of temperature are extant and that these can be conflated to produce a continuous record, one record not being sufficiently reliable on its own. Modern records from north-west England were also examined to determine climatic characteristics in different parts of the region to provide some standard against which to compare the patterns of earlier records. Manley examined each of the old records in turn and discussed the adjustment of values needed. He produced tables of monthly mean temperatures for every year since 1753. The pattern of fluctuations that he deduced accords well with the record elsewhere.

Wind and pressure records have also been used to reconstruct climatic conditions. Brooks (1926) considered that 'records of wind direction are probably most valuable of all', and in a study by Brooks and Hunt (1933) wind vectors are plotted from observations in diaries and other documents. Pressure patterns have been reconstructed from data in diaries in Japan (Maejima and Koike, 1976) and synoptic charts for individual days in the eighteenth century have been drawn from observations of wind in ships' logbooks (Oliver and Kington, 1970) (Figure 1.16). These charts have subsequently been used in analysis of frequency of weather types and of monthly and seasonal fluctuations (Kington, 1975a and b, 1976a and b). Once continuous quantitative data on climate are available these can be analysed for trends and cycles: for example, Kraus (1955a and b) has examined secular changes in rainfall in Australia over the last century. Extremes and other characteristics can also be analysed (e.g. Craddock and Weller, 1975).

In many cases, particularly in relation to rainfall, it is not only appropriate to examine totals and means but also frequency and variability. In analyses of rainfall characteristics in the south-west United States Leopold (1951a) found that the number and intensity of storms changed in the period since 1850, although the annual rainfall totals had not altered significantly.

An early theme in analysis of climatic change was the identification of spatial patterns and interrelationships to help understand the general circulation (Lamb and Johnson, 1959) and a further recurrent theme has been the examination of the basic and longstanding question of the causes of climatic change. Analysis of temporal patterns is also related to elucidating causes of variation by identifying time correlations in fluctuations of various climatic phenomena. Climatic data have, for example, been analysed for their relationship to sunspot cycles (Maunder, 1922; Clough, 1933; Brunt, 1937; Schneider and Mass, 1975) and to volcanic eruptions (Lamb, 1970), the latter study involving use of historical records of these events. A developing focus is the influence of human activities and urbanization on climate, though such studies have mostly used modern systematic records. Much work is also being done on the impact of climatic change on agriculture and human activities (e.g. Parry, 1978; Smith and Parry, 1981) and this has entailed analysis of the nature of climatic change and its effects.

Case study: the climate of early medieval times

Lamb (1965) in his paper on the early medieval warm epoch and its sequel noted how little was known when he began his work about climatic variations in the historical period. He used a variety of data—historical, physical and biological—to identify the 'Little Optimum' of the Middle Ages. He found evidence that ice limits in the Arctic in the period AD 1000–1200 were much farther north than later. In Norway, settlement and cultivation advanced up valleys from AD 800 and declined from the fourteenth century onwards. Abundant evidence is available that vine cultivation took place farther north in Europe during the Little Optimum and that tree limits in the Alps were higher. There is also archaeological and tree-ring evidence of different moisture régimes in parts of North America. Much other evidence from elsewhere is also cited by Lamb (1965) but this is mostly from secondary sources and from published analyses of original data. The evidence he accumulated suggests that temperatures were 1–2°C higher than present generally and possibly even higher nearer the Arctic.

Lamb considered that the most reliable surface weather indicators which can be derived from data in early manuscripts are mildness or severity of winter, and raininess or drought in the months of July and August. He calculated indices for these two characteristics and applied these to elucidate the characteristics of half centuries from AD 800 onwards (see Chapter 3). The results are indicated in Figure 4.2, on which the pattern of variations and shifts in the main climatic belts can be seen. Lamb (1965) assessed the meaning of these indices in actual temperature and rainfall terms. He tested for correlations between climatic statistics and the indices using mainly meteorological evidence, and then checked these against botanical and viticulture evidence of conditions. The evidence on incidence and location of vineyards in Britain and their later decline seems particularly reliable. Much progress in establishing conditions has been made since the publication of this paper, but Lamb's methodology was fundamental in developing this type of study.

4.2 GLACIER STUDIES

Much of the work using historical sources to assess glacier variations emanates from Scandinavia, Iceland and Greenland or is concerned with those areas. French and Swiss scientists have also investigated Alpine glaciers and much recent literature has been produced on fluctuations of glaciers in the United States and Canada. Almost all the glaciological studies using historical sources are concerned with establishing the extent and dates of variations in glaciers and ice sheets. These studies are undertaken for two main purposes, either for their value in indicating climatic variations, or for greater understanding of glaciological processes and glacier régimes. A few studies aim to establish the chronology of fluctuations in order to relate the impact of the ice sheets to human activities.

Different types of sources covering different time periods are available in the various countries that have formed the main foci of study. In Iceland, Greenland and Scandinavia sources such as sagas and Norse legends date back to the tenth century. However, the reliability of these may be questionable. Koch (1945) said that 'they must be characterised as historical novels, at times probably only little reliable historically'. This is probably also true of traditions and tales elsewhere (Grove, 1966). Another important source in relation to these northern lands are records of sea voyages and trade and of the difficulties encountered at various time periods due to ice-extension (Koch, 1945). Burrows (1976) has also compiled historical evidence on sightings of icebergs around Antarctica, including information derived from the voyages of Captain Cook.

The extension of ice sheets and glaciers particularly in the Little Ice Age caused a number of difficulties, including loss of farmland, destruction of settlements and travel in marginal areas (Table 4.2). These effects are recorded in many Scandinavian and Icelandic documents, for example, local registers show the location of farms and settlements at the ice-margin and give lists of abandoned farms (Thorarinsson, 1939). Similarly, tax records show where and when crop harvests diminished and enable ice fluctuations to be inferred. Some parish and ecclesiastical records include descriptions of ice-extent (Thorarinsson, 1943).

Table 4.2
ICE OFF ICELAND AND SOUTH GREENLAND, GLACIATION IN WEST GREENLAND AND ICELAND AND CULTIVATION OF CEREALS AND GRAIN IN ICELAND *C.* AD 865–1950
(after Schell, 1961)

Date	Ice off		Glaciation		Cereal and grain cultivation
	Iceland	South Greenland	West Greenland	Iceland	Iceland
(865)–900	0.1[1]		—	Probably less	
901–1000	0.7	—	—	severe than at	
1001–1100	0.2	Very occasional	Relatively light	other times	All districts
1101–1200	0.0			subsequently	
1201–1300	7.8	Generally moderate	—	—	Ceased first in
1301–1400	6.0	to light	—	—	north and east
1401–1500	2.8	Generally light	—	—	before end of 12th
1501–1600	3.2	—	—	Increase (2nd half)	century, then altogether near end of 16th century
1601–1700	22.6		—		
1701–1800	25.3	Generally severe	Relatively severe	Relatively severe	None
1801–1900	40.8				
1901–1950	8.6		Decrease	Decrease	Revived in 1920s

[1] Duration in weeks

Newspapers are not widely utilised in glaciological studies but reports of snowfall and depth were used by Curry (1969) in an analysis of glacier régimes. Oral evidence is quite widely quoted, particularly in relation to fluctuations of ice-dammed lakes (Thorarinsson, 1939; Lawrence and Lawrence, 1961) and some glaciers have changed sufficiently in an individual's lifetime for a person to recall former ice-margin positions. This oral evidence can also be used in association with rock-markings to date marginal positions.

An important source of information for the Arctic/sub-Arctic areas of the world and also for the North American cordillera are the records of explorers and travellers. Some expeditions were sent specifically to find ways across or through ice areas or to explore the extent of ice sheets and glaciers. In these and other travellers' accounts there are often quite detailed accounts of the route taken, descriptions of views of glaciers and valleys and notes of difficulties encountered in crossing or circumventing an area. Thorarinsson (1943) indicated how such descriptions can be compared with present conditions and several

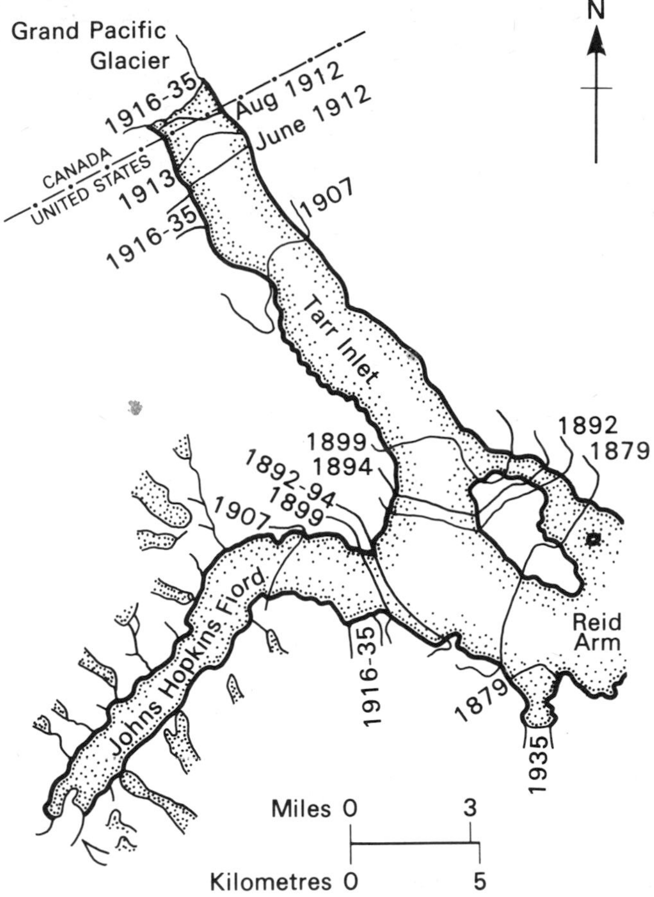

Figure 4.3. ICE-RECESSION IN GLACIER BAY, ALASKA, 1879–1935.
(Source: Cooper (1937))

workers have mapped glacier limits from such evidence (Cooper, 1937; Karlén, 1973) (Figure 4.3). In the late nineteenth century many expeditions were undertaken to study the glaciers themselves and their results have been compared with recent measurements (Carlson, 1939; Harrison, 1956). Some Himalayan glacier fluctuations have also been studied from explorers' and travellers' writings (Mayewski and Jeschke, 1979).

Maps are prime sources of evidence of glacier-extent and some very early examples have been used in the Iceland and Greenland studies. Thorarinsson (1943) stated that reliable mapping of Icelandic glaciers only began in 1902, but he did say of a 1570 map by Thorlaksson that it is 'extraordinarily good for its time', and of another which accompanies the travel descriptions of the men involved in compiling a land register in 1794 that 'the picture this map gives of Vatnajökull is amazingly correct for its time, although many details may be wrong'. He continued: 'Some older maps—even very primitive ones—may, however, give valuable information on glacier oscillations, but great care is essential in

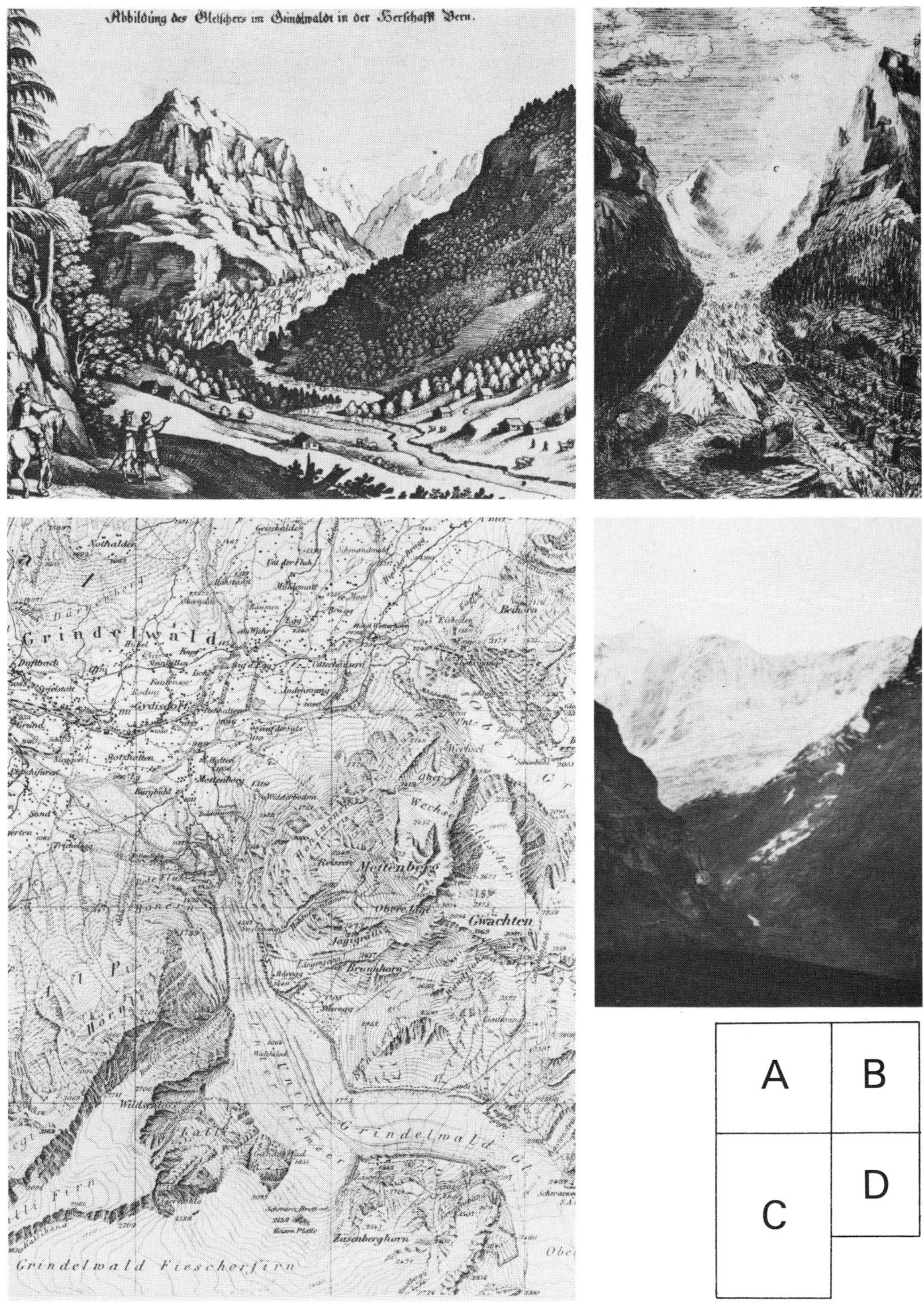

Figure 4.4. THE GRINDELWALD GLACIER IN THE SWISS ALPS FROM 1640 TO THE PRESENT DAY
A — In 1640. The lower of the two Grindelwald glaciers (see C) extended into the main valley at this date; B — in 1748. The lower Grindelwald glacier still stretched far beyond its present position; C — The Grindelwald glacier as portrayed on a late nineteenth-century 1:50 000 topographic map; D — in 1967. (Sources: Merian (1644), Altmann (1751), 'Siegfried Atlas' sheet 396 courtesy of Federal Office of Topography)

drawing any conclusions from them regarding glacier changes.' Thorarinsson corroborated this evidence by comparison with other maps and descriptions and also illustrated the difficulties of interpreting these older maps. In many parts of the world, particularly in North America, the earliest maps are systematic surveys by official expeditions or government agencies rather than the work of private individuals.

Other graphic sources such as paintings, engravings and photographs have been widely used, especially in European studies. Le Roy Ladurie's (1972) study of climatic fluctuations contains many such illustrations of Alpine glaciers and Messerli *et al.* (1978) have also used these sources extensively for the Bernese Oberland, and advised: 'For maximum value of the pictures, the date of the illustrated document must be precisely known, the glacier and the topography of the surroundings should be drawn exactly, and the location from which the artist made the picture must be established'. This evidence can then be used to map limits or to compare with present positions and so measure distances of recession (Figure 4.4).

One of the main features of the application of historical sources in glaciology and glacial geomorphology is the degree of integration attempted with other types of evidence. In North America, for example, historical evidence is usually not very abundant and most corroboration comes from morphological evidence of moraines and trim-lines, from biological evidence of tree-rings used to date old trunks and from living vegetation and evidence of pollen and lichens (e.g. Carrara and McGimsey, 1981). In many studies these are the primary forms of evidence and the historical evidence is corroborative (e.g. Lawrence, 1950; Sigafoos and Hendricks, 1961).

Many analyses involve the construction of chronologies of glacier fluctuations. These vary in level of detail and the degree of quantification which is is possible to achieve. Somc

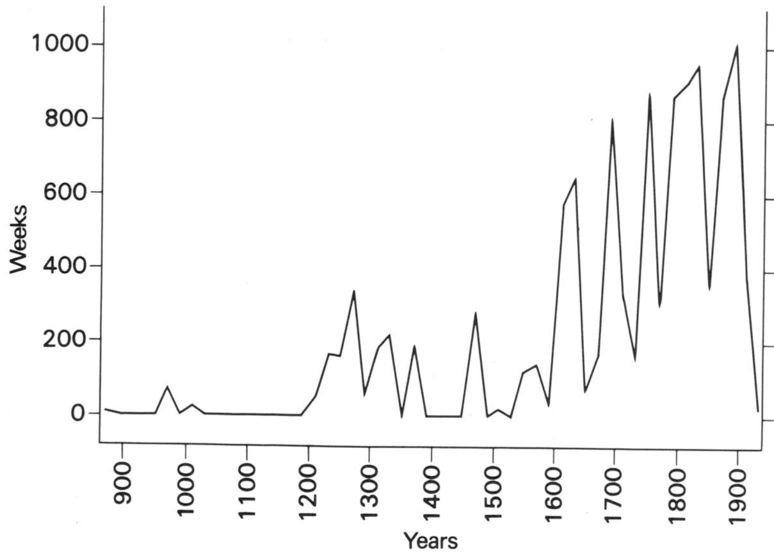

Figure 4.5. NUMBER OF WEEKS WITH ICE MULTIPLIED BY NUMBER OF AREAS WITH ICE AROUND ICELAND, AD 860–1939. (Source: Koch (1940))

consist of series of quotes and descriptions of states of glaciers at different periods while others also use this fragmentary type of information but provide in addition summaries of dates and fluctuations, as exemplified in Table 4.2. Yet others make measurements or estimates of changes in dimensions and position which can be tabulated or plotted on time-sequence graphs. The wealth of detail in some studies is outstanding, particularly in Thorarinsson's (1943) work on Iceland, Koch's (1945) and Weidick's (1959) work on Greenland and Le Roy Ladurie's (1972) work on the Alps (Figure 4.5).

Some authors discuss only single glaciers (Harrison, 1956) but others compare several within one region (Heusser, 1957) (Figure 4.6). Ahlmann (1953) provided a succinct synthesis of the knowledge of glacier fluctuations at the time he was writing and Figure 4.7 illustrates some of his world-wide information. Some of the studies which aim to elucidate climatic variations compare or correlate ice fluctuations with other climatic data. Bray and Struik (1963) correlated glacier variations with sunspot activity. Many studies also try to determine how climatic fluctuations affect glaciers, as Dightman and Beatty (1952) have done for glaciers in Montana. A major problem in assessing the mass balance of glaciers is that much of the information provided is on areal extent, whereas glaciers are three-dimensional and may alter in volume by thinning and thickening. Paintings and engravings, when available, may help to reveal this. In general, as Figure 4.7 shows, glaciers in

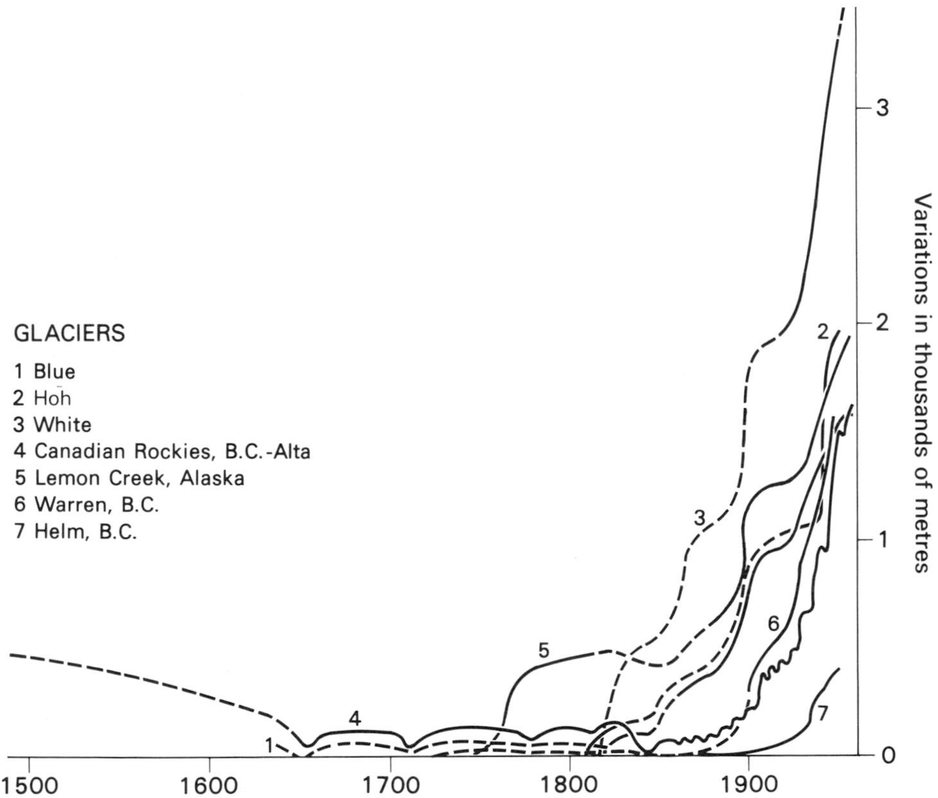

Figure 4.6. GLACIER MOVEMENTS IN NORTH-WESTERN NORTH AMERICA FROM 1500 TO 1950. (Source: Heusser (1957))

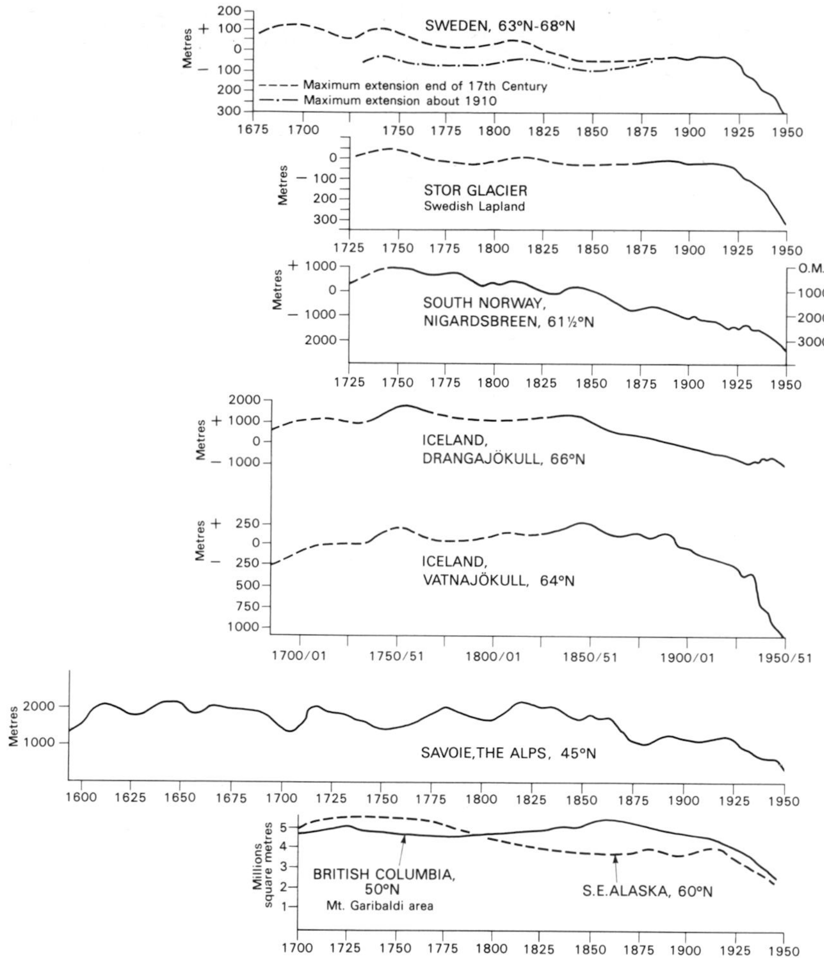

Figure 4.7. VARIATIONS IN GLACIER TERMINI IN THE NORTHERN HEMISPHERE *C.*1600–1950. (Source: Ahlmann (1953))

Scandinavia, Iceland, the Alps and North America have exhibited a general recession during the last two centuries from a maximum of advance in the middle of the eighteenth century or earlier. The pattern of individual glaciers varies because of their particular morphology and ice budgets, and some have shown minor readvances over the period.

Case study: the Mont Blanc Massif

J. M. Grove (1966) has examined fluctuations of glaciers in the Mont Blanc Massif. There is a great variety and rich volume of evidence for this area, although cartographic coverage is poor until the nineteenth century. Her study covers the period of the Little Ice Age and thus helps in our understanding of the extent of glacier fluctuations in that period and their impact on human activities. Some notion of the state of glaciers in the sixteenth century may be gained from local traditions and myths, and these can then be compared with later descriptions of the difficulty of travelling across certain areas. The first useful account was by an archdeacon who visited Chamonix in 1580 to sort out tax disputes and, incidental to this task, he described the number of glaciers he could see. By the early seventeenth century there is evidence that conditions were worsening; there were increased floods accompanied by supplications for tax relief. Land was overrun, houses destroyed and churchmen were called in to exorcize the glacier. Reports suggest that there was a major glacier advance in about 1605, another in about 1641–1643 and others later in the seventeenth century. The historical sources are corroborated by moraines and ^{14}C evidence. In the early eighteenth century supplications were still being made for tax relief and these give some details of the weather and the difficulties it caused. A major rock- and ice-fall recorded in 1717 which killed both cattle and people may have been associated with another glacier advance.

The frontal position of glaciers in the Arve valley are marked on a detailed 1728–1732 plan of land-holdings in the Chamonix area. By the mid-eighteenth century, travellers' accounts describe routes across glaciers and from then onwards writings by travellers and early glaciologists proliferate. The last three advances appear to have been in 1770–1780, 1818–1820 and 1850. From about this last date, information on the position and fluctuations of individual glaciers can be ascertained. Since the mid-nineteenth century there has been a general recession of the Mont Blanc glaciers, as evidenced by mountain huts left high above glaciers, by moraines which are well away from present ice-margins, travellers' descriptions and by photographs and systematic measurements.

4.3 COASTAL STUDIES

Some of the earliest applications of historical sources were in the field of coastal geomorphology probably because of the dynamic nature of coastal changes and the practical problems these caused. Much early work was undertaken by engineers like Redman (1852, 1864), who accumulated a great deal of historical evidence which is still widely quoted. Systematic study and explanation began during this century, notably by Steers (e.g. 1926) in Britain. Since the end of the Second World War studies relating changes in features to marine processes have proliferated. Most work has been concerned with the more mobile features such as spits and bars and with rapidly eroding coastlines. A developing theme has been the effect of man in building structures such as jetties and in altering sediment supply. Navigation problems have also provided an impetus for coastal studies with a time dimension.

In coastal studies much of the information derived from historical sources relates to the position of cliffs, beaches, spits and bars at particular times and the dimensions, shape, orientation and characteristics of these features. If these are then compared for different dates, the amount, rate and direction of change can be ascertained. The usual sources are maps of various kinds and hydrographic charts. For the more recent period, aerial photographs have proved even more important since they have the advantage of showing small features such as beach ridges, shallow-water offshore features and wave and sediment patterns (El-Ashry and Wanless, 1968; Stafford and Langfelder, 1971). Early maps have been used extensively at least to indicate the presence or absence of features (e.g. Nossin,

Table 4.3
SUMMARY OF MAP EVIDENCE USED TO TRACE CHANGES IN THE MALAYSIAN COASTLINE, 1640–1879
(after Nossin, 1965)

Year	Author	Map	Tingoram	Bay	River Pahang	Site of Pekan (Pahang)
c. 1640	Hondius	Asia recens summa cura delineata	Potigaram	No bay indicated	Connection of river Pahang with straits: Jor (Johore) shown as island	Pahang: north of river
1700	Eberard	Malaya	Trincora	Bay with two islands	Estuary, with island named Baham	Not shown (may be Baham)
1726	Velentijn	Nieuwe Kaart van het Eyland Sumatra	Tingaran	Large bay, far inland	River Pahang, island offshore	Pahang: north of river
c. 1740	Pierre Mortier	Le Royaume de Siam avec les Royaumes qui luy sont Tributaires	Tingoran	Bay far inland; three islands	Estuary; island. Also a river Pahang farther north off Pullo Capas	Pahang ou Paam: north of river
1752	d'Anville	Seconde Partie de la Carte d'Asie	Tingoran	Bay with two islands	Shown: short	Pahang: north of river
1753	Joannes van Keulen	Nieuwe Caart strekkende van Banca langs de kusten van Malacca, Siam, Cambodja, etc.	Tingorang	Bay far inland, three islands in southern part of bay	Short: island off mouth	Pahang: south of river (may refer to name of river)
1775	Munnevillette	Cote de Malaye	Tingoram	Bay with three islands	Short: island off mouth	Pahang: north of river
1778	Thomas Jeffreys	A Chart of the Straits of Malacca and Singapore	Tingoran	Bay far inland, three islands	Pullo Pahang, island off river mouth	Pahang: north of river
1800	Anonymous	Marine chart of Malaya (map bears no specific title)	Tincoram	Bay with three islands	Pullo Pahang, island off river mouth	Pahang: north of river
1825	John Cary	A new map of the East India Isles	Tingoram	Bay less far inland, no islands shown	River short: Pulo Pahang off coast	Pahang: north of river
1855	J. H. Colton & Co. (Publ.)	East Indies	Not shown	No bay indicated	River shown	Pahang: north of river
1876	Edward Stanford (Publ.)	Malay Peninsula	Not shown	No bay indicated	Two distributaries, river Pahang and S. Pantoh (= S. Pahang Tua)	Pekan: south of river
1879	Robinson	Malaya	Not shown	No bay indicated	Sungei Pahang Tuah (north) and river Pahang (south) as distributaries	Pekan: south of river

1965) and often the orientation and rough shape and position of any features (e.g. Steers, 1926; Carr, 1969), if not the exact dimensions. Table 4.3 summarizes the map evidence used by Nossin (1965) to trace the changes in a bay on the coast of Malaysia.

A number of sixteenth- to eighteenth-century maps have been used in coastal studies in Britain. Steers (1926) used maps by Saxton, Norden, Speed, Bryant and others to trace the growth of Orford Ness. Lovegrove (1953) has used several sixteenth-century maps as well as later ones to date beach ridges near Camber Castle in Sussex. A number of county maps by surveyors like Yeakell and Gardner, Armstrong, Budgen and Greenwood have also been used (e.g. Redman, 1852; Brookfield, 1952; Owen, 1952) to provide at least an indication of general characteristics at a particular time. Some of these, such as Norden's maps of Orford Ness, are of large enough scale to provide quite detailed information, though many others are of dubious or variable accuracy. Estate, enclosure and tithe maps have similarly been used to reconstruct the position and nature of coastal features by Kidson (1950), Kestner (1962) and Carr (1969) (Figure 3.6).

Hydrographic charts are a most important source for coastal studies. By the mid-nineteenth century revisions of many charts are frequent enough to permit construction of a detailed picture of changes, as Robinson (1960, 1964, 1966) has shown (Figure 3.6). These charts have the advantage that they show depths in some detail and thus offshore changes can be examined. Published topographic maps have also proved valuable but perhaps users should note Williams' (1956) cautions:

> More often than not the only reliable measurements are the widely spaced (in time) surveys of the Ordnance Survey and the Hydrographic Department of the Admiralty. While these surveys are of greatest value they are, for obvious reasons, insufficient for complete understanding. They occur too rarely; the scale, especially of the hydrographic surveys, is often too small, and the practical difficulties of making precise surveys are considerable where the foreshore is regularly covered by the advance of the tide. These surveys can be regarded, therefore, as occasional (though authoritative) glimpses in a long story.

Documentary sources useful for coastal studies include manorial, estate and parish records, tax, port and trade records, and explorers' and travellers' writings and chronicles. This kind of material is particularly useful for early periods when maps are unavailable and also to corroborate cartographic evidence. But, as Homan (1938) said, these documents are 'fragmentary and frequently unintelligible' and require much search and interpretative work to be useful. Estate records, particularly deeds, have been used to establish the extent of land at particular dates and have been compared with maps (e.g. Cozens-Hardy, 1924) and other records including Domesday Book (e.g. Pickwell, 1878; Sheppard, 1909) and surveys (Steers and Jensen, 1953) to show changes. Loss of land is occasionally mentioned in tax records where there has been a remission of taxes or dispute over payment (e.g. Owen, 1952). Manorial records have provided evidence on a variety of topics, ranging from the date of construction of buildings (Carr, 1969) to information on sea-flooding (Brookfield, 1952) and evidence of siltation of estuaries (Homan, 1938). Drainage records can also provide evidence on reclamation and the state of marshes (Sheppard, 1912; Owen, 1952). Much indirect and contextual information and corroboration on matters such as the state of ports and settlements, the presence and construction of sea walls and jetties and incidence of major storms is provided by annals and chronicles and by local histories and area descriptions. These sources are especially valuable where there have been major changes causing destruction of settlements, as on the east coast of Britain (Sheppard, 1912) or silting up of harbours, as on the south coast of England (Redman, 1852). In areas of rapid change, local people can often provide some useful oral evidence, particularly those involved in coastal management such as harbour masters and local authority engineers and surveyors. Some authors have used drawings, illustrations and photographs to examine changes in seafronts and beaches (e.g. Valentin, 1954). Sheppard (1912) also examined illustrations of former churches and villages which have been destroyed by coastal erosion.

There are a number of sources unique to coastal studies. These include pilot books and navigation manuals, particularly for major estuaries (Robinson, 1951) and records of lighthouses (de Boer, 1964). Many government and official documents such as reports of harbour commissioners and engineers and records of local authorities can provide valuable information to locate shorelines and features at specific dates (So, 1974). The Royal Commission on Erosion of Coasts (1907) collated information and opinion on this subject in Britain at that date.

The accuracy of source material has received a variety of assessments in coastal studies ranging from vague comparative statements of 'more' or 'less' accurate than another source to actual measures of accuracy for specific maps. Carr (1962, 1980) has discussed accuracy in some detail and Bird (1974) identified sources and variability of error in his study of the Australian coast:

> early maps are more reliable in showing the alignments of well-defined cliffed and rocky shores than the configuration of swampy coasts, especially where the tidal range is large, the inter-tidal zone broad and mappable features intricate and diffuse.

The use and interpretation of all these sources requires careful corroboration, and consideration of the theoretical likelihood of the different situations conveyed. Brookfield (1952) put all this very neatly in his report on the estuary of the Adur river in Sussex, England:

> The Armada Survey of 1587 showed a lagoon behind the beach at Southwick and also a series of lagoons extending west of the Adur past Worthing to Goring. The first minute book of the Shoreham Harbour Commissioners (1760–1812) has a number of references to areas of salterns and reclaimed land on the north bank of the harbour, much of which was overwhelmed by the increased tidal flow that resulted from the opening of the new entrance in 1763. One parcel of land at Southwick was referred to as the 'Old Salts'. It is unlikely that it would have this name if 150 years before the site had been open sea, as suggested by Cheal and Morris. More probably it represents the lagoon (and adjoining marsh) at Southwick of the map of 1587, then separated by a shingle bank from the sea, but by 1760 fronting on the river.

Corroboration of historical evidence in coastal studies is often provided by a variety of field evidence. Morphology is very important for confirming positions of old shorelines in the form of beach ridges (e.g. Lewis, 1932) as is the nature of deposits (e.g. Barnes and King, 1957). Kestner (1962) has combined evidence of salt-marsh discontinuities with documentary evidence of their development. Others have used evidence of structures such as old bridges (Kidson, 1950; Brookfield, 1952), Martello towers and pill-boxes (Hardy, 1966) and lifeboat houses (Athearn and Ronne, 1963; Hardy, 1966) to prove erosion or deposition. Archaeological evidence has also been widely used (e.g. Akeroyd, 1972) to investigate both vertical changes in sea level and horizontal changes in the position of the shoreline.

Although historical sources are used in coastal studies for reconstructions at particular times they are more often employed to elucidate changes, often with the aim of predicting future changes or solving practical problems. Much work has concentrated on establishing sequences of development and change in coastal spits and offshore bars, both because they

Table 4.4
THE OPENING AND CLOSING OF INLETS IN THE SNOW MARSH AREA OF NORTH CAROLINA, 1760–1962 (after El-Ashry and Wanless, 1968)

Year	Corncake inlet		New inlet		Northern inlet	Southern inlet
	Width (ft)	Migration direction	Width (ft)	Migration direction	Width (ft)	Width (ft)
1760	—		Opened		—	—
1880	?		Closed		?	?
1926	1200		950		—	—
1935	950	2000 ft northward	Closed	—	—	—
1950	1450	400 ft southward	Closed		500	400
1953	650		300		Open	Closed
1954	1600		800		120	220
1958	Closed		350		Closed	250
1962	Closed		600	$\frac{1}{2}$ mile southward	Closed	Closed

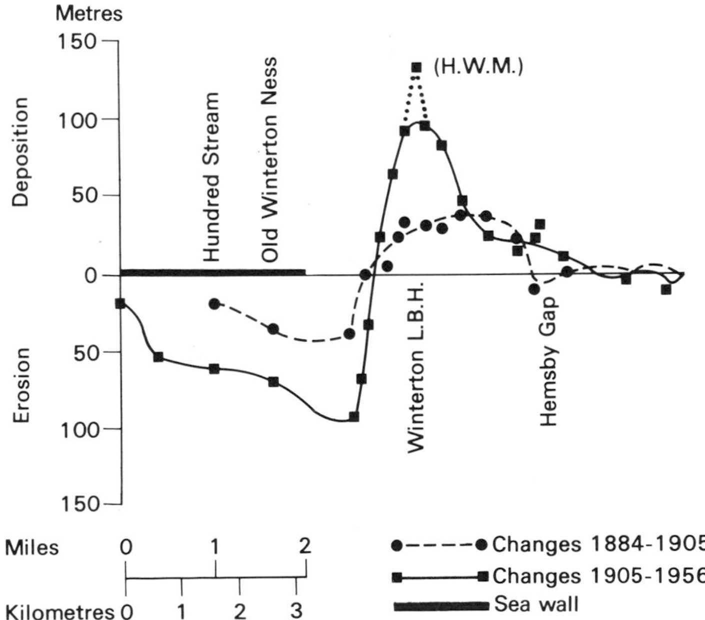

Figure 4.8. COASTAL CHANGES IN THE WINTERTON AREA OF NORFOLK.
(Source: Hardy (1966))

are noticeable landscape features and also because their changes have practical implications. Thus spits such as Spurn Head (de Boer, 1964), Gibraltar Point (Barnes and King, 1957), Orford Ness (Steers, 1926; Carr, 1969), Dungeness (Lewis, 1932), Dawlish Warren (Kidson, 1950) and many others in Britain and elsewhere have been subjected to close historical analysis. Similarly the movement of offshore bars is illustrated in Robinson's (1960) study of Hallsands. Such studies have often been closely linked with the investigation of processes and the influence of currents, tides and winds. Others have elucidated phases of different activity, the appearance and disappearance of features such as inlets, as shown in Table 4.4, and the variation in rates of processes over time (Figure 4.8). Rates of erosion derived from maps and other documents are widely quoted, as in Cocco's (1976) study of the Italian north Ionian coast, Valentin's (1954) study of the east coast of Holderness, Lincolnshire, Rude's (1923) investigation of Cape Hatteras and Seeling and Sorensen's (1973) work on the Texas shoreline.

Reconstruction of conditions before and after European settlement has been a major theme in areas such as Australia (Bird, 1974). Historical sources have played a major role in dating features such as old shorelines (e.g. Lovegrove, 1953) and in dating man-made structures (Homan, 1938) from which rate and mode of subsequent development can be inferred. The effects of major storms have also been widely studied, and these often involve investigations of the historical background (e.g. Stoddart, 1971).

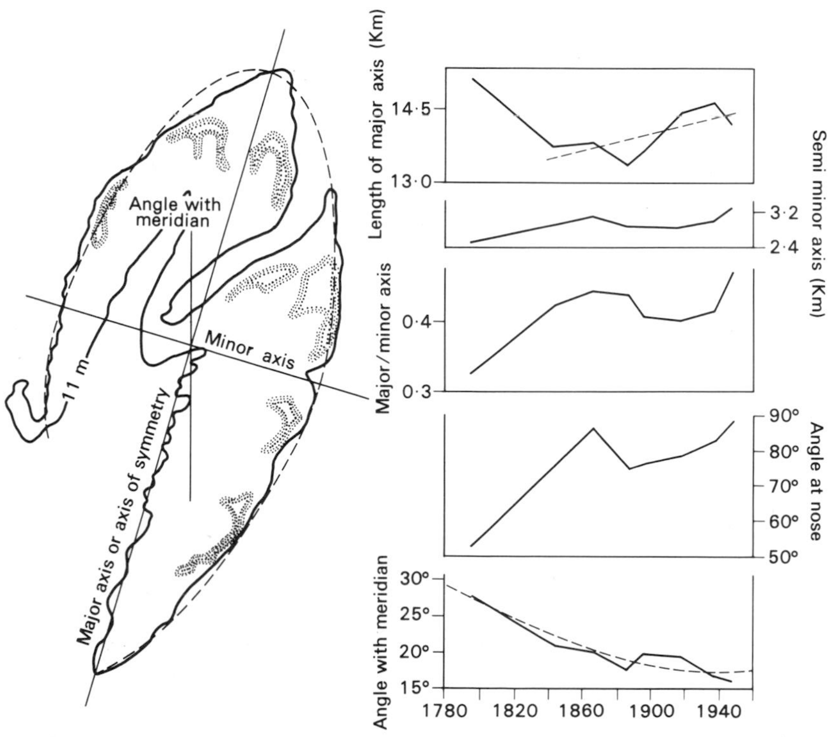

Figure 4.9. MEASUREMENTS OF CHANGE IN THE GOODWIN SANDS, 1780–1960. (Source: Cloet (1954))

The amount of information that can be obtained on the location, nature and time of occurrence of change depends largely on the detail of reconstruction possible from the historical sources. Ideally, a sequence of maps or superimposed outlines may be obtained (Figure 3.6) and maps derived from these (Robinson, 1966). The distribution and patterns of change may also be described, measured and analysed. Figure 4.9 shows measurements of change in the Goodwin Sands over the period 1780-1960 (Cloet, 1954). Cloet (1963) and others have derived maps of minimum and maximum extents of shoals and of thickness of zones of movement by comparison of isobaths on successive hydrographic surveys. This can be automated by digitizing and computer-aided plotting.

Case study: Spurn Head, Humberside

One study in which a variety of historical sources are used is now examined in more detail. De Boer's (1964) study of the evolution of the spit at Spurn Head on the east coast of Humberside developed from a desire to understand its history since there was some evidence that it had been breached several times and there was a possibility of this happening again in the near future. He made extensive use of records of legal disputes over

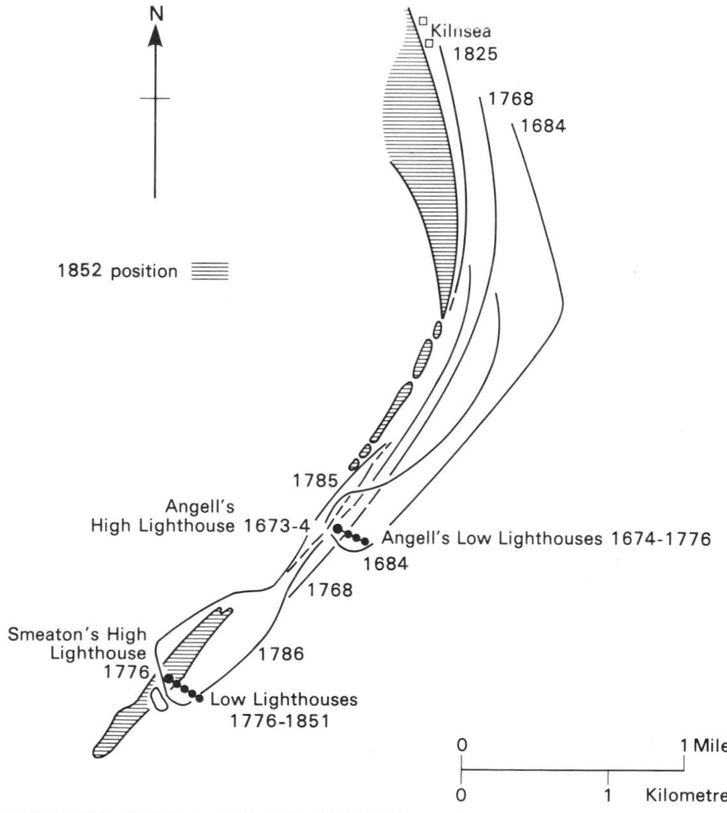

Figure 4.10. THE EVOLUTION OF SPURN POINT, HUMBERSIDE, FROM 1684 TO 1852.
(Source: de Boer (1964))

lighthouses which arose because their location had to be changed a number of times as the coastline moved. Such legal evidence is very precise, and in some cases gives details of the position and form of the spit. Books of navigation describing the channels and navigation charts are also cited and maps were used for reconstructions in the more recent period. From this evidence a picture of the development of the spit over the last three centuries or so was derived (Figure 4.10). De Boer then tested this model to see if it fitted earlier phases of development. For the thirteenth to sixteenth centuries he used the evidence of place-names, manorial rolls which indicate the existence of habitation on the spit at one time, and Abbey chronicles which describe the settlement. Continual damage to the settlement from the fourteenth century onwards is evidenced by depopulation, tax-remittances and decline of the port. There is also some fragmentary evidence for an earlier phase of development of the spit. Thus a cyclic model of development appears to apply but, as de Boer himself acknowledged, much of the reconstruction must remain somewhat speculative.

4.4 RIVER STUDIES

The spectacular and rapid movements of rivers such as the Mississippi were instrumental in drawing attention at an early date to the dynamic nature of some streams and provided the stimulus for early studies such as those of Fisk (1944) on alluvial features and Matthes (1947) on Mississippi cut-offs. Other very mobile channels which have caused damage or navigation difficulties such as the Brahmaputra and Ganges systems have also long been studied (e.g. Login, 1872; Wood, 1924). Figure 4.11 shows the extensive nature of channel changes on the Kosi river in northern India. Historical sources, particularly maps, have been widely used in the last decade or so to elucidate the spatial distribution of changes in channels and to understand the controls on channel movement. They have also been used to investigate the nature and rates of channel changes and to understand the relation of channel changes to fluvial processes and sediment dynamics. The impact of human activities on river channels has also been a major theme of study. A selection of recent applications based on the use of historical sources is to be found in *River Channel Changes* (Gregory, 1977) and *Timescales in Geomorphology* (Cullingford, Davidson and Lewin, 1980).

Some of the studies using historical sources aim simply to establish whether a stream is stable in position (e.g. Alexander and Nunnally, 1972). Much fluvial work, though, focuses on the chronology and sequences of changes exhibited by a series of maps and analyses the characteristics and nature of the changes. By far the greatest number of studies are of changes in channel pattern, for which maps provide the prime source. For example, Mosley (1975), Hooke (1977b) and Lewin and Brindle (1977) have studied planform changes in rivers in Britain. The mechanisms and types of changes in river bends have been studied by Carey (1963), while Brice (1974) produced general models of change from such analyses. Handy (1972) has developed a method of dating cut-offs from analysis of the subsequent development of the river channel by relating the rate of movement to distance from the edge of the meander belt. Historical sources have also been used to date old river courses either directly (e.g. Davies and Lewin, 1974) or by relative age and form (e.g. Schattner, 1962; Klimek and Trafas, 1972) and to elucidate, or at least corroborate, the development of sedimentary features (Fisk *et al.*, 1954; Bluck, 1971). The development of flood plains has

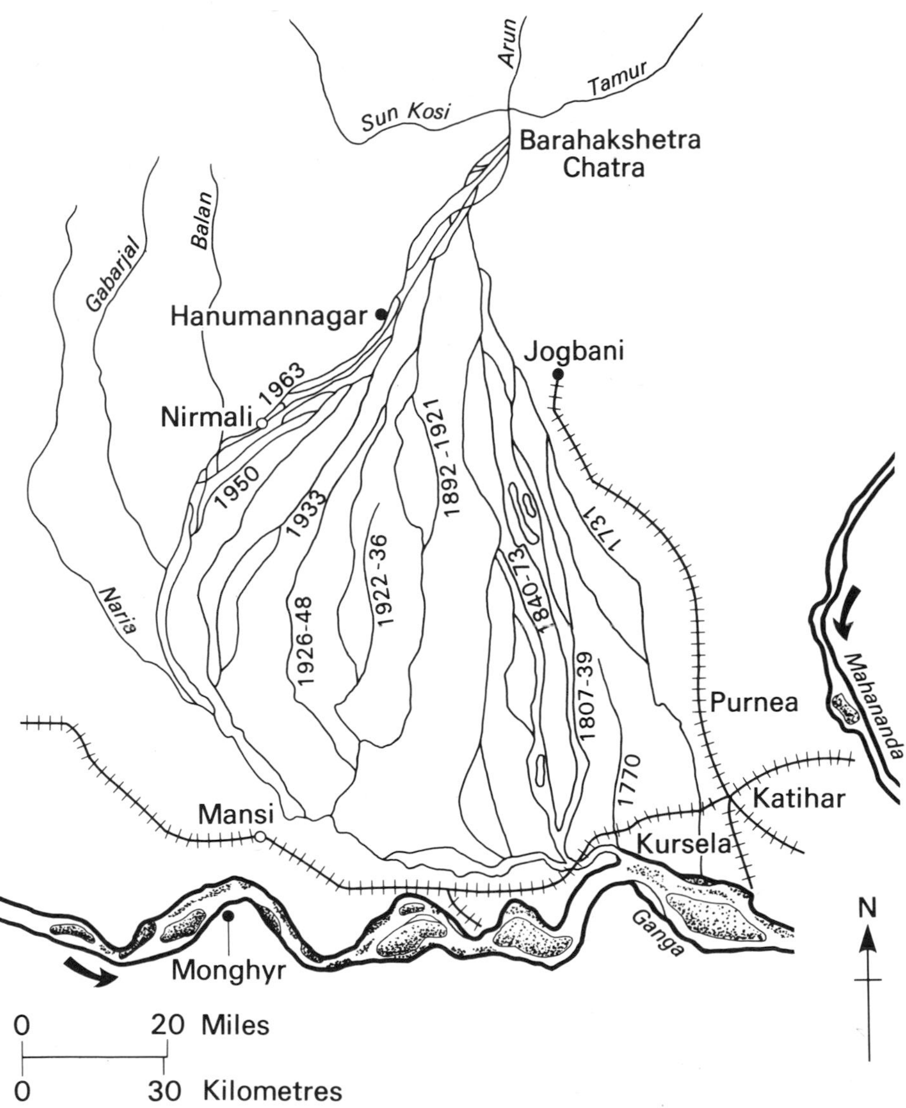

Figure 4.11. SHIFTING COURSES OF THE KOSI RIVER, NORTH INDIA, 1731–1963.
(Source: Gole and Chitale (1966))

also been investigated by studying the nature and rate of changes in stream courses (Schmudde, 1963; Carey, 1969).

Sources for study of cross-sections are more limited in availability, but Daniels (1960) examined changes in the form of a drainage ditch in Iowa using drainage records, and many studies have been made of erosion and deposition in channels resulting from land-use changes (e.g. Piest *et al.*, 1976). Both Burkham (1972) and Schumm and Lichty (1963) have studied channels in the United States in which very large changes in width and other characteristics have occurred in the last 100 years or so.

Historical studies of changes in channel networks are also few in number. The works of Edgar and Melhorn (1974) and Piasecka (1974) are examples. A recent study by Ovenden and Gregory (1980) has illustrated the potential of maps for studying changes in networks in Britain. In spite of the problems of scale and varying criteria for portraying stream lines on maps, they have shown that comparison of networks portrayed on large-scale Ordnance Survey maps indicates significant network expansion, and their study demonstrates the impact of human activities on the stream network.

The solution of practical and engineering problems is also an important field of study and much Russian work has been concerned with this aspect (e.g. Kondrat'yev and Popov, 1967). There are studies of rivers undercutting structures (e.g. Duncanson, 1909), or devastating land (e.g. Gole and Chitale, 1966). The effect of straightening and channelization of rivers has been usefully investigated with the aid of historical documents, especially in relation to the Mississippi (Matthes, 1947; Winkley, 1972; Stevens *et al.*, 1975)

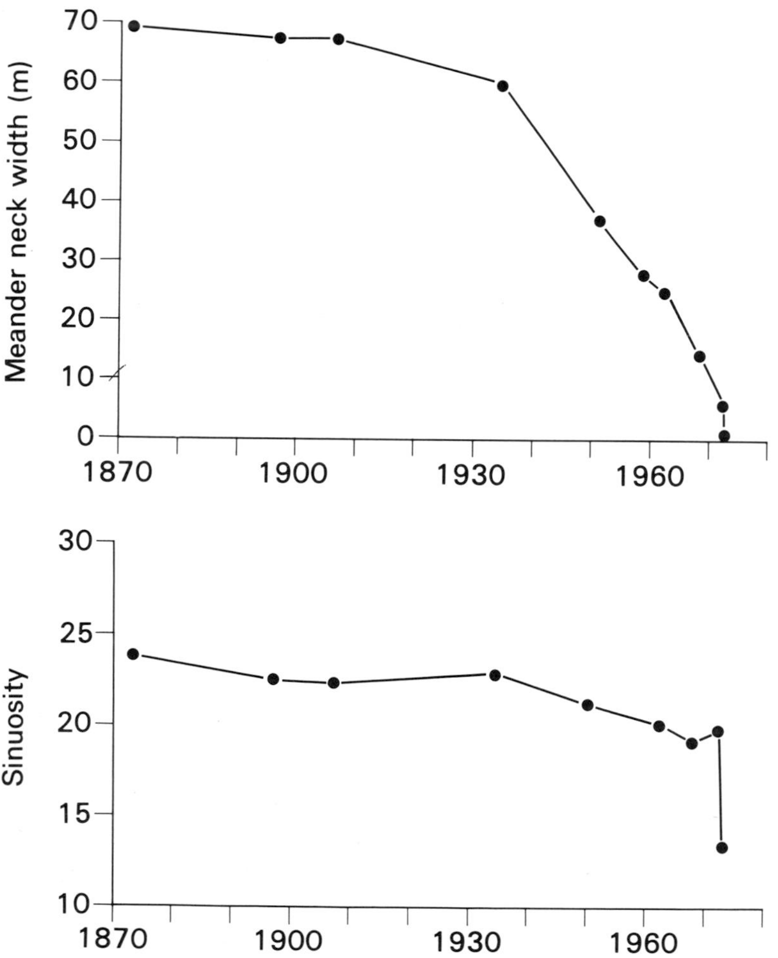

Figure 4.12. CHANGES IN MEANDER NECK-WIDTH AND SINUOSITY ON THE RIVER BOLLIN, CHESHIRE, 1872–1973. (Source: Mosley (1975))

and also other rivers such as the Hernad in Hungary (Laczay, 1977). The sequence of meander development after channel-straightening has been examined with aerial photographs by Hussey and Zimmerman (1953) and Noble and Palmquist (1968). Historical sources have been used to provide some 'background' context to events and to establish the situation obtaining prior to a particular change such as a cut-off (Johnson and Paynter, 1967).

In many river studies, maps have proved to be the most important source. These vary in date, scale and coverage and thus in amount of information they can provide but generally they can show by comparison whether or not there has been a change in river planform. With the aid of a series of detailed maps the sequence of changes in particular parts of a river course can be established, measurements of direction, distances and rates of change can be made and processes of movement and change analysed, as shown by Mosley (1975) in Figure 4.12 and Sundborg (1956) in Figure 4.13. The scale of map required will depend on the purpose of the study and the size of the river and is often a limitation on detailed analysis. For detailed analysis of planform changes, scales of 1:10 000 or larger are

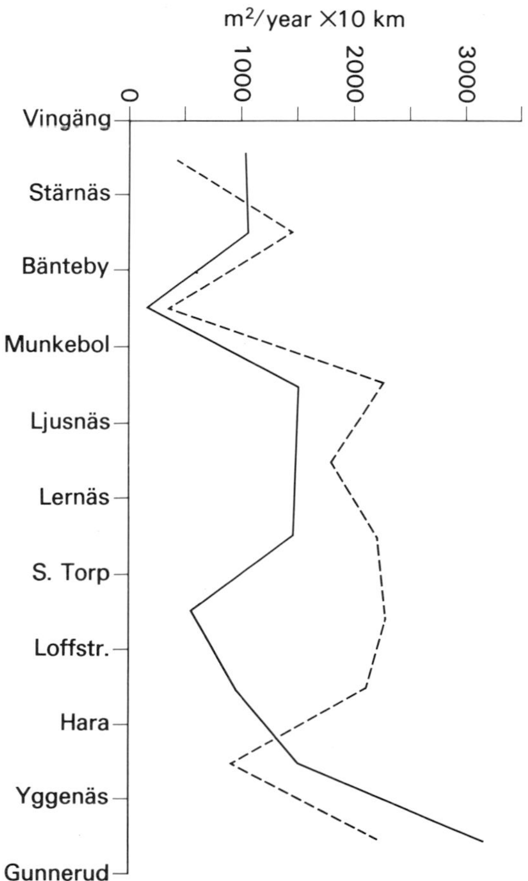

Figure 4.13. RATES OF EROSION ON THE RIVER KLARÄLVEN, SWEDEN.
The solid line relates to the period 1850–1950 and the dashed line to 1800–1850. (Source: Sundborg (1956))

advisable. The other main limitation is that maps are two-dimensional whereas changes may be in three dimensions. The methods available for analysing changes are described by Dietz (1952), Sundborg (1956), Koc (1972), Lewin and Hughes (1976) and Hooke (1977b).

There are some specific problems in the use of maps for river studies particularly where accurate measurements are required; these stem from factors such as the methods of survey used and the manner of representing channels. Various criteria for mapping the level of water have been used and conventions vary. For example, sometimes gravel bars are mapped, sometimes they are not. As much background as possible on survey methods and mapping conventions should be obtained and small changes investigated fully to check that they are real. (See above 3.2–3.6). Partly because of the rapidity of change, braided channels have received less attention than meanders, but studies using map evidence have been conducted on Scottish rivers (Werritty and Ferguson, 1980) and in Norway (Nordseth, 1973). Lewin and Weir (1977) have derived indices of braiding from maps of different dates for the river Spey in Scotland, and some of their results are summarized in Table 4.5.

Table 4.5
BRAIDING INDICES FOR SIX DATED MAPS OF THE RIVER SPEY, SCOTLAND
(after Lewin and Weir, 1977)

	1967	1889*	1887	1882*	1876	1760
Map scale	1:10 000	1:2500	1:2500	1:2500	1:11 000	1:21 000
Braiding index (Brice, 1960)	2.26	6.50	5.43	5.32	5.00	7.40
Number of complete links	56	62	86	26	95	59
'Islands'	22	21	29	7	29	19
'Islands' with 2 links	4	1	6	3	3	1
3	5	2	6	3	—	—
4	5	7	4	1	14	7
5	3	7	4	—	3	4
6	—	1	4	—	2	4
7	3	3	2	—	4	—
8+	2	—	3	—	3	3

*Maps cover only half the length covered in other surveys; the rest cover the same 3 km of valley.

Another valuable source often useful in close association with maps is the aerial photograph (Figure 4.14). These can show clearly traces of old stream courses in flood-plains as well as considerable detail of the contemporary channels, and for this reason they are frequently used for the analysis of braided channels. Aerial photographs may be used to corroborate historical sources as well as providing evidence of a course at a particular date. Ground photographs are occasionally useful for examining historic changes. For example, Schumm (1971) compared two photographs of the North Platte river taken in 1895 and 1967. Other illustrative and graphic sources have not been widely used, though Towl (1935) used a painting of the river Missouri from a century ago to show how heavily wooded its valley was then.

There are also some sources specific to fluvial applications. These include various river surveys and technical data produced by government or public organizations or by engineers. Some valuable records, including maps and sections, may still be held by private organizations such as mining or industrial companies, and are data sources which remain

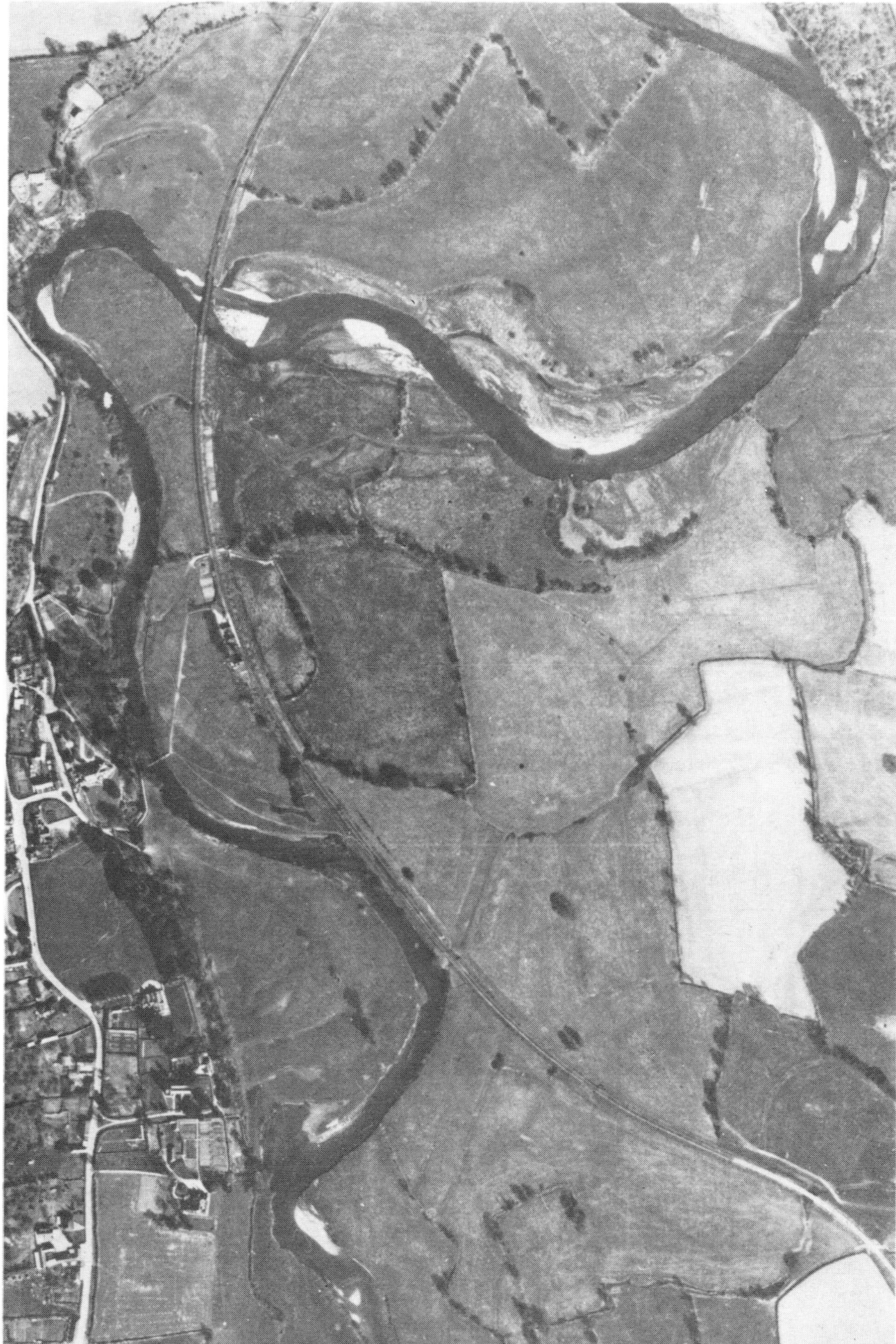

Figure 4.14. VERTICAL AIR-PHOTOGRAPH OF PART OF THE RIVER EXE NEAR EXETER, DEVON.
(Source: Department of the Environment CPE/UK/1995)

largely unexploited. Drainage and bridge records exist for many areas and for some counties in England there are books containing vertical and horizontal bridge sections dating back to the early nineteenth century on which details of streams are shown. Pilot books are available for some of the largest rivers of the world, and Kondrat'yev and Popov (1967) highlighted their value: 'In compiling pilotage charts one uses field data on the channel which far exceed navigation needs. These data are an extremely valuable source for obtaining much information used in a wide range of hydrological computations.' Early Mississippi surveys have likewise been used (Fisk, 1944; Matthes, 1947).

As in many other areas of study, the diaries and journals of expeditions and travels in the New World and recently settled countries are a useful source. These have been used by Towl (1935), Eardley (1938), Trimble (1970) and Burkham (1972) in various parts of North America and by Eyles (1977a and b) in Australia to ascertain conditions before and during the settlement period and to corroborate information from other sources such as maps. Often these accounts provide only qualitative descriptions but occasionally they include estimates of the dimensions of, say, a river channel. Closely related to personal written observations is oral evidence which has been widely used in studies of the early settlement period in the United States. It has not been much used elsewhere but its value should not be underestimated. One of the present authors gained considerable insights by talking to a local resident and water bailiff with long familiarity of the river Exe in Devon. This oral evidence confirmed the movements of the river discovered from nineteenth-century maps but also told of unsuspected works on the river, including use of explosives early in the twentieth century to blow up obstructed pools.

In India and other areas of early civilization, ancient writings do provide some evidence, though it is usually rather scanty. Some examples of the use of this kind of material are the work of Russell (1954) in Anatolia, Turkey, Schattner (1962) in the Jordan Valley and Wood (1924) in India. General histories have also been used (Wilhelmy, 1966; Taher, 1974) and occasionally other types of writings provide information; for example, Stevens *et al.* (1975) have referred to the work of Mark Twain and Charles Dickens. Records such as medieval manorial rolls, chronicles and estate and ecclesiastical papers have not been widely used in British river studies. This may be because of the abundance of information provided by maps but it is more likely because of the fragmentary, qualitative nature of these sources. Nevertheless, they remain to be more fully exploited, as coastal and other geomorphological studies have demonstrated their potential. Sundborg (1956) has used some such records; for example, a seventeenth-century tax return describing the destruction of land by a river in Sweden. He added, however, that it is only with the help of old maps that the course of the river can be traced in any detail. Legal records can also prove a valuable source because disputes occasioned by loss of land and moving boundaries often reach the courts. It was a legal case in Des Moines, for example, which first drew the attention of Handy (1972) to the extent of river movement since a cut-off in 1880. Legal disputes over boundaries often fully document all the evidence of river movement (e.g. Bowman, 1923).

Examination of administrative and similar boundaries on a map can often provide a clue to possible channel changes as where a sinuous boundary runs adjacent to, but not coincident with, a river. Dating these boundaries is not easy. Most parish boundaries in England, for example, were established by the late Middle Ages but were much altered in the nineteenth century to accommodate changing distributions of settlement and the growth of towns.

There is perhaps a keener awareness in fluvial geomorphology than in some other fields of physical geography of the need to assess the accuracy of historical documents. This is probably because of the quantitative nature of analysis in fluvial geomorphology and because much of the work is very recent and related to process studies. It is probably also due to the considerable use of maps which are amenable to relatively straightforward tests of accuracy. Actual measures of accuracy are presented by Sundborg (1956) and Mosley (1975), while Lewin and Hughes (1976) give quantitative estimates of errors involved in the use of small-scale maps. Field reconnaissance is the main method of confirmation used. It involves checking the positions and forms of old channels for correspondence with documentary evidence and distances from fixed objects can also be measured in the field. Other corroborative evidence employed includes the age of trees (e.g. Bowman, 1923; Eardley, 1938; Trimble, 1970), the presence of datable remains (e.g. Alexander and Nunnally, 1972) and the evidence of deposits, sedimentation patterns and stratigraphy (Fisk, 1944; Nilsson and Martvall, 1972; Brice, 1974).

Case study: channel changes on Devon rivers

The purpose of this study was to investigate where, in what way and at what rate rivers had changed, over as long a period as accurate evidence allowed (Hooke, 1977a and b). It was designed also to complement analysis of present processes and rates of erosion. An initial survey was made to identify unstable streams and reaches by comparing courses on 1:10 560 Ordnance Survey maps of different dates. Several mobile streams were located in south-east Devon and detailed evidence was then compiled at a scale of 1:2500 using County Series and National Grid Series Ordnance Survey maps and tithe maps (Figures 4.15 and 4.16). The map courses were superimposed on a base tracing and then digitized. The data were used to analyse the meander characteristics at each date, in spectral analysis of the patterns and to produce graph plots of the courses. Meander forms and changes were also measured directly from the superimposed map traces. Considerable lengths of these streams were found to have changed course in a 130-year period. Rates of movement of up to 3 metres per year were found and the type of change showed a certain consistency—a tendency for the bends to become more sinuous, to increase in meander-amplitude and for them to shift downstream, as can be seen in Figure 4.17. There were few natural cut-offs and the changes were consistent enough in most places for maps and aerial photographs to give a reasonably representative picture of changes over time. The influence of man could be seen in formerly straightened reaches which had resumed meandering. Rates of movement for different periods were compiled and showed an apparent increase: historical material was used to investigate possible causes (see section 6.3). The historical map evidence was corroborated by field evidence of old channels. In general the study indicated widespread instability, rapid and long-continued changes and also demonstrated the nature of the changes. The results have implications for interpreting flood-plain history, prediction and control of channel movement and illustrate the kind of findings emanating from other recent fluvial studies which show that changes are much more rapid than formerly thought.

Figure 4.15. PART OF THE TITHE MAP OF BRAMPFORD SPEKE PARISH, DEVON.
Showing the same part of the river Exe as in Figures 4.14 and 4.16. (Source: Devon Record Office)

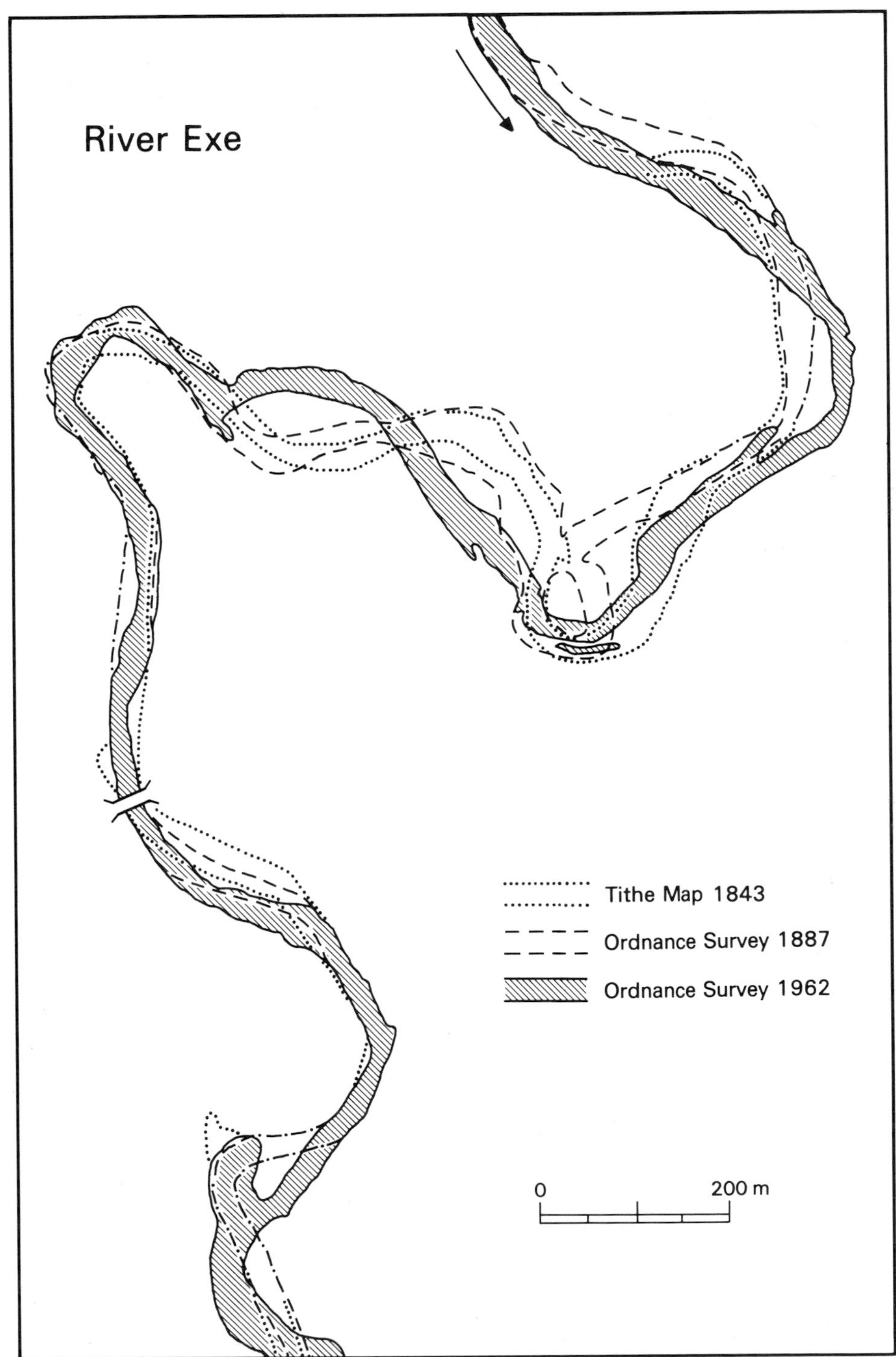

Figure 4.16. THE COURSE OF THE RIVER EXE ABOVE EXETER, DEVON, IN 1843, 1887 AND 1962

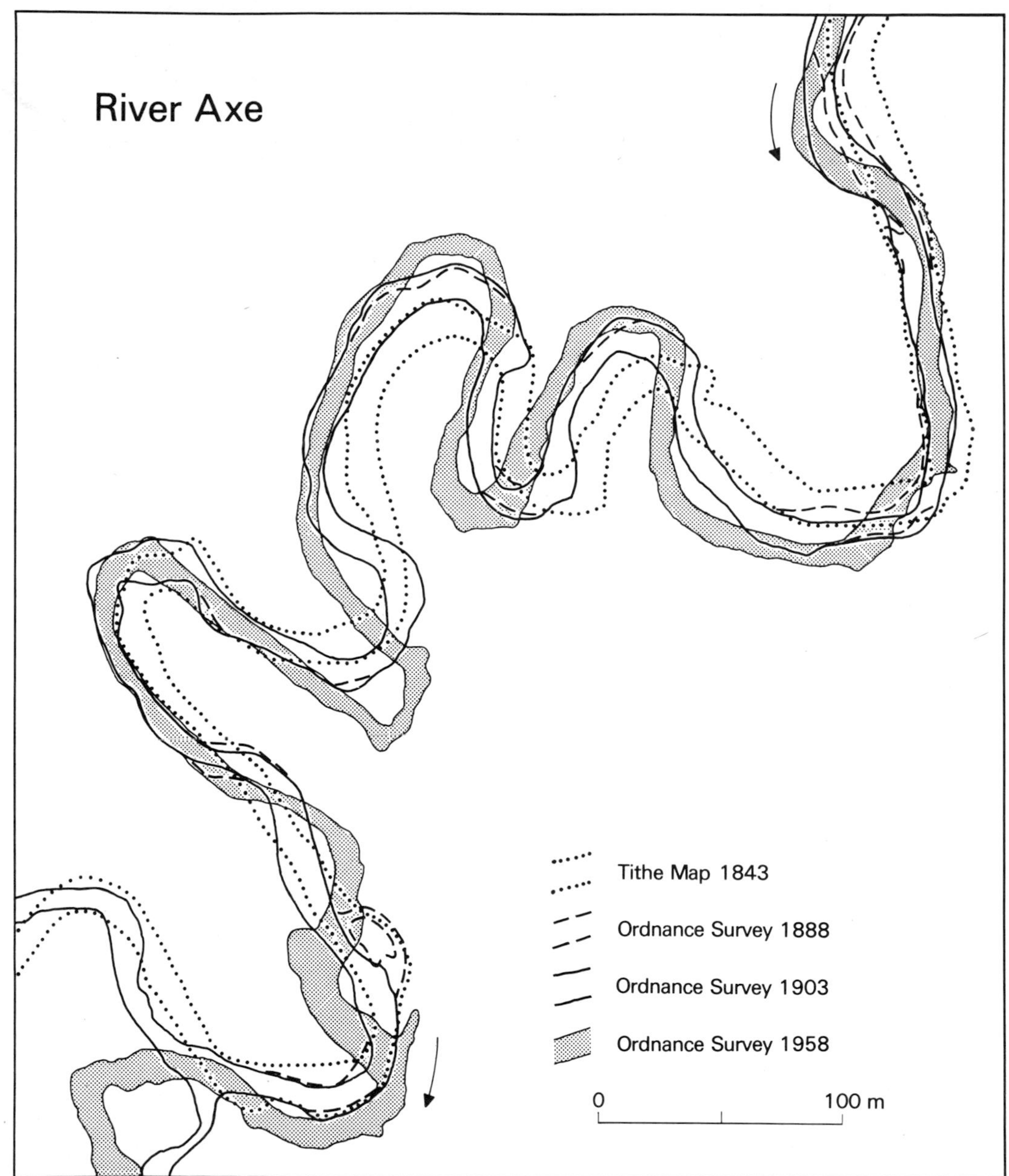

Figure 4.17. THE COURSE OF THE RIVER AXE NEAR AXMINSTER, DEVON, IN 1843, 1888, 1903 AND 1958.

Work in the fluvial field is continuing to elucidate the types of changes, their variability and distribution, and these longer-term changes are being related to detailed studies of processes and magnitude–frequency of events.

4.5 VEGETATION STUDIES

The impact of man's activities on the vegetation patterns of countries which underwent European settlement during the last two centuries has been a theme which has been followed profitably with the assistance of historical documents. The general approach has been to reconstruct the pre-settlement vegetation and to assess the causes of change and the effects of particular types of land use and land management (e.g. Cottam and Stewart, 1940; Burcham, 1956). Much work has focused on the most sensitive marginal areas such as the semi-arid south-west of the United States (e.g. Humphrey, 1958; Johnston, 1963; Buffington and Herbel, 1965). An important impetus for this work was the need to assess the potential of land for different types of land use. In this context Stearns (1949) said,

> It was evident that some knowledge of the history of the stands would be necessary before the relative persistence and reproductive capacity of the various species could be understood.

Another major theme in these studies of vegetation is related to the more academic question of vegetation successions—whether certain associations are climax vegetation, the progress of invaders and exotic plants and assessment of whether certain plants are on the increase or decrease (e.g. Chavannes, 1941; York and Dick-Peddie, 1969).

Travellers' and explorers' descriptive accounts are important sources of information on New World environments. They vary in value depending on the writer's perception and previous experience, contemporaneity of the records, the season of travel and other potential causes of bias discussed in Chapter 3. The most useful descriptions are those which can be relocated and compared with later periods, as Mason (1963) has suggested. A major problem with many of these early descriptions is that they lack details of species and vegetation composition. They usually allow the presence of woodland or grassland to be identified, but Burcham (1956) commented in a North American context that,

> while accounts of contemporary travellers are of great value in giving us an appraisal of the general nature of the forage cover at the time California was being settled they afford few details of its botanical composition or floristic characteristics. It is to early botanical collections that we must turn for these details.

Some early military expeditions did include trained botanists. Although some of their surveys are more systematic than the travel records, many suffer from similar problems of perception and variability. Johnston (1963), in his study of grassland in south Texas and north-east Mexico, is very sceptical of the value of many of these sources, pointing out confusions over plant-names. The problem of perception of vegetation and the image presented by reports is exemplified by the 'myth' of the Great American Desert east of the Rockies which prevailed in the public mind for much of the nineteenth century. Many early explorers and settlers were unfamiliar with extensive grassland areas of the type found on the plains of the American West and thought these a barren, alien and hostile environment (Lewis, 1962, 1965a and b; Watson, 1967).

It is from the plats and survey notes of the United States Federal Land Survey and to a lesser extent boundary, railway and settlement surveys that most useful detail has been derived. The practice of identifying trees at section corners and along certain lines means that some idea of the species composition and density of woodland can be obtained. York and Dick-Peddie (1969) have summarized the method:

> The use of trees to witness a section corner constituted a one mile square grid sample. A witness tree was selected in each quadrant as near the corner as possible and the species, trunk diameter, and distance and azimuth from the corner were recorded. In the New Mexico surveys the surveyor noted the vegetation on each line of all sections in every township. He then also summarised the vegetative and topographic features of the entire township.

Stearns (1949) has also tested the accuracy and consistency of these surveys, on which further detail has been provided in Chapter 2 above.

Sears (1925) used the plats and notes to construct a map of the natural vegetation of Ohio.

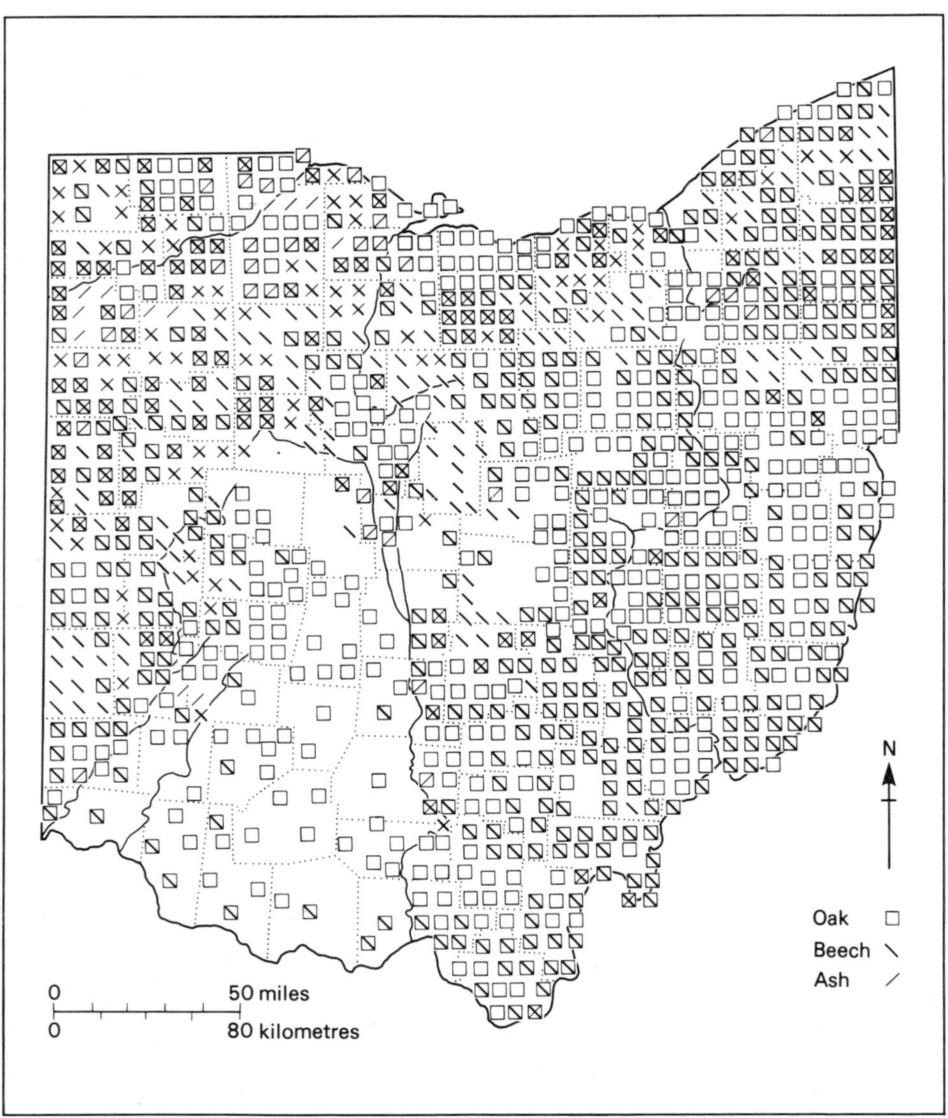

Figure 4.18. MAP OF THE VIRGIN FOREST OF OHIO.
(Source: Sears (1925))

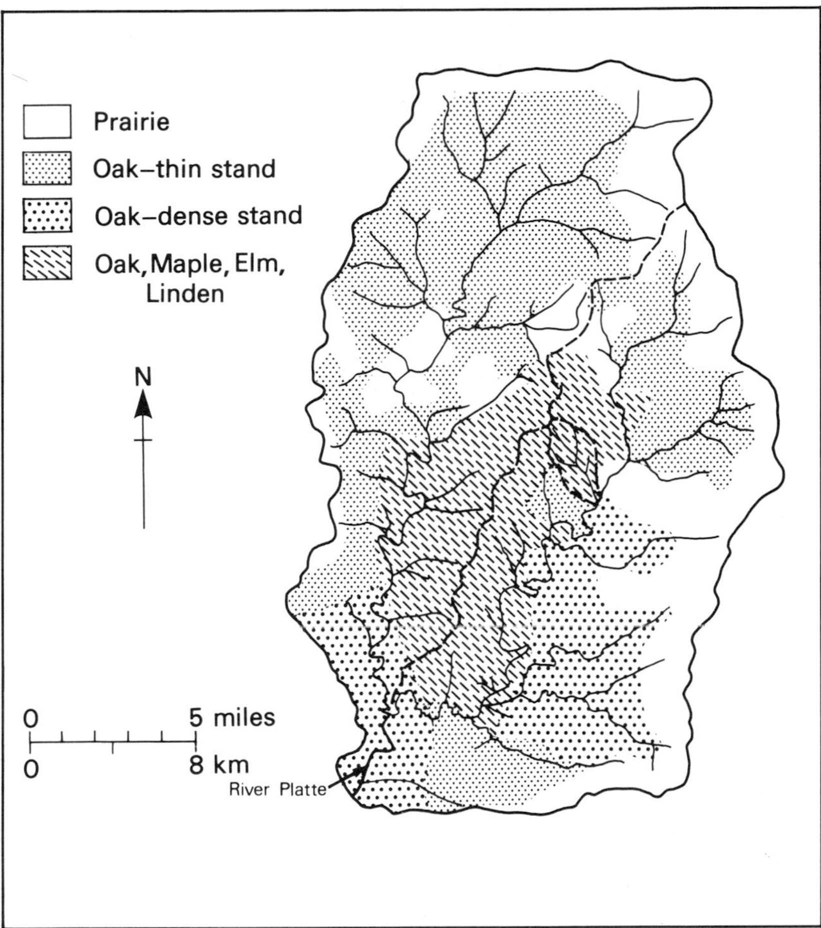

Figure 4.19. THE NATURAL VEGETATION OF THE PLATTE RIVER BASIN, WISCONSIN, 1832–1833. (Source: Knox (1977))

He generalized the forest type in each township but indicated the dominant species, whether oak, beech or ash (Figure 4.18). Knox (1977) also based a reconstruction of pre-settlement vegetation in the Platte river basin, Wisconsin, on these surveys and Trewartha's (1940) transcription of them (Figure 4.19). Similarly, Murton (1968) has reconstructed the vegetation of an area of North Island, New Zealand, from early land surveys (Figure 2.4).

Ground photographs have been quite widely used to reconstruct vegetation cover, especially in North America. For example, Graf (1979b) has mapped the progress of deforestation associated with gold-mining in Colorado from photographs (Figure 4.20). Several authors have tried to photograph present vegetation from the same location for comparison with old photographs (e.g. Shantz and Turner, 1958; Hastings and Turner, 1965). Shantz and Turner encountered a number of difficulties when attempting to re-photograph sites in Africa due to factors such as urban growth and agricultural development. Bahre and Bradbury (1978) have assessed the effects of different

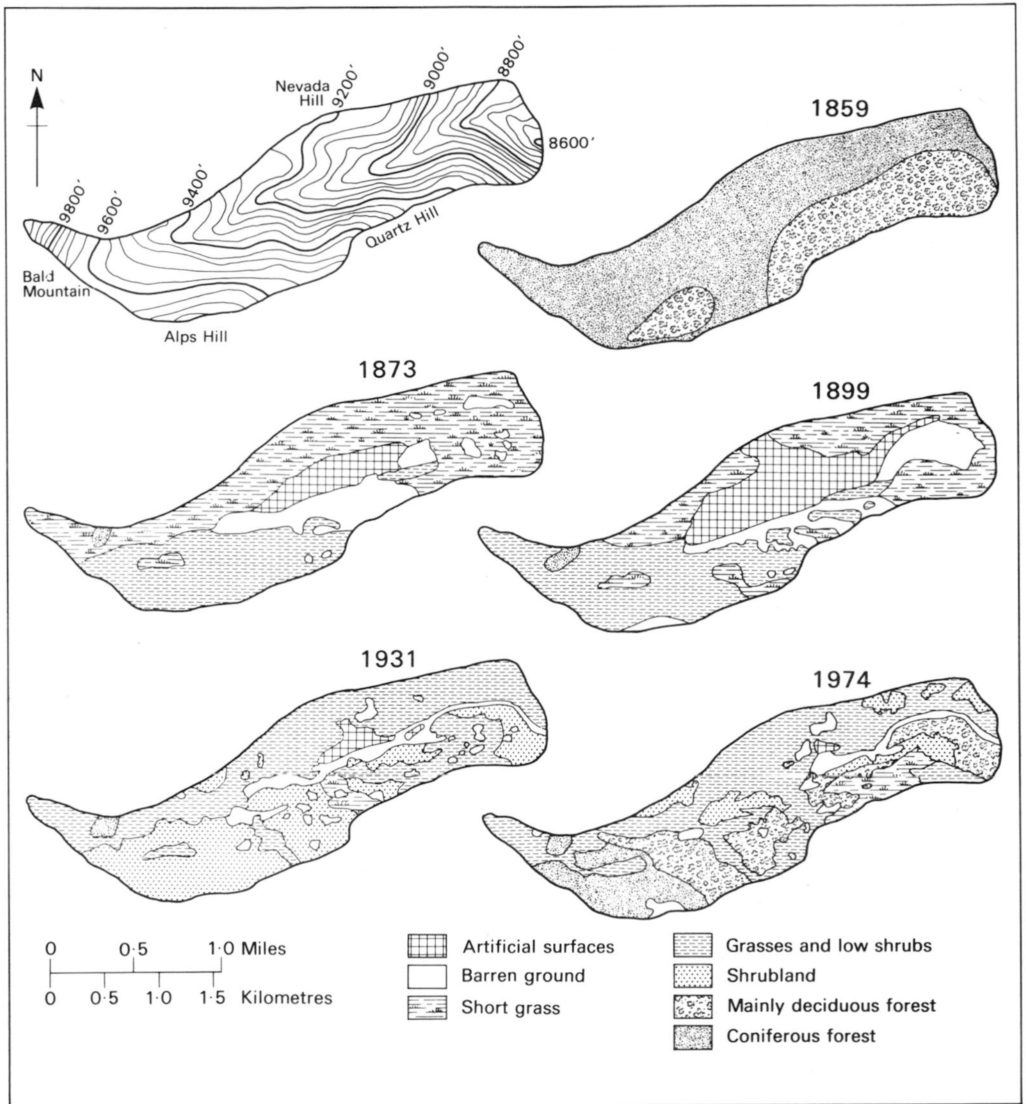

Figure 4.20. VEGETATION OF THE NEVADA GULCH BASIN, COLORADO, 1859–1974, RECONSTRUCTED FROM PHOTOGRAPHS. (Source: Graf (1979b))

management on either side of the Mexico/United States border by repeat photography of boundary posts. Other graphic sources have been used in vegetation studies, including sculptures in ancient India (Randhawa, 1952), bas-reliefs showing transport of timber in the Lebanon (Mikesell, 1969) and eleventh-century woodcuts in Britain showing swine-feeding and wood-cutting (Darby, 1956).

In European and other long-settled countries it becomes difficult to separate strictly biogeographical studies from studies in agricultural history, which has long been an important focus in historical geography and economic history (e.g. Darby, 1951, 1956).

Two quotations can perhaps demonstrate the increasing awareness of the value of historical sources to ecological studies:

> The existing state of vegetation in a particular site may be regarded as a product of interaction between its natural development and its use and exploitation by man. The former can be deduced from knowledge of the ecology and growth characteristics of the dominant species but for the latter one is dependent mainly on documentary sources. (Peterken, 1969)

Sir Harry Godwin, in the preface to Rackham (1975), said:

> Not least, ecologists, like local historians and economists, will note the new dimension provided by the combination of ecological expertise and effective use of documentary history.

Some records relating to ancient civilizations can provide information on vegetation and some of these are described by Mikesell (1969) for the Lebanon and Randhawa (1952) for India. Most of the earliest documentary sources in Europe date from medieval times. Wightman (1968) has reconstructed the vegetation of the Vale of Pickering for *c.* AD 1300 (Figure 4.21), and said of his sources:

> For most of England a largely unexploited supply of information on local vegetation is preserved in legal, ecclesiastical, royal and manorial documents of early medieval date. Most of the useful data appear in connection with three common situations: the transfer of land, the settlement of legal matters and the evaluation of manorial assets. Formal land transfers appear most frequently in cartularies, deeds and fiens which often include brief descriptions of the property being conveyed and of appurtenant rights in meadow, marsh, moor and woodland . . . Less numerous are manorial extents or inventories . . . which, while varying in their degree of detail, list and describe specific manorial assets including woodland and moorland.

Wightman (1968) also noted some of the problems of these early sources such as their uneven geographical distribution and their lack of quantitative values. Indirect evidence of

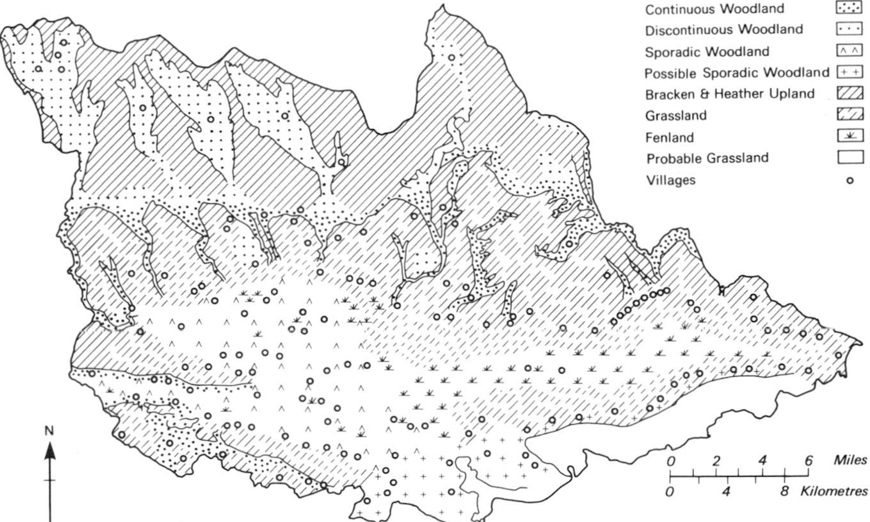

Figure 4.21. THE PROBABLE VEGETATION PATTERN OF THE VALE OF PICKERING, YORKSHIRE, *C.* AD 1300. (Source: Wightman (1968))

activities and products of an area can also give an indication of the vegetation as, for example, Darby (1956), Oldfield (1963) and Peterken (1969) have demonstrated. Nicholls (1972) and Tinsley (1976) have used records of legal disputes over rights to woodland and timber in their studies of the forest-cover of Staffordshire and north Yorkshire respectively. Rackham's (1980) study of the history of woodland in Britain provides invaluable information on sources as does Sheail (1980) in a review of sources for historical ecology.

In the Old World, some specific taxation and land-use surveys are available at very early dates: probably the most notable is the Domesday survey of eleventh-century England. Oldfield (1963), Peterken (1969), Rackham (1975) and others have used early maps to reconstruct the extent of woodland at various times and review some of the difficulties of using these sources (Figure 4.22).

Much corroboration of historical evidence comes from the present vegetation itself; the presence of relict plants and the present distribution of different associations and communities. Palynological evidence and historical sources are also frequently used to corroborate each other and historical documents may be used to extend pollen sequences forward. A selection of such studies are contained in a theme volume of the *Transactions of the Institute of British Geographers* (1976). Place-names can also be used, while an unusual

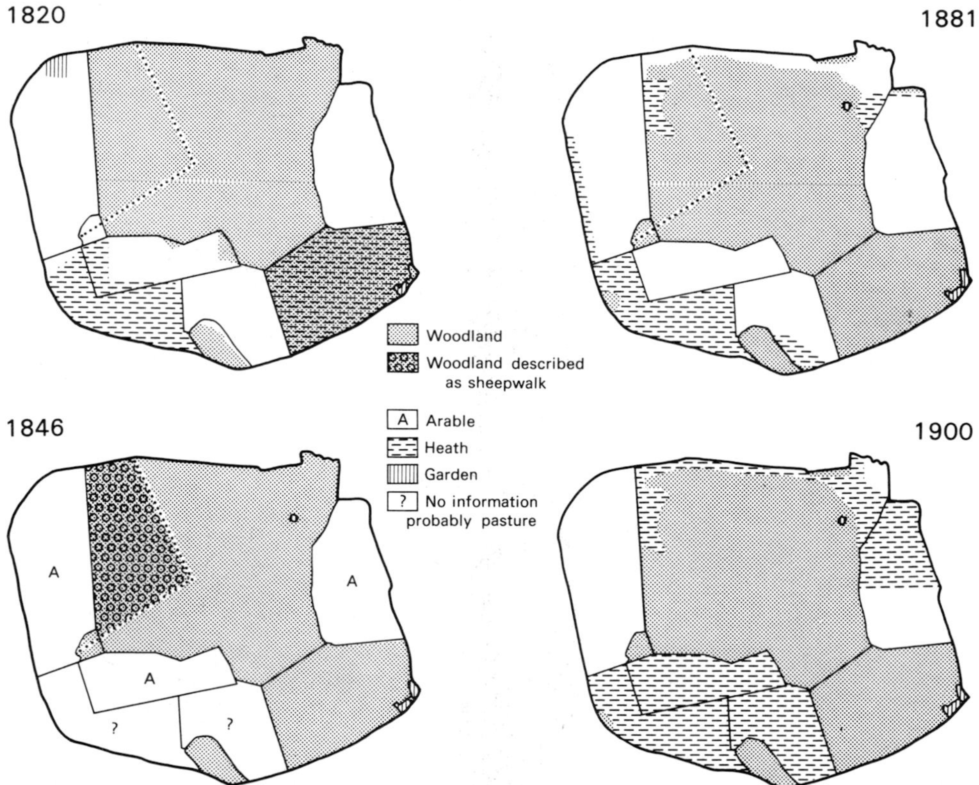

Figure 4.22. VEGETATION AND LAND USE OF STAVERTON PARK, SUFFOLK, IN THE NINETEENTH CENTURY. (Source: Peterken (1969))

Table 4.6
POLLEN-ANALYSIS OF SITES IN THE SOUTH-EAST LAKE DISTRICT OF ENGLAND
(after Oldfield, 1969)

Ecological phases determined by pollen analysis	Land-use history from documentary evidence
Forest regeneration and little agricultural activity	(1) The obscure post-Roman period which appears to have been a retrogressive and unsettled phase. (5th and 6th centuries AD)
Subdued farming activity with probably some lowland clearance	(2) The Anglo-Saxon settlement of the area (6th to 8th centuries AD), which was concentrated on a few lowland centres and was of only slight importance in the area
Oak wood clearance, probably on uplands and mainly for pasture	(3) The period of Norse colonization (9th and 10th centuries AD), an important episode locally and one involving much upland settlement
Brief forest-regeneration phase with ash	(4) The Norman Conquest and Domesday survey (late 11th century AD) are poorly recorded for this area, though contemporary sources suggest that most of the reclaimed parts of this area were laid waste
Resumption of mainly pastoral farming at a rather subdued level	(5) With the establishment of Furness Abbey in 1127 a period of clearance and enclosure, mainly for pasture, began
Period of extensive pastoral farming	(6) During the 14th and 15th centuries the creation of sheep pasture throughout Furness reached its peak, under monastic influence
Extension of cleared land for both arable and pasture	(7) From *c.* AD 1500 onwards enclosure began to impinge on the Abbey lands . . . a wave of forest clearance and settlement by yeoman farmers . . . most of the common pasture land was enclosed and turned over to a more mixed system of farming, whilst new clearings were created in the woodland
Greatly increased cereal cultivation in an almost completely cleared and drained landscape, 19th century	(8) The writings of the late 18th and early 19th-century agriculturalists record the gradual adoption of improved farming methods . . . Enclosure awards involving the local moss and marsh lands were passed, and their partial reclamation and drainage led to a further extension of both arable and pasture lands . . . from the middle of the 18th century onwards, creation of artificial parklands and plantations, including pine and other conifers took place

form of corroboration used in the south-west of the United States is the presence of plant-remains in adobe bricks (Hendry, 1931).

Information can be obtained from all these sources to reconstruct the vegetation pattern at a particular period. This may take the form of a map, as in Wightman's (1968) study, or a description, often with extensive use of quotations, as in the American studies of desert grassland by Humphrey (1958), and of nineteenth-century vegetation watersheds by Leopold (1951b). Most studies have as an aim the assessment of vegetation change or, at least, how an earlier period differed from the present. In some cases a general chronology is established in which periods of differing land use and vegetation are identified (e.g. Oldfield, 1963) (Table 4.6).

The level of detail which can be obtained from historical sources varies greatly. Many of the examples cited are concerned only with general categories of vegetation such as woodland and grassland. Other studies attempt to analyse the species mentioned in exploration journals and early surveys (Buffington and Herbel, 1965; York and Dick-Peddie, 1969). Some authors have tried to identify associations and communities and whether these have changed (e.g. Sears, 1925; Bryan, 1928a, Watts, 1960), while Mason (1963) has identified invaders, increasers and decreasers. Cottam and Stewart (1940) have mapped the area of invasion by juniper from explorers' and settlers' accounts (Figure 4.23).

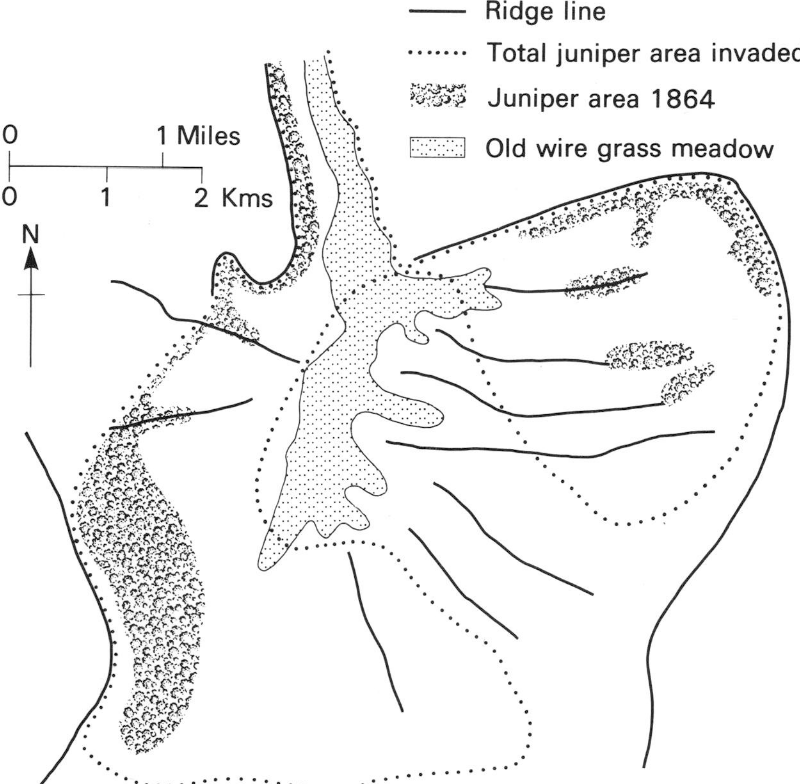

Figure 4.23. JUNIPER INVASION OF THE MOUNTAINS MEADOW VALLEY, UTAH. (Source: Cottam and Stewart (1940))

Where detailed information on coverage or composition can be obtained it can form the basis for some quantitative assessment. It may be possible to tabulate areas or proportions of different types of vegetation and, if information is available for two dates, to calculate changes (e.g. Rackham, 1975). From sources such as the United States' land surveys it is possible to estimate densities of trees, as Kenoyer (1933) showed for areas around Lake Michigan (Table 4.7). Buffington and Herbel (1965) have coded the presence and

Table 4.7

PERCENTAGE DISTRIBUTION OF PLANT ASSOCIATIONS RECONSTRUCTED FROM FEDERAL LAND SURVEY PLATS AND NOTES

Averages for the three counties bordering Lake Michigan are compared with those for three counties remote from the lake (after Kenoyer, 1933)

	Beech–maple	Beech–maple with hemlock	Oak–hickory	Oak–pine	Dry prairie	Lake and swamp
Allegan	51	18	8	16	0	7
Van Buren	40	15	33	3	0.2	9
Berrien	62	0	32	0	1	5
Cass	28	0	59	0	4	9
Kalamazoo	23	0	56	0	8	13
St Joseph	5	0	76.5	0	11.5	7
3 near lake	51	11	24	6	0.4	7
3 remote	19	0	64	0	6	10

associations of three plants and assessed their positions and densities, and Stearns (1949) has compared the size of major tree species present in 1857–1859 in Wisconsin with those in 1946 (Table 4.8).

Some information derived from historical sources has considerable practical significance. As Oldfield noted in 1969,

> there are many situations especially in fairly wild 'semi-natural' plant communities where the determination of long-term planning and conservation aims and policy must take into account historical factors. In the first place historical knowledge enhances our appreciation of the site and its ecology and so increases the scientific and educational value of the locality. In the second place historical evidence of this type forms part of the framework within which the extrapolation of trends and choice of alternatives must be made.

Case study: the English Lake District

The use of the historical–ecological approach to the analysis of land use and vegetation in settled areas can be exemplified by the studies of Oldfield (1963, 1969). In the opening paragraph of his 1969 paper, Oldfield stated that,

> in trying to understand the nature and status of present land use patterns, especially where the direct and over-whelming influence of intensive, contemporary farming methods is absent, historical questions may be very significant.

He discussed the values and limitations of pollen in indicating land use in the recent period and analysed the sequence shown by pollen-analysed sites in the south-east Lake District.

Table 4.8
PERCENTAGE DENSITY OF MAJOR TREE SPECIES IN A NUMBER OF SIZE CLASSES IN WISCONSIN IN 1857–1859 AND 1949
(after Stearns, 1949)

	Sugar maple		Hemlock		Yellow birch		Basswood		White pine		Totals	
	1857–1859	1946	1857–1859	1946	1857–1859	1946	1857–1859	1946	1857–1859	1946	1857–1859	1946
Size classes												
7–12.9 inches	53.8	40.4	55.1	48.8	38.7	0	40.0	58.3	16.7	0	49.1	39.3
13–18.9 inches	40.4	30.7	29.1	39.0	45.0	20.0	40.0	16.6	16.7	0	36.3	30.3
19–24.9 inches	4.8	21.1	14.9	4.8	8.7	53.3	20.0	0	33.3	0	10.9	17.2
25 inches and over	0.9	7.7	0.8	7.3	7.5	26.6	0	25.0	33.3	100	3.6	13.1
Number of trees	104	52	127	41	80	15	5	12	12	2	328	122

At the bottom of the sequence a phase of late Roman forest-clearance is dated by ^{14}C and an upper limit is provided by the clay from a haematite mine which was dated from documentary evidence to the first few decades of the nineteenth century. He demonstrated how the succession of land-use changes deduced from fossil pollen can be correlated in detail with documentary, place-name and cartographic evidence of changing culture, settlement and economy during the period (Table 4.6). For example, the decline in hemp cultivation (1750–1800) can be seen, there is a fall in mugwort pollen as a result of the introduction of deep ploughing in 1810, and the fall in alder and other tree pollen can be interpreted by reference to parliamentary enclosure. Oldfield has suggested how pollen evidence may usefully supplement historical evidence: by illuminating contrasts in type and intensity of farming within an area at any given time and at any locality through time; by substantiating points not completely resolved by documentary records; and by furnishing a context within which the recent history and present status of significant local plant species may be evaluated.

Oldfield (1969) then discussed two case studies in detail. One was a bog and fen area in which the effects of peat-cutting and drainage-alterations were apparent and pollen evidence could be related to documentary records of these activities. In the second example, Roudsea Wood, different phases of land use were recognized from the pollen sequence preserved in a nearby bog and these phases were supported by the historical sources which enabled him to suggest the vegetation chronology.

5
Events and Processes

In Chapter 4 we examined sources which can be used in studies of morphology and the assessment of change. Here we review some of the rather different evidence which can help to reconstruct and understand events and processes. The division of these chapters is somewhat arbitrary, since changes are brought about by the operation of processes, and processes are themselves simply series of events of various magnitudes and frequencies. However, the methods of analysing individual events are rather different from the analysis of morphological changes. The easiest processes to investigate with historical sources are those in which changes come about through infrequent large events, which are often very well documented.

The reasons for studying events using historical information vary, although often an initial aim is to establish the dates of particular incidents or to date the features produced by an event. Other work is concerned with reconstructing the chronology of events and calculating the frequency of occurrence. This has practical implications for risk-assessment, land-use planning and prediction. In some studies the aim has been to identify the conditions under which particular events took place and to elucidate general principles of cause, characteristics and effects. Others have investigated the perception of events and reaction to them which provides data for designing effective warning and alleviation procedures.

Several important types of historical source can provide information on physical landscape events but the main sources used are those which incorporate personal observations. These include newspaper reports, articles and descriptions in journals, magazines, diaries and letters, accounts in local histories and oral evidence. Pictorial material can be useful but maps are relatively unimportant sources for documenting events unless produced at the time to show the area affected. Newspapers are extremely valuable, but their content can be biased and sometimes exaggerated and sensational. Reports of events occur in a variety of magazines and journals, ranging from the technical and professional to the lay and popular. Some of the most useful for descriptions of events are the journals of local history and scientific associations. Stories of major events are often passed down in history, tradition and folklore and, though subject to distortion, usually have some elements of truth (Grove, 1966). It is often difficult, however, to corroborate such reports and cite specific documentary evidence. Personal papers such as diaries and letters can prove invaluable but may be difficult to procure and may contain much irrelevant material which has to be sifted through. Several researchers have collected material by placing requests and advertisements in newspapers and magazines (e.g.

Louderback, 1947; Pitts, 1973; Stanford, 1975). Associated with disastrous events are records of pleas for aid from administrative authorities, for relief of taxes from government, and supplications for 'divine intervention' from ecclesiastical authorities. With almost all these sources search of the material is difficult and tedious if no prior knowledge of dates of events can be obtained.

There is likely to be bias in the frequency and coverage of reports of events for several reasons. Only events which have an effect on human activities are commonly reported, so that small events and those occurring in sparsely populated areas tend to be neglected. Frequency of reports also depends on availability of suitable reporting media while the nature of reporting will depend on the authors' particular perceptions and attitudes (see Chapter 3). Estimates of magnitudes are not always to be relied on and are often given in relative terms like 'biggest flood since . . .' or 'most severe . . . in living memory'. Such assessments are prone to exaggeration, but can be useful if corroborated by other sources. There may be problems in isolating individual events in a sequence, and definition of terms will probably need some clarification. Malin (1946), for example, found that different observers used different terms to describe dust-storm phenomena which made it difficult to distinguish the type of event.

As in Chapter 4, the use of historical sources to help understand environmental processes and events is discussed in this chapter by taking examples from some major fields of application. These are: meteorological events, hydrological events, soil-erosion and gullying, mass movements and tectonic activity.

5.1 METEOROLOGICAL EVENTS

There are many meteorological phenomena on which data can be obtained from historical sources including severe frosts, heavy snowfall, intense rainstorms, tornadoes and hurricanes. Some information on individual events can be obtained from the records on climate mentioned in Chapter 4. For early periods, state and ecclesiastical chronicles and annals can provide some information, though the events need to be carefully corroborated, as Buchinsky (1963) has shown for the Ukraine. Later, personal diaries provide a major source and general as well as meteorological diaries are valuable since they highlight the most significant events, though they are not reliable for assessing frequency. In the nineteenth century, newspapers are the main source of material and often give quite detailed reports. Gregory and Williams (1981), for example, have constructed maps of Derbyshire showing weather and other hazards reported in the *Daily Evening Telegraph*, 1879–1978. Other journals such as the *Gentleman's Magazine* also carried relevant articles. By the end of the nineteenth century, specialist periodicals such as the *Meteorological Magazine, Quarterly Journal of the Royal Meteorological Society* and *Weather* had begun publication and contain reports of significant weather occurrences. *British Rainfall* also carries reports of the most remarkable storms of each year. Local histories and parish records often contain mentions of the most notable weather events; stories of events are also transmitted and carried down by word of mouth. Illustrations and photographs can provide some indication of conditions, particularly of heavy snowfalls and severe winters. The other main source of information is in records of activities which would have been affected by the events, such as agriculture, land and sea travel and military activity. For example, Meaden (1977) obtained information on weather in 54 BC from Caesar's writings on the *Gallic Wars*. For more recent times, instrumental records provide direct information on the duration, intensity and magnitude of events.

Details of weather events have been compiled in various chronologies (e.g. Lowe, 1870; Britton, 1937), though some early publications should be used with caution. (See above 4.1). These chronologies mostly list descriptions of events or conditions for particular days. Information on individual types of phenomena has been collated in works such as Andrews' (1887) book *Famous Frosts and Frost Fairs in Great Britain*, in which he used magazines, newspapers, parish registers and accounts, chronicles, diaries, poems and illustrations to compile evidence of the occurrence and characteristics of severe frosts. Two 'frost fairs' are illustrated in Figures 5.1 and 5.2. Individual seasons have also been analysed. Manley (1975) discussed the cold winter of 1684 in Britain with the aid of diary evidence and the records of activities such as shipping which were affected by the abnormal weather. Another example of work in this vein is Hoyt's (1958) discussion of the cold summer of 1816 in the United States based on newspaper reports, diaries and published articles.

Analyses of contemporary events or extreme conditions have often prompted investigation of the occurrence of similar conditions in the past. For example, the severe

Figure 5.1. FROST FAIR ON THE RIVER THAMES IN THE REIGN OF CHARLES II. (Source: Andrews (1887))

Figure 5.2. FROST FAIR ON THE RIVER THAMES IN 1814.
(Source: Andrews (1887))

winter of 1962–1963 in Britain occasioned analysis of past records, and Lamb (1963) remarked that up to that time a description of snow in *Lorna Doone* had been thought to be wildly exaggerated, but 1962–1963 demonstrated that thick snow on Exmoor was possible.

Evidence of events has been subjected to analysis of frequency and return periods. Pearson (1976, 1978) analysed the frequency of snowstorms in Scotland from newspaper reports and included details of depth and duration of snow-cover and attendant loss of animals. The importance of determining the frequency of natural hazards for land-use planning is emphasized by Carrara (1979). He used a combination of historical records and tree-ring analysis to analyse the frequency of snow-avalanches at Ophir, Colorado. Malin (1946) analysed dust-storms occurring in Kansas in the second half of the nineteenth century; he used newspaper reports extensively and evidence of failure of crops to reconstruct the conditions and assess effects and frequency.

The effects of severe storms and the incidence of coastal flooding are frequently mentioned in coastal studies, particularly the effect of storms in breaching spits and barriers or causing other devastation (e.g. Martin, 1872; de Boer, 1964). Brookfield (1952) referred to medieval flooding recorded in the Nonae Rolls, and Owen (1952) cited thirteenth-century coastal storms recorded in abbey chronicles. Erosion events on the Australian coast over the last 100 years and damaging storms which have contributed to coastal recession at Aberystwyth are the subject of studies by Thom (1974b) and So (1974) respectively. A phenomenon occasionally analysed is the occurrence of very low ebbs in estuaries. Reports of these in the Thames have been examined by Aranuvachapun and Brimblecombe (1978), but they cast doubt on some reports since even in the severe drought of 1976 it was not possible to wade across the Thames.

The incidence of hurricanes and tornadoes in the south-east United States and the Caribbean has been intensively analysed. One of the main purposes of such analyses is to assess storm-frequency and pattern of occurrence. Most technical data and standardized observations of hurricane-occurrence date from the late nineteenth century in the United States. Elsewhere they are not available until the early years of the present century. Dunn and Miller (1960) have compiled a table of hurricane frequency in different areas of the world. Prentiss (1952) has tried to assess the frequency of hurricanes in Texas and compiled lists of those that have occurred for each century since the fifteenth. These lists show increasing frequency of hurricanes but indicate some problems which arise from the type and availability of evidence. Likewise, Tannehill (1938) found references to sixteen hurricanes in the sixteenth century and thirty-three in the seventeenth, although he estimated that about 600 would have occurred. He has listed all known hurricanes and tropical storms from 1494 onwards. Much of the evidence available arises from reports of ships sunk and towns destroyed.

The effects of hurricanes on coasts have been examined by Brown (1939), Howard (1939), Stoddart (1971) and others. Many of these are studies of recent hurricanes but often they include an historical perspective. Hurricanes can cause coastal flooding, and Tannehill (1938) gives historical examples of storm-waves inundating coastal areas, particularly around the Bay of Bengal. He noted the case of one storm-wave at the mouth of the Hooghly river in 1737 which killed 300 000 people.

Giles (1927) has compiled data on tornadoes in Virginia between 1814 and 1925 using newspapers and personal correspondence as well as technical sources, and emphasized the difficulties, particularly in the earlier years, of recording information because of poor communications and the transient nature of the phenomenon. More recently, Galway (1977) has analysed the number of outbreaks, areas and effects of tornadoes in the United States and highlighted problems of perception and reporting of such events. Stanford (1975) collected a number of photographs of tornadoes in Iowa by placing requests in local newspapers and farm magazines, and from these photographs, dating back to 1899, he analysed the varying structures and effects of tornadoes. The literature on hurricanes and tornadoes is predominantly North American, but Meaden (1975) discussed a tornado which struck Scarborough, England, in 1975 and compared it with one in 1165 which was described by a monk in the Chronicle of Melrose Abbey.

5.2 HYDROLOGICAL EVENTS

Probably the most important area in hydrology in which historical sources have been used is for the purpose of extending flood- and flow-records back in time. Accurate gauging records are often limited to the period since 1950 in Britain and do not, therefore, provide a sufficiently long data sequence from which to calculate flood-frequencies or recurrence-intervals. Benson (1962) analysed the factors influencing occurrence of floods and explained the need to use historical sources to extend the length of flood-records. The use of such material increased the time base for his analysis in the United States from fifty to 200 or 300 years. He said that the 'importance of such an investigation for extending flood knowledge and hence the period of time on which the frequencies of floods are determined cannot be overemphasised'. Cruff and Rantz (1965), in comparing several methods available for calculating flood frequencies, likewise explain how historic data can be incorporated. Much of the information available in historical sources is in the form of, or is reducible to, relative statements such as 'highest flood since . . .'. The values and dates can then be ranked and frequencies and recurrence-intervals calculated, though allowance must be made for likely bias in reporting and exaggeration of levels and effects.

Some information on floods and droughts can be obtained from sources on weather and climate outlined in Chapter 4. Weather and other diaries may include observations on floods and prolonged periods of dry weather, as may chronicles and annals. Newspapers are a useful source of information on floods, though news articles usually focus on the most dramatic human effects and any damage or loss of life. Newspaper offices often hold many more photographs of floods than they publish, and these may be worth examining. Reports of floods were also published in the *Gentleman's Magazine* until the 1830s, and Lamb (1966) cited a description of a bog-burst in Scotland in 1772 reported in this periodical. Records from factories, mills and other buildings affected by floods are often available and it may be possible to relate these to marks on walls and bridges. Local authorities may also have records of 'clearing-up' operations after floods and reports were often made by local surveyors and engineers. Reports and descriptions of major floods are frequently published after the events as a way of raising money for relief and to attract help, as happened in Exeter in 1960, and there may be administrative records of aid and payment for flood relief. The records of the commissioners of sewers and subsequent organizations concerned with rivers and drainage can provide much valuable information, as can government and parliamentary papers relating to drainage schemes, which often include evidence of floods and their effects. Professional reports are to be found in engineering and other journals such as the *Proceedings of the Institution of Civil Engineers* and the *Transactions of the Institution of Water Engineers.* Water authorities produce reports after major floods and some of these are deposited in the library of the Institution of Civil Engineers. Major floods are also described in articles published in meteorological magazines. Occasionally maps showing areas of inundation are available (Daniels, 1960; NERC, 1975) especially if associated with a drainage scheme or problem.

Potter (1978), in association with the Institute of Hydrology, has published a guide to use of historic records for augmenting hydrological data. This primarily covers floods and river flows but includes information on sources relating to rainfall and storms and to drainage alterations. He discusses instrumental and official records, newspapers, reports and histories, diaries, medieval manuscripts and said of historical sources on river-flooding:

> It is possible to acquire information on the circumstances leading to the flood, the occurrence of surface runoff, the time of the flood arrival at a particular place, the peak height reached, the period of inundations and the rate of its recession; also details of the damage to crops and buildings, the effect of flood defence banks, and any breaches or overtopping that might have taken place, damage to weirs, sluices, roads and bridges, mills and occasionally changes in river channels.

A major application of historical sources is in the Institute of Hydrology's Flood Studies

Report (NERC, 1975) in which historical records, mostly collected by the water authorities in Britain, are used to calculate flood statistics. Sources used include newspapers, a map of a 1771 flood, records at works and mills, plaques and marks on bridges and structures, and technical papers and articles in journals and magazines. Jones (1975) has assessed these historical records, and says:

> The accuracy of such records is likely to be less than for recent records using modern instruments, flow measurements and critical appraisal; channel changes may have introduced errors of unknown magnitude. Nevertheless, it is considered that the value of the additional information outweighs these disadvantages.

Much early hydrological information is in the form of reports of individual floods. Brooks and Glasspoole (1928) discussed problems of collecting accurate information in their book *British Floods and Droughts*. They stated that before measurements of rainfall become available we have to rely on 'more or less vague descriptions of diarists and annalists' and that before 1650 there are mostly only references to great floods and droughts. They referred to several publications in which dates of major floods have been compiled though cast doubt on the authenticity of some of the dates. They mentioned the difficulty of identifying even the year of floods from early documents. A considerable amount of information on flood events can be obtained from descriptions of their effects on agriculture, as Brandon (1971a) showed in his study of agriculture in Sussex in the Middle Ages, and on military and other activities (Brooks and Glasspoole, 1928).

Important past flood events have been reassessed, as, for example, in Anderson and Calver's (1977) study of the present state of features created by the 1952 Lynmouth flood. They compared published and unpublished contemporary observations and photographs with present conditions. This showed that deposits and channel-deepening persisted longer than the other flood effects. The geomorphological role of floods is often considered in

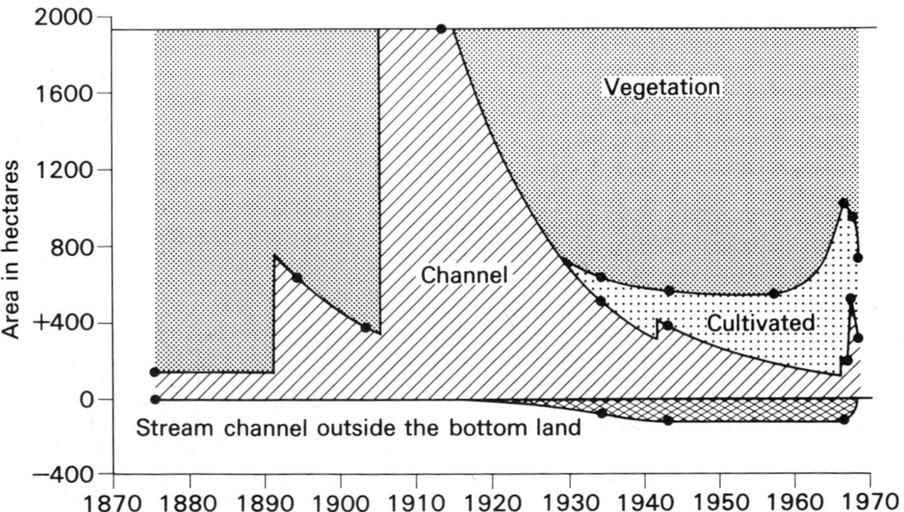

Figure 5.3. CHANGES IN AREA OF PART OF THE CHANNEL OF THE GILA RIVER, ARIZONA, 1870–1970. (Source: Burkham (1972))

relation to river channel changes, particularly cut-offs (e.g. Towl, 1935; Eardley, 1938; Carey, 1969). Schumm and Lichty (1963) and Burkham (1972) have related periods of channel-widening to increased incidence of large floods (Figure 5.3).

Woolley (1946) has analysed cloudburst floods in Utah in the period 1850–1938. His major historical sources are newspaper reports incorporating eye-witness descriptions. Many of these are quoted verbatim and include information on damage and the height and passage of the flood-waters. The distribution and frequency of the floods were analysed; Woolley (1946) also noted dates of large mudflows on alluvial fans. He suggested that the increase in number of floods reported, as indicated in Figure 5.4, was related to increased settlement and population and also the increased efficiency of reporting media.

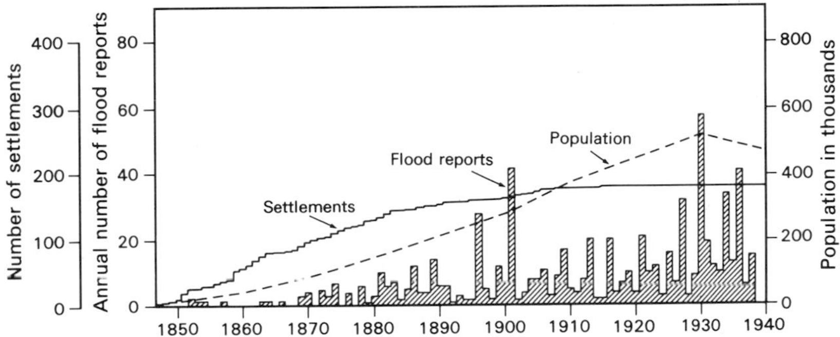

Figure 5.4. NUMBERS OF FLOODS REPORTED IN UTAH FROM 1837 TO 1938 IN RELATION TO INCREASES IN POPULATION AND THE NUMBER OF SETTLEMENTS.
(Source: Woolley (1946))

Drought and floods are often considered together in the literature but in analysing droughts from historical sources the difficulty, as Brooks and Glasspoole (1928) have indicated, is that there is much less information because of the less drastic effects of drought and its more gradual onset. Droughts do affect agriculture, fishing, mills, waste-disposal and navigation, and valuable information on their occurrence and characteristics can be found in the records of these activities. More cognizance is taken of droughts in areas where water-supply is a problem. Thomas *et al.* (1963) analysed droughts in the south-west of the United States and mentioned mid-nineteenth-century occurrences; they used mission notes and diaries to estimate rainfall between 1769 and the beginning of systematic measurements. Lawson *et al.* (1971) have analysed the spatial and temporal distribution of droughts in Nebraska in the period 1931–1969 and also briefly discussed instrumentally measured droughts since the 1890s.

Historical technical records and direct measurements of discharge arc rarely available, but include twelfth-century BC records of the Nile carved on stone and measured in units of handspans (Bell, 1970). The records of the Nile floods have been subject to much analysis of ranges, periodicities and heights and to comparison with present values (Brooks, 1928; Pollard, 1968). Chu (1973) analysed old Chinese records of floods and droughts and found that the early records even discriminate between floods caused by excessive rainfall and those due to inundation. A 100-year record of lake-levels in Nevada was used by Hardman and Venstrom (1941) to calculate variations in discharge of the Truckee river; the lake-records were corroborated by photographic and descriptive evidence. Modern discharge-

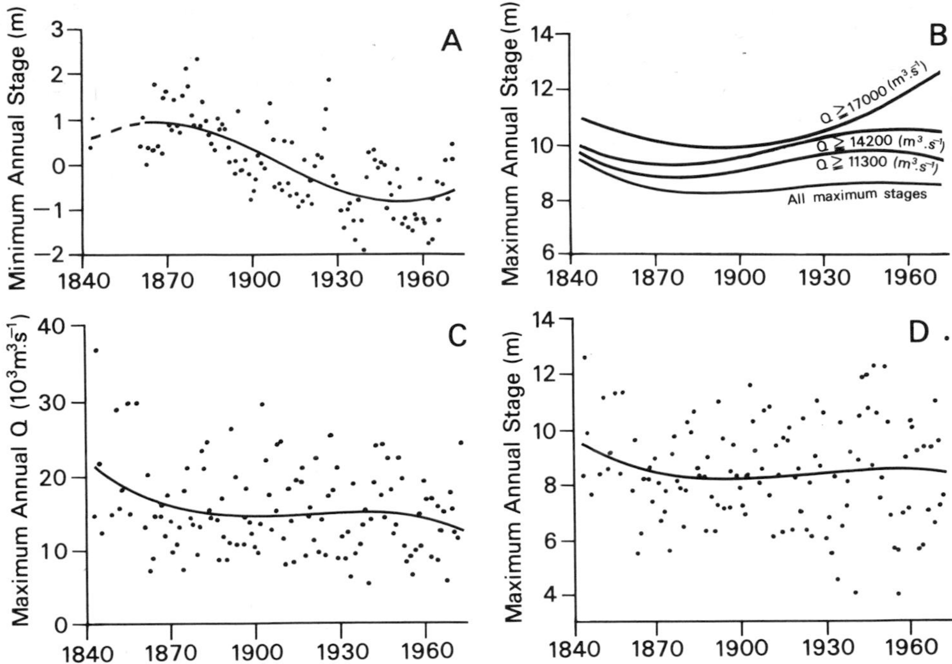

Figure 5.5. FLOWS ON THE RIVER MISSISSIPPI AT ST LOUIS, 1840–1975.
A — minimum annual stages; B — maximum annual stages at different values of discharge (Q); C — maximum annual discharges (Q); D — maximum annual stages.
(Source: Belt (1975))

records have been used to examine the effects of channel-changes on the passage and height of flood waves; Belt (1975) studied changes in flood-levels in relation to man's constriction of the Mississippi river since the mid-nineteenth century (Figure 5.5) and Happ (1944) and Burkham (1976) have assessed the effects of channel-changes on floods. In much work the

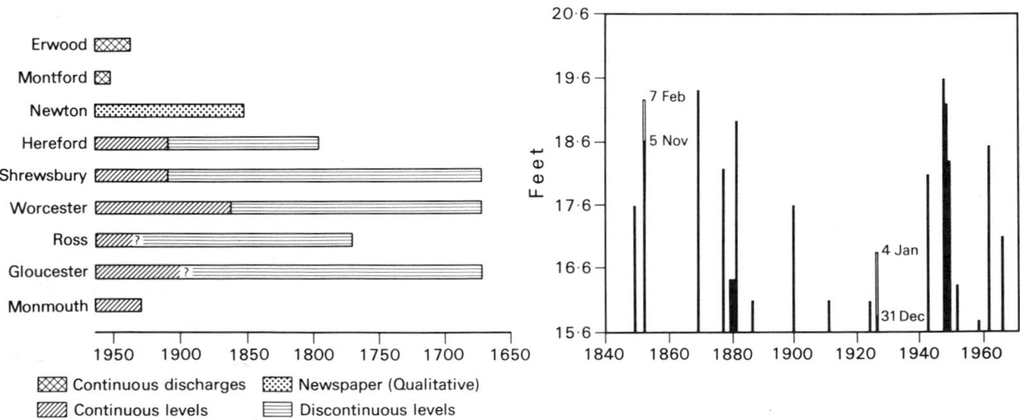

Figure 5.6. LENGTH AND RELIABILITY OF THE FLOOD RECORD AT SELECTED STATIONS ON THE RIVER SEVERN AND RECORDED FLOOD LEVELS AT THE WELSH BRIDGE, SHREWSBURY.
(Source: Howe *et al.* (1967))

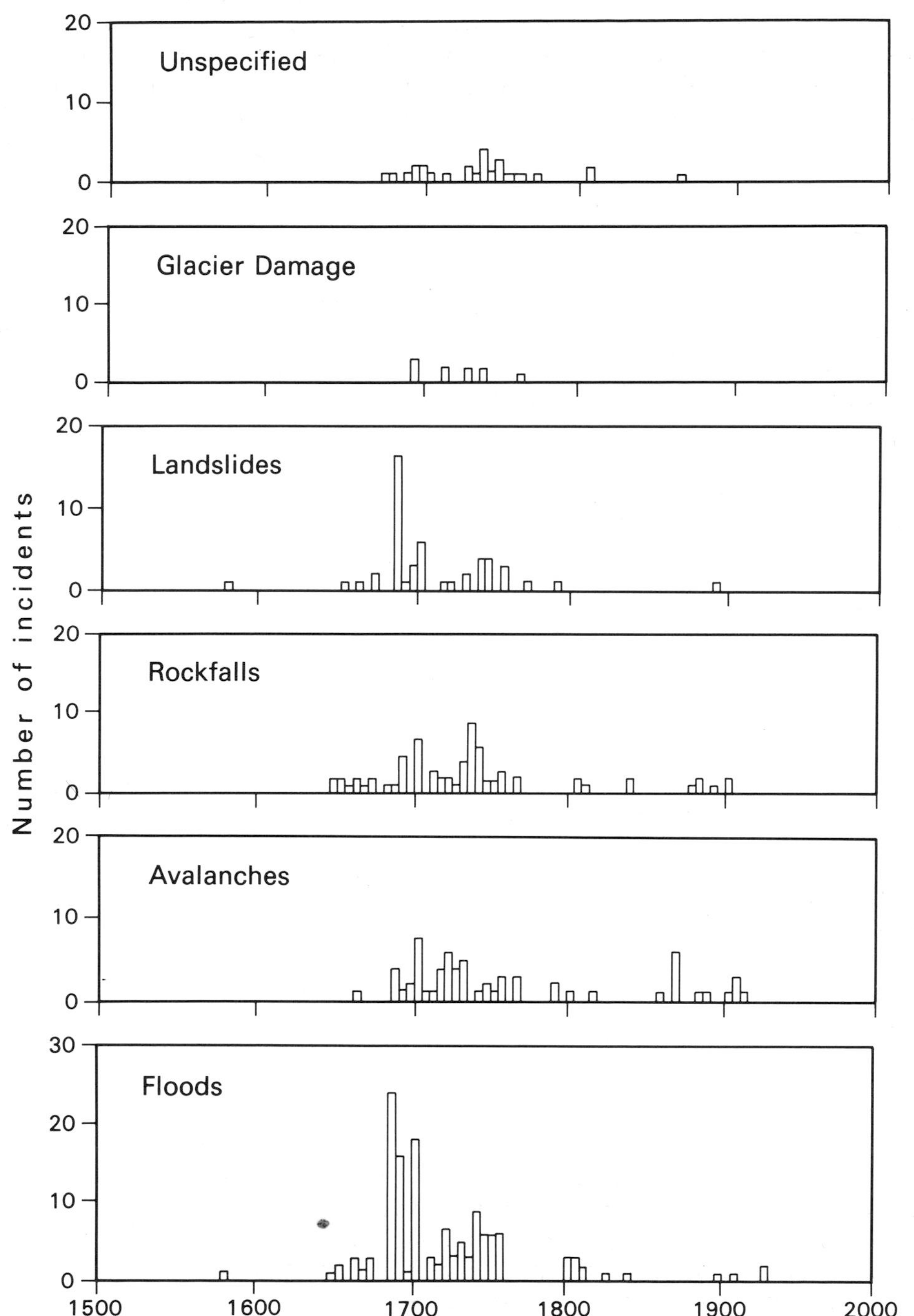

Figure 5.7. THE INCIDENCE OF MASS MOVEMENTS AND FLOODS IN FOUR NORWEGIAN PARISHES AS REVEALED BY *LANDSKYLD* RECORDS FROM THE SIXTEENTH TO THE TWENTIETH CENTURIES.
(Source: Grove (1972))

aim has been to identify whether changes in hydrological régime have occurred and to assess the effects of such changes. Many studies are restricted to recent periods because of lack of earlier discharge-records as, for example, in Pickup's (1976) study of runoff changes in the Cumberland Basin, New South Wales.

Analyses of the hydrological effects of changes in factors such as land use often makes use of runoff data as in Knox *et al.*'s (1976) investigation of the hydrological response to land use and climate change in the Mississippi valley since the mid-nineteenth century. Howe *et al.* (1967) related changes in flow-régime of the river Severn to land-use changes and reconstructed former flow-levels and floods from newspaper reports, height observations and measurements (Figure 5.6): they suggested these show periods of high river-levels in the mid-nineteenth century and since 1944. Grant (1965) used evidence from explorers' writings in New Zealand together with sedimentary and botanical evidence to establish that a greater incidence of mass movements, forest damage and other processes occurred in the past and to infer changes in the régime of the Tukituki river since 1650. Grove (1972) has analysed the increased frequency of floods and mass movements during the 'Little Ice Age' in Norway (Figure 5.7). Changes in environmental factors such as the elements of climate have implications for hydrological processes and runoff, but as Lockwood (1979) has discussed for Britain over the last 50 000 years, historical applications are still somewhat limited and further reconstructions could be achieved.

5.3 EROSION AND GULLYING

The problems of soil-erosion on slopes and of erosion and deposition in river valleys are most widely reported in semi-arid regions of the world. Much of the English-language literature on this subject concerns the Mediterranean and the semi-arid areas of North America. Widespread soil-erosion on slopes is difficult to detect over a short time period, and thus most of the evidence on this is obtained from morphological or vegetation indicators or from evidence of deposition lower in the drainage system. However, both Bennett (1931) and Happ *et al.* (1940) have cited some documentary evidence of changes in soils in the United States, particularly loss of upper layers. Much more obvious is the rapid development of gullies or enlargement of stream courses.

Active soil-erosion has been a problem in the Mediterranean region for many centuries. Classical writings, travel accounts and histories provide some evidence of its occurrence and the problems caused. For example, Butzer (1974) has quoted Plato on soil-erosion but he also warned that some discussions of soil-erosion in the Mediterranean have exaggerated conditions. Van Zuidam (1975) discussed erosion in the Zaragoza area of Spain and cited evidence from classical writings and chronicles. However, archaeological and stratigraphical evidence tends to be more valuable for analysis of soil-erosion, as Vita-Finzi (1969) and Butzer (1974) have shown in their work on the history and chronology of erosion and sedimentation in Mediterranean valleys. They have used such evidence to date deposits and to assess rates and distribution of erosion and deposition. Other areas of the world have, of course, also experienced severe soil-erosion; Butzer (1974) noted the long-standing nature of this problem in south-east Asia and the lack of historical investigation there. Soil-erosion in the Machakos Hills of Kenya has been investigated by Moore (1979) partly using records of European explorers and early colonial writings and reports.

There is a growing realization that soil-erosion in Britain is greater than was thought traditionally, and in a recent article reviewing the situation Morgan (1980) has provided some historical perspective and cited descriptions from Defoe, from a natural history and from newspaper articles. He also suggested that local authority records of clearing sediment from roads after storms and water-authority records of clearing silted reservoirs may provide useful information.

In many parts of the American South-west several sequences of cutting and filling can be traced, many having occurred within historical time, as Womack and Schumm (1977) have shown in their analysis of terrace sequences of Douglas Creek, north-western Colorado. It was, however, the widespread appearance of extensive and rapid channel-trenching in the late nineteenth century which caused severe problems for settlers and farmers and stimulated study of the processes and causes. Much of the evidence and many previous studies are reviewed by Cooke and Reeves (1976). The extensive literature on arroyos (gullies) focuses on several major themes including the date of initiation of gullying (which is particularly important in identifying cause), the rate of gullying and its progress, the mechanics and processes of channel-cutting, the conditions, frequency and distribution of occurrence and the effects of particular factors such as agricultural methods.

In areas of recent gullying and soil-erosion, oral evidence has been widely used. Much channel-development had taken place within the living memory of residents at the time of some of the most important studies on channel-erosion such as those of Bryan (1925, 1928b). Settlers were able to give him information on the situation prior to trenching and often to date its initiation and progress. Of course, such information has to be treated with the caution indicated in Chapter 3.

Travellers' accounts, records of military expeditions, early surveys and diaries and journals of early settlers also provide much information on the early state of valleys and

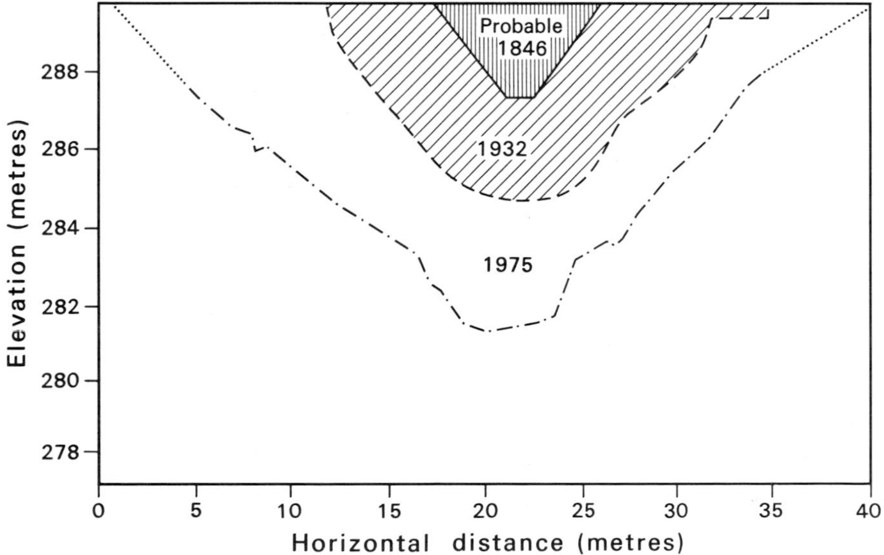

Figure 5.8. CROSS-SECTIONS OF THE EAST FORK LITTLE TARKIO CREEK, MISSOURI, 1846–1975. (Source: Piest *et al.* (1977))

channels especially the situation prior to arroyo-cutting, as shown by Thornthwaite *et al.* (1942) and Hastings (1959). This information may be direct, as in a description of a river valley, or may be indirect in that problems of crossing what is now a deeply incised stream are not mentioned. Early land surveys can be used in conjunction with accompanying notes or other descriptions which may also include cross-sections (Happ *et al.*, 1940; Knox, 1972; Piest *et al.*, 1977) (Figure 5.8). The period of channel-cutting in the south-west United States also coincided with pioneering photography and, although few photographs exist of channels prior to trenching, many are available which show the progress of erosion and the

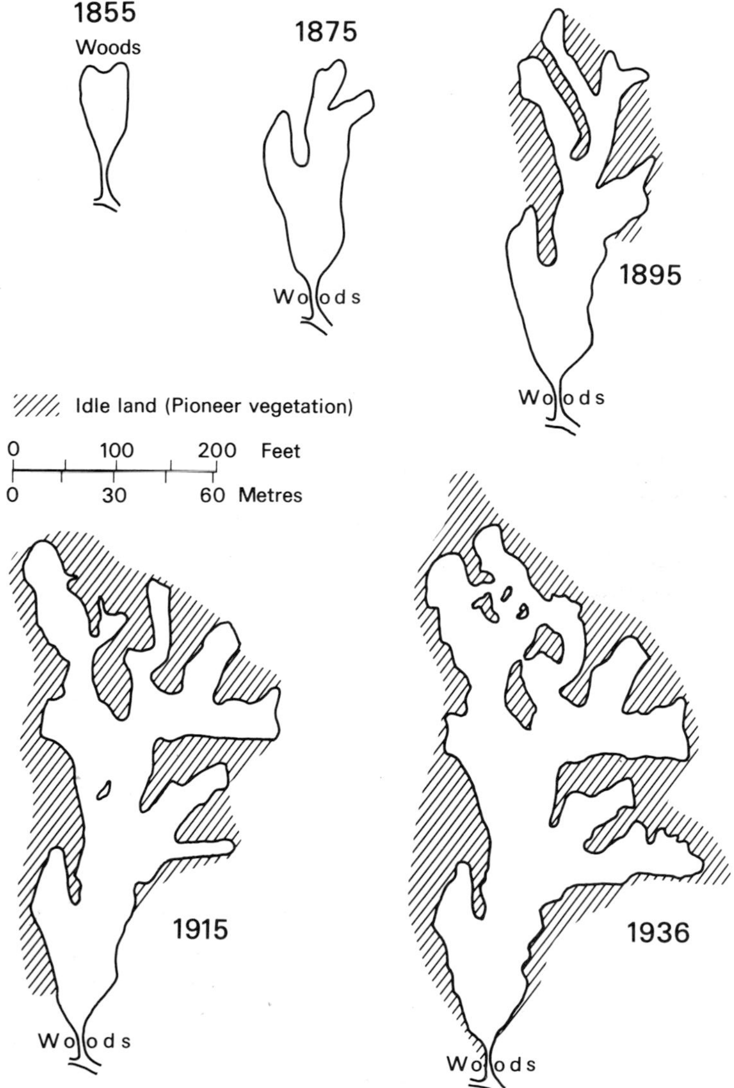

Figure 5.9. RECONSTRUCTED HISTORY OF WALDEN'S GULLY NEAR MOORE, SOUTH CAROLINA, FROM 1855 TO 1936. (Source: Ireland *et al.* (1939))

relation of the channels to fixed structures such as houses and roads (e.g. Duce, 1918; Ireland *et al.*, 1939).

Such rapid erosion obviously caused problems for settlers by undermining and destroying structures such as bridges and in causing loss of agricultural land. Bridge records, engineers' records and legal disputes have occasionally been used to analyse the channel-changes (Cooke and Reeves, 1976). Erosion was also so rapid and spectacular and had such effects on homes and livelihoods that there were reports and articles in local newspapers; these have been used by Thornthwaite *et al.* (1942) and Hastings (1959), for example. All these sources are of limited value unless they can be tied to location and date.

Most studies of soil-erosion and gullying involve extensive comparison of sources and use of non-documentary corroboration, because much of the historical evidence is vague or circumstantial and because much analysis is aimed at establishing the cause of increased erosion, mostly through time-correlation of activities (see Chapter 6). It is often only possible to elucidate approximate dates for the initiation of landscape features but in some cases a detailed chronology of arroyo development may be built up, as Cooke and Reeves (1976) have done for a number of arroyos in southern Arizona and coastal California. They used sources such as land surveys and notes, maps, government reports, newspaper and travellers' accounts and oral evidence. Ireland *et al.* (1939) were able to reconstruct the growth of a set of gullies in South Carolina from field, historical and oral evidence, as shown in Figure 5.9. Much work has been concerned with rates of gully development because this has practical implications for prediction of future erosion activity and its prevention. Much gully extension is by headward erosion, and thus information on position of the headcut at different dates is extremely important. Again only vague indications may be obtainable from historical sources, but in some instances, as Bryan (1928b) showed, it may be possible to plot or map the position of the headcut at several dates and calculate rates of erosion.

Many descriptions by travellers, explorers and settlers give at least approximate dimensions of channels, and these can be compared with later measurements, as

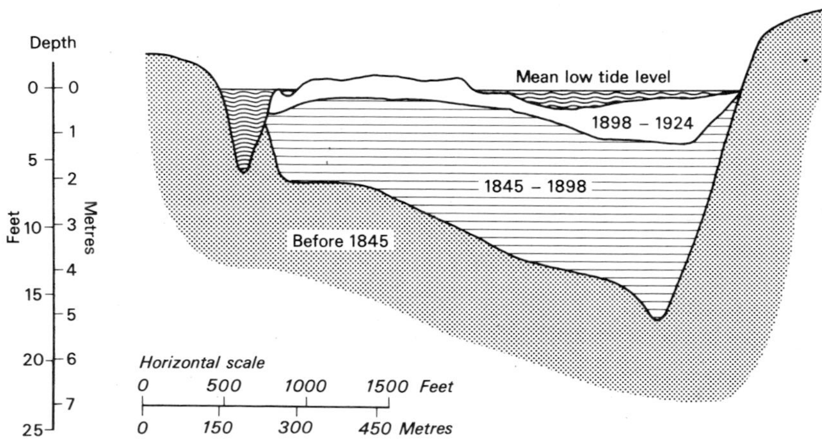

Figure 5.10. SEDIMENTATION OF THE PATAPSCO RIVER, BALTIMORE HARBOUR, MARYLAND. (Source: Gottschalk (1945))

Cooperider and Hendricks (1937) and Piest *et al.* (1977) demonstrated. Occasionally, soil-erosion has been studied historically through analysis of sediment loads in rivers, as Meade and Trimble (1974) have done for the Atlantic streams of the United States, but usually lack of long sediment records precludes this.

Many studies of deposition lower down in basins are based on sedimentary evidence, but the dating of sediments can sometimes be corroborated by historical evidence as, for example, in Knox's (1972) study of valley-alluviation in Wisconsin and Trimble's (1970) study of the Alcovy river swamps in Georgia, in which he used a variety of sources to show that the swamps were formed from the late nineteenth century onwards and that their formation can be related to increases in soil-erosion in contributary basins. Gottschalk (1945) demonstrated by use of historical sources how upland erosion caused sedimentation in Chesapeake Bay which interfered with the operation of newly created ports in the eighteenth and nineteenth centuries (Figure 5.10). The head of navigation near Baltimore has been pushed seven miles downstream since 1608. Siltation of ports and coasts has also been recorded in the Old World, for example in Tunisia (Vita Finzi, 1969) and in ancient Mesopotamia (Jacobsen and Adams, 1958).

5.4 MASS MOVEMENTS

Most slope studies using historical sources have focused on unstable slopes and features resulting from rapid mass movements such as landslides. These studies have had two general purposes: to promote scientific understanding of slope forms, processes and changes and also in more practical terms to provide information on stability in relation to engineering construction on slopes. Many of the more technical analyses date from the last twenty years or so, but prior to that time individual occurrences of movements were studied and significant developments made towards the general understanding and classification of slope movements.

As with all events, the spectacular tend to be best reported and most studied, but in a few instances historical sources have helped in the analysis of more gradual changes. For example, Brunsden and Kesel (1973) have studied the degradation of a Mississippi river bluff over time; they used maps to date successive migrations of the river away from the base of the slope and from this information they analysed rates and characteristics of the sequence of cliff-deformation. Similar types of study have been made of moraines dated from historical and other evidence (e.g. Welch, 1970).

Landslides are probably the most frequently studied slope features, and their occurrence is usually the subject of some kind of contemporary report, even if in early reports the events are not fully understood and are described in terms such as 'act of God'. Some of the reports in newspapers, chronicles, parish and local histories, magazines and technical publications have been used in subsequent analyses but many remain to be examined. Extracts from an example of one such description in a nineteenth-century local history (Randall, 1880) are given here to illustrate the type of information these can provide. The description is of a landslide which occurred at Buildwas on the banks of the river Severn in Shropshire in 1773, and the words are those of a witness, the Reverend Mr Fletcher.

'When I went to the spot', says Mr. Fletcher, 'the first thing that struck me was the destruction of the little bridge that separated the parish of Madeley from that of Buildwas, and the total disappearing of the turn-pike road to Buildwas bridge, instead of which nothing presented itself to my view but a confused heap of bushes, and huge clods of earth tumbled over one another. The river also wore a different aspect; it was shallow, turbid, noisy, boisterous, and came down from a different point. Whether I considered the water or the land the scene appeared to me entirely new, and as I could not fancy myself in another part of the country, I concluded that the God of nature had shaken his providential iron rod over the subverted spot before me . . . I climbed over the ruins and came to a field well grown with rye-grass, where the ground was greatly cracked in several places, and where large turfs some entirely, others half turned up exhibited the appearance of straight or crooked furrows, imperfectly formed by a plough drawn at a venture. Getting from that field over the hedge, into a part of the road which was yet visible, I found it raised in one place, sunk in another, concave in a third, hanging on one side in a fourth, and contracted as if some uncommon force had pressed the two hedges together . . . Between a shattered field and the river there was on that morning a bank on which besides a great deal of underwood grew twenty-five large oaks, this wood shot with such violence into the Severn before it that it forced the water in great columns a considerable height, like mighty fountains, and gave the over-flowing river a retrograde motion. This is not the only accident that happened to the Severn; for near the Grove the channel which was chiefly of a soft blue rock burst in ten thousand pieces, and rose perpendicularly about ten yards, heaving up the immense quantity of water and the shoals of fishes that were therein. Among the rubbish at the bottom of the river, which was very deep in that place, there were one or two huge stones and a large piece of timber, or an oak tree, which from time immemorial had lain partly buried in the mud, I suppose in consequence of some flood; the stones and tree were thrown up as if they had been only a pebble and a stick, and are now at some distance from the river, many feet higher than the surface of it . . . The tossing, tearing and shifting of so many acres of land below, was attended with the formation of stupendous chasms above. At some distance above, near the wood which crowns that desolated spot, another chasm, or rather a complication of chasms excited my admiration; it is an assemblage of chasms, one of which that seems to terminate the desolation to the north-east, runs some hundred yards towards the river and Madeley Wood; it looked like the deep channel of some great serpentine river dried up, whose little islands, fords, and hollows appear without a watery veil. This long chasm at the top seems to be made up of two or three that run into each other, and their conjunction when it is viewed from a particular point exhibits the appearance of a ruined fortress whose ramparts have been blown up by mines that have done dreadful executing and yet have spared here and there a pyramid of earth, or a shattered tower by which the spectators can judge of the nature and solidity of the demolished bulwark.'

Descriptions such as that quoted above give some idea of the nature of slope movement and its effects, though the evidence should not be accepted uncritically but should be corroborated wherever possible.

Case studies have been widely used to elucidate the general principles of mass movements and the conditions under which they occur. Some of the early collections of case studies such as those of Collin (1846) and Sharpe (1938) can now be regarded as historical sources themselves. More recent technical studies, such as that by Skempton and Hutchinson (1969), include stability-analyses of particular cases, though they mostly use recent data.

Geologists and engineers often reported on landslides shortly after they occurred and published their observations in technical journals. These often incorporate the comments of other observers and information published in newspapers at the time. The studies by Morse (1869) of a landslide in Maine and by Newland (1916) of landslides in the Hudson valley are examples. Early geological surveys can often provide useful background and may themselves include descriptions of major landslides (Howe, 1909).

Similarly, the nature and characteristics of mudflows have been studied from case histories and remarkable events such as the Wrightwood mudflow (Sharp and Nobles, 1953) are frequently quoted as examples. Azimi and Desvarreux (1974), when studying a special type of mudflow in the French Alps, referred to a 1914 book on Savoie which mentions thirteen mudflows between 1732 and 1908.

A number of old landslips which occurred in the historical period have been reappraised because the slips still form significant features in the landscape that require explanation and dating and because they are of technical interest and require analysis using more modern techniques and knowledge. Thus, for example, Heim's (1882) description and analysis of the great Elm rockfall of 1881 has been subjected to simulation analysis by Hsu (1975) to assess the nature of the movement. Both Arber (1941) and Pitts (1973) have examined the Bindon landslip which occurred on the Dorset coast in 1839 to assess the type and nature of the movement. Sources used include newspapers, technical and lay articles, local histories and engravings (Figures 5.11 and 5.12). Sections have been constructed from these, the prior conditions analysed and the exact nature of the processes discussed. The observations of geologists who were present at the site have proved particularly valuable. Such information also permits examination of subsequent changes such as removal or degradation of fallen blocks, as Pitts (1973) has shown using a series of photographs. Movements can be corroborated by the morphological and sedimentological characteristics of the slipped material (see Chapter 2).

Information on conditions relating to slope movement, such as moisture content and water levels, has occasionally been provided by the historical record. For example,

Figure 5.11. A VIEW OF THE 1839 BINDON LANDSLIP, DORSET, LOOKING WEST TOWARDS THE EXE ESTUARY. (Photograph courtesy of Dorset County Library)

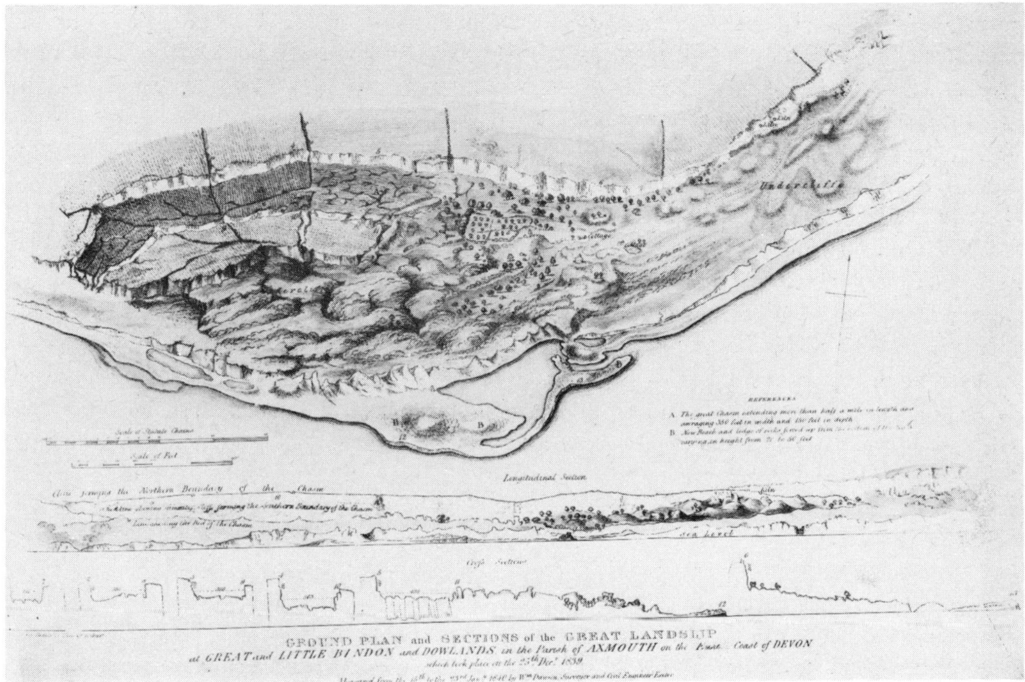

Figure 5.12. PLAN AND SECTIONS OF THE BINDON LANDSLIP, DORSET, DRAWN TO ACCOMPANY CONYBEARE'S CONTEMPORARY ACCOUNT.
(Photograph courtesy of Dorset County Library)

Chandler (1970b) found a description of the flooding of a brickyard which suggested high water-pressure in a nearby slope. Letters and drawings of a mill were used by Skempton (1971) in a geotechnical analysis of slope stability, and Hutchinson (1961) calculated the safety factor of a slope in the year 1877 from contemporary maps and oral evidence. Cracking of ground prior to a slip has been described, for example by Suklje and Vidmar (1961) at Gradot Ridge in Macedonia. Railway archives have been used to analyse movement at Folkestone Warren, England (Toms, 1953; Hutchinson, 1969) and in Norway (Eide and Bjerrum, 1954).

A further use of historical sources is in dating slope features. The accuracy with which this is possible varies considerably; there may be direct evidence on movement, or only indirect comments on damage or effects. Often only a maximum or minimum date can be deduced. For example, Hutchinson and Gostelow (1976) dated movements at Hadleigh Castle in Essex by the way the castle and slopes are indicated in various illustrations and documents. These are listed in Table 5.1. Bromhead (1978) has analysed movements on slopes at Herne Bay, Kent, as indicated by various editions of Ordnance Survey maps:

> Between the first (1872) and second (1898) series of the Ordnance Survey, great changes appear to have taken place along this stretch of the coast line at Herne Bay, the earlier map showing the slope to be fairly even and well vegetated but breaking up into land slips at the time of the later map . . . the cliffs at the position of the Miramar landslide are shown considerably to seaward of their present position, and at some time between the two series, a major cliff failure took place [Hutchinson

Table 5.1
EVIDENCE FOR SLOPE MOVEMENTS AT HADLEIGH CASTLE, ESSEX, *C.* 1230–1975
(after Hutchinson and Gostelow, 1976)

Event and location	Date	Cause
Possible collapse of 'phase I structure', sited up to 22 m E of tower F	1230–1260	Probably 'local movements in the underlying clay'
Castle gate and other buildings in need of repair	1240	Not stated
Collapse of western curtain wall between towers F and G, the adjoining 'Phase II hall' and probably tower G	Towards end of 13th century	Movement 'towards the SW due to slumping of the underlying clay'
Collapse of part of curtain wall, probably near the barbican	1317–1318	Not stated
Landslide to the S which carried away foundation of tower D, leaving the bulk of the S curtain wall intact	Post 1738, pre-1863	Continuing degradation of Hadleigh Cliff
Major landslide to the S which carried practically the whole S curtain wall downhill for about 12 m	Post 1881 pre-1920 some movement by 1895	Continuing degradation of Hadleigh Cliff
Landslide to the NE which back-tilted NE half of tower B	Post 1930, pre-1955	Slipping at crest of the north-eastern side valley
Collapse of most of remaining eastern part of tower B	29–30 Jan. 1965	Continued ditto
Minor landslides to the S in the W part of rear scarp, roughly 20–30 m E of tower F: the largest was 6–7 m wide and sank 3 m	On and around 31 March 1969 following snow-melt	Continuing minor degradation of rear scarp left by major late 19th-century landslide
Minor landslides to the S in the W part of rear scarp, roughly 0–30 m E of tower F	Winter 1969–1970	Ditto
Continuing minor landslides to the S on rear scarp, generally involving the spoil bulldozed there in 1971 and 1972	1975	Softening of spoil, and renewals of movement in earlier slides due in part to loading of spoil

1965b, on the basis of local records, places this in 1883] with the loss of about 20 m along approximately 100 m of cliff. The O.S. second edition clearly shows a ridge and graben structure, similar to that of the second landslide . . . By the time of the third series (1906), the ridge moved seawards some 12 m and had moved more at its western end than elsewhere. Later, at the survey for the fourth series (1933) no traces of this landslide were mapped.

Hutchinson (1965b), in his comprehensive report on the coastal landslides of Kent, used many historical sources including newspaper and technical reports. In some cases it may be possible to trace a sequence of movements, as Bjerrum (1971) did in his study of quick clay flows in Norway. The detailed nature of movements on a complex unstable slope near Charmouth, Dorset, has been documented, in part from maps and aerial photographs, by Brunsden and Jones (1976).

Historical sources may occasionally corroborate other evidence of movement. Legget and Lasalle (1978) found some organic material buried by clay in the St Lawrence river valley of Quebec and dated this to about 400 years old, much younger than previously thought. Letters and travellers' reports confirmed that an earthquake and landslide which produced the overlying clay occurred in 1663. The occurrence of a landslide at Rubble Creek in British Columbia in 1855–1856 has been corroborated by reports from military personnel and Indians (Moore and Mathews, 1978). However, the age of features cannot always be placed so precisely. Chandler (1971), using ridge-and-furrow and map evidence, can simply ascribe the occurrence of slight movements on an escarpment near Rockingham, Northamptonshire, to the broad period since 1615 (Table 2.2).

In some cases, slope movements and signs of instability may have become obscured, but Chandler (1970a) has said that 'the importance of recognising old landslips so that they may be considered at an early stage in the design of engineering works cannot be over emphasised'. Historical sources can also provide useful background information on the conditions contributing to slope movements, for example, on the undermining of slopes by rivers, by coastal erosion or on the addition or removal of man-made structures on slopes. Hutchinson (1965a) used an early map and oral evidence to determine the influence of a river on a slope at Vibstad in Norway and recently Hutchinson *et al.* (1980) have analysed the influence of the construction of a harbour pier on the coastal landslides at Folkestone Warren.

Another use for historical sources is in the assessment of the frequency of mass-movement events. Frequency-analyses are mostly obtained by collating case studies and reports: for example, Seed (1968) analysed landslides occurring from liquefaction associated with earthquakes and listed events from which frequency can be calculated. The number of landslides in a century in Norway are noted by Bjerrum (1954), though with no indication of sources. Grove (1972) has analysed the incidence of landslides, avalanches, rockfalls and floods in western Norway and has identified their varying frequency during the 'Little Ice Age'. Allowance was made in land-rent assessments for the effects of such natural disasters and Grove (1972) demonstrated the concentration of these events in the period 1650 to 1760 and corroborated the rent-records with eye-witness accounts. She has also discussed the difficulties of assessing frequency because of sequences of events, the difficulty of their separation and lack of exact dates in some sources. For these reasons frequencies were calculated for five-year periods (Figure 5.7).

5.5 TECTONIC ACTIVITY

Many texts on earthquakes and volcanoes describe significant and spectacular occurrences in the past. For example, Rittmann and Rittmann (1976) discussed the history and dates of the last eruptions of many of the world's major volcanoes and Bullard (1976) included many case histories in his book on volcanoes. The longest historical records emanate from Europe and the Far East. For example, Turner (1925) has postulated a 284-year cycle in Chinese earthquakes using records dating back many centuries. Rittmann and Rittmann (1976) stated that the oldest record of a volcanic eruption is probably that of Etna, described in one of Pindar's odes from the fifth-century BC.

Lyell, in his *Principles of Geology* (1867 edition), discussed earthquakes and volcanic activity in a number of areas using detailed descriptions of events by eye-witnesses, written accounts and local histories. Often specific reports of volcanic eruptions and earthquakes, as with other events, were written in order to obtain relief and help, and intercessionary pleas were often also made to church leaders. The texts of these can be useful. Lyell (1867 edition), Rittmann and Rittmann (1976) and others have demonstrated the value of illustrations, drawings and paintings. The spectacular nature of eruptions and earthquakes have made them interesting subjects for artists whose work has proved useful in indicating the nature and path of lava flows, although allowance must be made for embellishment and artistic licence. Historical sources have also proved useful in helping to date different lava flows and fault scarps and can be corroborated by field evidence of the position and nature of the features.

A major purpose of analyses of earthquake and volcanic activity has been to assess the liability of an area to such events. This is generally achieved by accumulating evidence of activity and information on intensity and area of effect in order to estimate frequencies of occurrence for risk-assessment. This has practical implications for land-use zoning and building design. As Ambraseys (1979) said:

> To reduce earthquake losses in an area of little-known seismicity it is necessary to determine in advance what is an acceptable risk. To do this a critical evaluation of all existing seismological, tectonic and historical data should be made.

As might be expected, much of this work is restricted to specific regions, most notably California and Japan, and to a lesser extent the Euro-Asian belt of tectonic activity. The literature is now vast and only a few case studies can be cited here. The major developed regions experiencing tectonic activity have institutions and journals devoted to seismology and vulcanology which produce a wealth of studies including historical investigations; for example, Hatori and Usami have published many articles using historical sources in the *Bulletin of the Earthquake Research Institute* of the University of Tokyo (Hatori, 1976, 1977, 1978; Usami, 1976, 1977, 1978).

Much work has been done to establish the dates of earthquakes in the western United States, and California in particular. Townley and Allen (1939) produced a catalogue of earthquakes for the Pacific coast between 1769 and 1928. This was based on two earlier catalogues and also on printed lists in scientific journals, accounts in books, magazines and newspapers, lists from manuscript records of individuals, oral accounts, and the earthquake records of various institutions. In their work they discussed the varying coverage of information and the problems of estimating intensity and timing before the introduction of Standard Time. Byerly (1951) has used this catalogue alongside diaries, oral evidence and mission reports to investigate the history of earthquakes in the San Francisco area. The occurrence of earthquakes in central California in the 1830s was investigated by Louderback (1947) because of general confusion over the number and strength of events. His article contains interesting examples of how the veracity of reports can be investigated and assessed. Huber's (1930) study is concerned with the San Francisco earthquakes of 1865 and 1868, which are important because they appeared to equal the intensity of the devastating earthquake of 1906. The history of earthquakes on the San Andreas fault is also described by Oakeshott (1976), who noted that the earliest known report of an earthquake was made in 1769 by the explorer Gaspar de Portola. It was only in 1887 with the installation of the first seismograph in the United States that instrumental records became available.

The problems resulting from early, inaccurate compilations of earthquake dates have been emphasized by Ambraseys (1979), who investigated two areas in Iran. One of these has a long and well-documented history and was thought to have experienced much earthquake activity; the other is more remote and was considered seismically inactive. He found by careful investigation of travel accounts, histories and letters that almost the reverse is, in fact, true. Davison (1924), in a comprehensive review of British earthquakes, has also indicated the problems of using unverified annals and chronologies and of distinguishing between real and spurious earthquakes.

Some individual past earthquakes have been reassessed using scientific methods. For example, Reid (1914) has re-examined the reputedly severe Lisbon earthquake of 1755 to establish the sequence of events and the nature of the movement. He warned of the way that the embellished nature of writings at the time (see Chapter 3) distorts the truth and demonstrated how some phenomena have later been falsely ascribed to the Lisbon earthquake. Nuttli (1973) has examined the Mississippi valley earthquakes of 1811 and 1812, which are the largest earthquakes to have occurred in North America since European settlement. Using newspaper accounts and other descriptions, Nuttli deduced the earthquake-intensities and constructed an isoseismal map. Contemporary accounts described lakes disappearing and water being uplifted as well as damage to buildings. There were several principal shocks and aftershocks, though their number and exact timing are difficult to ascertain. Similarly, Leblanc (1981) has reassessed the intensity of a 1732 Montreal earthquake from additional documentary evidence.

Landslides and mass movements are often associated with earthquakes, and many authors have investigated the relationships between these. Tamura (1978) has examined the relationship of areal distribution of landslides to earthquake magnitude in Japan and Youd and Hoose (1978) have studied historic ground-failures in northern California. In a more unusual application, historical evidence of earthquake-frequency has been used by Healy *et al.* (1968) to prove that underground disposal of waste at Denver, Colorado, has caused earthquakes.

Past volcanic eruptions have also been investigated by a number of authors who have found eye-witness accounts especially valuable. With the help of such sources Thorarinsson (1970) has reconstructed the Lakigigar eruption of 1783 in Iceland; Williams (1955) that of Coseguina, Nicaragua, in 1835; Horner (1847) the eruption of Monte Nuovo, near Naples, in 1538; and Snead (1964) the mud-volcanoes of Baluchistan.

As with earthquakes, many investigations of volcanic activity are concerned with identifying areas liable to eruption, the assessment of risks and the prediction of future activity. Duncan *et al.* (1981) have analysed historical sources to produce a predictive model of the volcanic activity of Mount Etna, Sicily. An Icelandic geothermal power station, for example, has been built alongside what was thought to be an extinct volcano, but the volcano recently started to show activity very similar to that described by a pastor in his diary in 1724–1729. According to Fairhall (1978) this had previously been thought a rather embellished account, but geologists are now studying it carefully. In fact, the whole range of Icelandic volcanic activity in the past has been much studied. Thorarinsson *et al.* (1959) stated that 'Icelandic written records provide detailed and rather reliable accounts of volcanic eruptions for a comparatively long period of time'. Barth (1942) has discussed water- and fissure-eruptions at Myvatn in Iceland and noted the eruptions of 1725 and 1783.

Wadell (1935) has examined the relationship between ice-floods and volcanic activity in Iceland and has identified the sequence of events from 1681 onwards, mostly using the evidence of travellers' reports.

In reviews of recent volcanic events, some historical perspective is often provided; for example, Trask (1973), in his discussion of the Mexican volcano Paricutin, recorded that although there had been no volcanic activity there within living memory, a volcano had erupted fifty miles away in 1759. Moore *et al.* (1973) noted that since 1572 the Taal volcano has erupted explosively twenty-six times. The recent volcanic activity of the western United States, particularly of Mount Rainier and Mount Baker, has been investigated by Crandell and his collaborators (Crandell and Waldron, 1973; Crandell and Mullineaux, 1967), and these include some historical data. Oakeshott (1976) listed the dates of the then last eruptions of these Cascade Mountain volcanoes.

6
Explanation of Changes in Historical Time

Three main forms of causal inference are commonly employed in the earth sciences; first, cause may be inferred on the basis of temporal coincidence of phenomena, second, from the determination of spatial association, and third, by direct observation or experimentation (Harvey, 1969). Time-association is probably one of the most important methods in historical studies. Broadly speaking, if change in one factor or condition is coincident in time with a change in another, then a causal link may be suggested. However, as Hastings and Turner (1965) commented in relation to arroyo development, matters can be much more complicated:

> The principal historical and photographic evidence bearing on the question of cattle as the agent of change is equivocal; there is evidence that overgrazing acted as a primary cause, there is evidence that it did not. On both sides of the question, the timing of pairs of events is important. What is not clear in any single case is the degree to which coincidence is involved; a cause and effect relationship need not follow from a temporal correlation.

The precision with which a time-association can be established is very important. This can vary from a detailed reconstruction of continuous temporal variations, which can be tested by statistical correlation, to single and approximate dates that appear to show coincidence but are based on scanty evidence. Conversely, lack of coincidence in time can be used to eliminate a possible cause and refute an hypothesis, but allowance has to be made for time-lags and for any inherent variability in a phenomenon.

A second way of identifying cause and effect is to test for spatial association. It is known from the nature of distributions and processes that effects from certain causes are likely to be widespread whereas others may be much more localized. For example, the effects of a climatic change on vegetation are likely to be general over an area but the effect of cattle grazing and trampling is likely to be more isolated. The combination and interrelations of time- and space-associations should also be examined, for some changes may be progressive in space and time whereas others may be simultaneous over a whole area.

The third main form of evidence of causation derives from known processes and observed effects. If the effects of a certain influencing factor are known, and there has been a change in the factor in the past, then the environment is likely to reflect this change. The reverse is not necessarily true because given effects may arise from a number of causes. Much knowledge of processes and effects of individual factors is derived from recent studies, and three main approaches may be followed to identify the effects of particular factors: the analysis of a 'before and after' situation, analysis and comparison of affected and unaffected areas and the measurement of effects as a factor is altered or applied, sometimes in controlled experiments.

There are many problems associated with identification of cause and effect in environmental systems because of the interrelationships, links and feedbacks within systems and because the direction of operation or causation may depend on the timescale of change (Schumm and Lichty, 1965). A basic division is often made between natural and anthropogenic causes of change, although this is in many ways arbitrary since man can be considered part of the whole ecosystem. It is also often difficult to distinguish between natural and human effects since the latter may only modify the rates of processes which are inherently variable. However, the natural/human division can be important and useful since human activities are tending to have an increasing impact on the natural environment, and in many cases they do significantly alter processes and disturb equilibria, as is explained in Gregory and Walling (1979), *Man and Environmental Processes.* From the point of view of environmental management it is important to be able to identify the methods and extent of human interference and to understand its effects in order to predict future conditions and to take preventative action if so desired. Historical investigation can extend the scope of such studies by helping to understand the background to present conditions, by highlighting the progressive development of certain phenomena and by demonstrating the effects of activities no longer in operation.

6.1 NATURAL CAUSES

The major natural causes of change are mostly those phenomena already discussed in Chapters 4 and 5. They will be discussed here only briefly since the main sources have been referred to in the appropriate sections above and are quite familiar to earth scientists.

Change is inherent in many phenomena as periodic, cyclic or random variations, trends or migrations. Such changes do not necessarily require an alteration in any other factor or external conditions. For example, de Boer (1964) and Kestner (1970) have postulated cyclic variations in coastal processes and the development of river meanders and cut-offs has been similarly considered. Vegetation assemblages vary naturally due to competition and succession, and some workers have suggested that a sequence of slope processes also occurs naturally over time. It is thus important to consider autogenic models of explanation in the search for causes of change. Formerly, many of these models have assumed continuous, linear change but more recent theories such as Schumm's (1973) proposal of complex response and thresholds and Graf's (1979a) use of catastrophe theory incorporate elements of discontinuous change.

Geophysical causes relevant to the historical timescale include earthquakes and volcanic activity. These can have a direct effect on topography and materials but can also trigger landslides and mass movements and alter stream-courses. Tsunamis produced by earthquakes can have profound effects on coastal areas. Other similar causes of change which may be included here are base-level and sea-level changes; in some coastal areas, as around the Baltic, for example, these have had an effect within historical time (Jones, 1977). At the Royal Society in 1972 evidence for the subsidence of south-east England was discussed (Kelsey, 1972) and others, using levelling, tidal and similar evidence, have also suggested that relative sea-levels have changed in the historical period, particularly on the east coast of Britain (e.g. Valentin, 1953; Green, 1961).

Climatic change is a major cause of change in other phenomena though to establish it as a cause in any particular case the amount, type and timing of any change or variation must be carefully examined. Some changes may be quite subtle; Schumm and Lichty (1963) found that change in the magnitude and frequency of rainstorms while total rainfall remained the same had affected the vegetation and, in turn, the Cimarron river channel in Kansas. Changes in all aspects of climate must be considered, including, for example, the combined effects of temperature and precipitation on evapotranspiration. Probably the most important effects of changes in climate are on vegetation and runoff but these have repercussions on other parts of the environment. Areas of marginal vegetation growth are particularly sensitive.

Relief, topography and slope morphology change little on the historical timescale but any alterations can modify sediment-supply, drainage, vegetation and the stability of slopes. The sediments and materials of slopes may alter naturally during historic time through weathering, erosion and deposition, and alteration in the properties of materials may affect slope processes and sediment-supply. Sediment characteristics also vary in space and thus, as different areas are acted upon by processes such as fluvial or coastal erosion, the rates and nature of these processes may alter.

Changes in soil characteristics as by leaching and progressive loss of nutrients may cause changes in vegetation and ecology, but such changes under natural conditions are difficult to detect from historical sources. Vegetation is inherently more dynamic and varies with seasons, succession and competition and as a result of changes in other factors such as climate.

Drainage and runoff are altered by many of the above factors but are themselves also causes of change. As with climate, changes may be in the magnitude–frequency of régimes. Alteration of the hydrological processes may affect sediment- and solute-removal and the regolith, vegetation and ecology of an area; alteration of streamflow directly affects channel morphology and sediment transport.

6.2 THE EFFECT OF HUMAN ACTIVITIES

The increasing influence of human activities on the natural environment means that analysis of their mechanisms, rates of operation and their effects is extremely important. Many recent studies in physical geography have focused on this problem, though it is by no means a novel field of inquiry (Marsh, 1864; Sherlock, 1922; Thomas, 1956). Historical studies have the advantage of allowing examination over a long period of time, and can perhaps include a 'before and after' situation, a less complex set of circumstances or may allow conditions to be investigated which cannot now be replicated. Jennings (1965), referring to man-made landforms, said:

> more significant for geomorphological investigation than the huge increase in area so affected has been the increasing recognition, in this period, that unrecorded direct human interference in the past has left behind more effects to complicate interpretation than had previously been realised.

The main human activities which have been identified in the literature as causes of environmental change are land use and agricultural practices, urbanization, drainage, engineering construction and industrial activity. Both direct and indirect effects can be distinguished; for example, a change in land use is a direct change in vegetation but can also have some indirect effect on runoff. Likewise, drainage of marshland is a deliberate change in hydrological régime but indirectly it affects ecology. Constructional and excavation activities have a direct effect on slope morphology but may also have indirect effects on drainage and sediment-supply. Sometimes deliberate actions are taken which do not have the desired effects; for example, erosion-control structures may fail to prevent erosion. Likewise, an activity may have detrimental consequences which were not anticipated. A review of the impact of man on environmental processes is provided in Gregory and Walling (1979).

Land use

Various changes in agricultural land use can affect elements of the natural environment. The fundamental change is from natural to cultivated vegetation, which may disturb natural equilibria as cultivation often involves exposure of the soil, making it vulnerable to erosion and generating runoff. Susceptibility to erosion varies with climate, topography, soil type and other factors and the problem may also be aggravated by the nature of agricultural practices. A useful review of the general principles of soil-erosion is that of Butzer (1974). The results of the change from natural to cultivated land use can be seen particularly clearly during the period of initial settlement in the New World countries. Cumberland (1941) has examined such changes in New Zealand.

The clearance of forest has major repercussions on runoff, erosion and sediment-yield from slopes. These effects have been investigated in a number of studies (e.g. Ursic and Dendy, 1965; Clarke and McCulloch, 1979). The removal of trees decreases interception and evapotranspiration producing increased runoff which is able to transport greater amounts of sediments, thus increasing erosion on slopes and sedimentation in valleys and estuaries. Wolman (1967) estimated that the sediment-yields from forest areas in the eastern United States could be about 100 tons/square mile/year compared with 300-800 tons/square mile/year for cultivated land. Also referring to North America, Striffler (1964) produced figures for the Michigan area of 100 pounds/square mile/day for 'wild vegetation', 400 for forest, 1800 for pasture, and more than 2000 for cultivated land. Curry (1971) and Bailey (1973) have shown how certain forest-management and logging practices have a detrimental effect on the chemical and physical characteristics of soils, and Binns (1979) has examined the impact of afforestation in Great Britain, showing that land-acquisition policy is important in determining water-yield. The type of vegetation cover is also an important factor affecting slope stability (e.g. So, 1971).

A number of studies of the precise impact of deforestation have been made, particularly in those most recently deforested areas of the New World, where extant documentation is good. Studies by Gottschalk (1945), Nelson (1966), Trimble (1976) and Knox (1977) have shown that increased erosion on slopes and sedimentation in river valleys coincided with the periods of deforestation and early agriculture in the United States. Much of their information on land use is derived from Federal Land Survey plats and notes and travel and expedition descriptions. Knox (1977), for example, found that settlement and agricultural

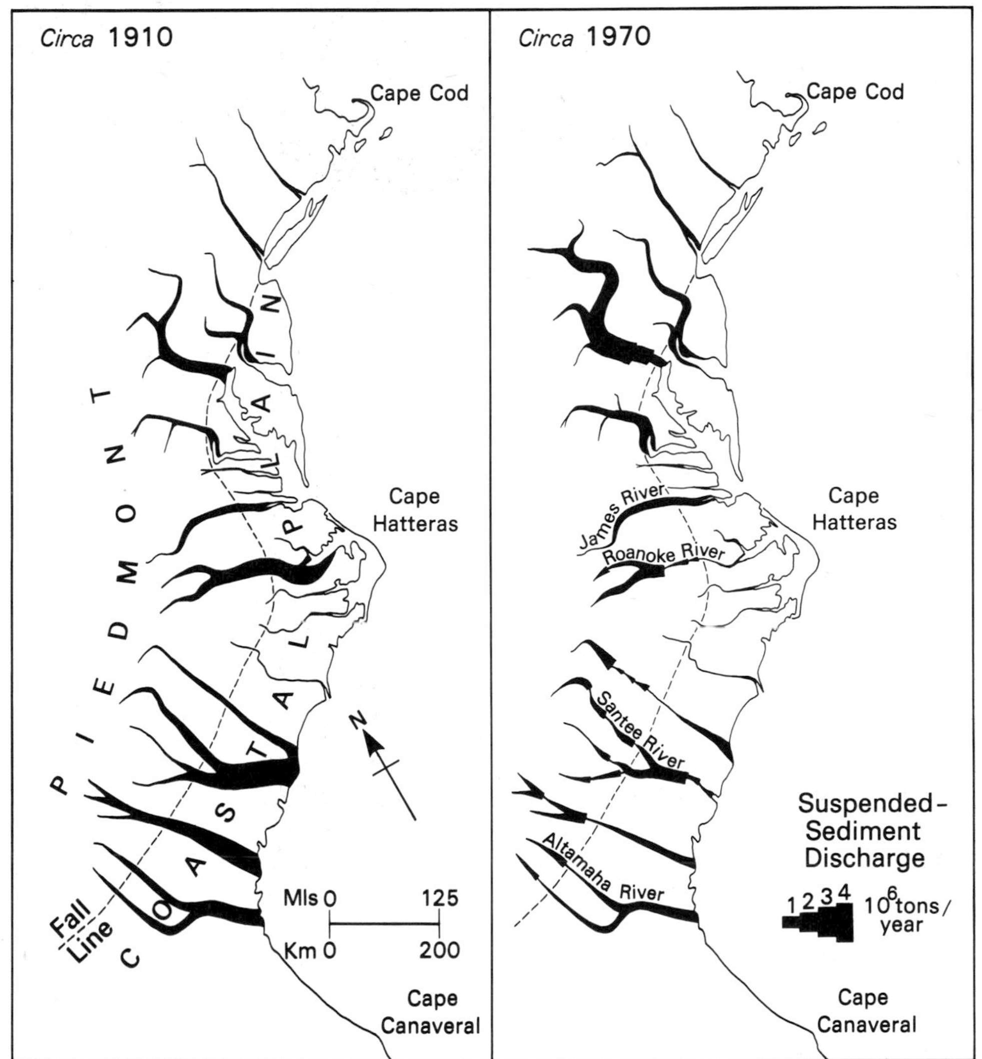

Figure 6.1. AVERAGE ANNUAL LOADS OF SUSPENDED SEDIMENT CARRIED BY RIVERS OF ATLANTIC DRAINAGE IN THE UNITED STATES *C.*1910 AND *C.*1970.
(Source: Meade and Trimble (1974))

development caused changes in flood-magnitude, sediment-yields and channel-morphology and that the impact varied between the upper and lower parts of watersheds (Figure 4.19). Towl (1935) and others have suggested that clearance of vegetation adjacent to river channels has caused channel changes. Many studies, particularly in Europe, have used pollen and sedimentary evidence of changes and related these to land use. They concern mainly prehistoric times but some enter the documentary period (e.g. Oldfield, 1963; Roberts *et al.*, 1973; Davis, 1976).

Striffler's 1964 data show that the change from grassland to crop-cultivation can also cause an increase in sediment-yield. Methods of cultivation such as contour ploughing can also have a considerable impact on physical systems. Piest *et al.* (1976) have postulated from a number of plot and small watershed studies in the United States that surface runoff from unprotected row-cropped fields is two to three times surface runoff from natural prairies, and that peak runoff rates from such fields are often ten to fifty times the pre-settlement rates. Bennett (1931) and others have also discussed soil changes and increases in erosion due to cultivation. Meade and Trimble (1974) have examined the possible causal link between the decrease in row-crop agriculture in the southern Piedmont, United States, 1919–1967, and the associated decrease in suspended sediment-loads of streams (Figures 6.1 and 6.2).

Comparatively little attention has been paid to the effects of the cultivation of particular crops on erosion, though factors such as the proportion of bare ground, rooting depth, water and nutrient demands, and returns to the soil are obviously significant. Klimek and Trafas, (1972) in one study of a basin in the Carpathians, suggested that change from corn to potato growing accompanied by poor farming practices in nineteenth-century Poland led to soil-erosion and channel changes which can be detected on topographical maps. Likewise, Costa (1975) has demonstrated the demanding nature of tobacco-cultivation and its effects on soils in nineteenth-century North America.

The effects of animals, particularly cattle, include compaction of the ground, trampling of vegetation, wearing of tracks, spread of seeds and selective grazing. By depleting vegetation it is suggested that livestock increase erosion in the same way as other factors that reduce vegetation cover. These effects of livestock husbandry have been studied particularly in relation to the formation of arroyos (e.g. Duce, 1918; Hastings and Turner, 1965; Denevan, 1967). Hadley (1977) has suggested that a reduction in cattle numbers was the cause of a decrease in the sediment load of the Colorado between 1926–1941 and 1942–1960.

Although the major destructive effects of land use arise from increased cultivation and deforestation, reverse changes do occur. The effects of afforestation and reclamation have been examined with historical documents, for example by Thorpe (1957) in Jutland, Kestner (1962) in the Wash and Bird (1974) in coastal Australia.

As noted in the introduction to this section, it is not only the type of land use but also the nature of agricultural practices that affect soil-erosion, though their exact influence is difficult to detect. Certainly, conservation techniques such as contour ploughing, terracing of slopes and mulching and maintaining a vegetation cover tend to reduce erosion. Ireland *et al.* (1939), in their discussion of gully-erosion, stated:

> The rate and manner of soil erosion, however, and the localisation of its action depend largely on the treatment that the land receives. Improper methods of farming, construction of roads and railroads, digging of ditches and throwing up of poorly planned terraces share the blame for the destructive erosion of the soil.

Reagan (1924) and Evenerai *et al.* (1961) have shown how the use of check dams and irrigation once enabled agricultural exploitation in two semi-arid areas, the south-west United States and the Negev, where there are now problems of erosion. Aghassy (1973) has suggested that the development of some Middle East badlands is partly due to ploughing techniques. Some authors have also hypothesized that soil-exhaustion and erosion were

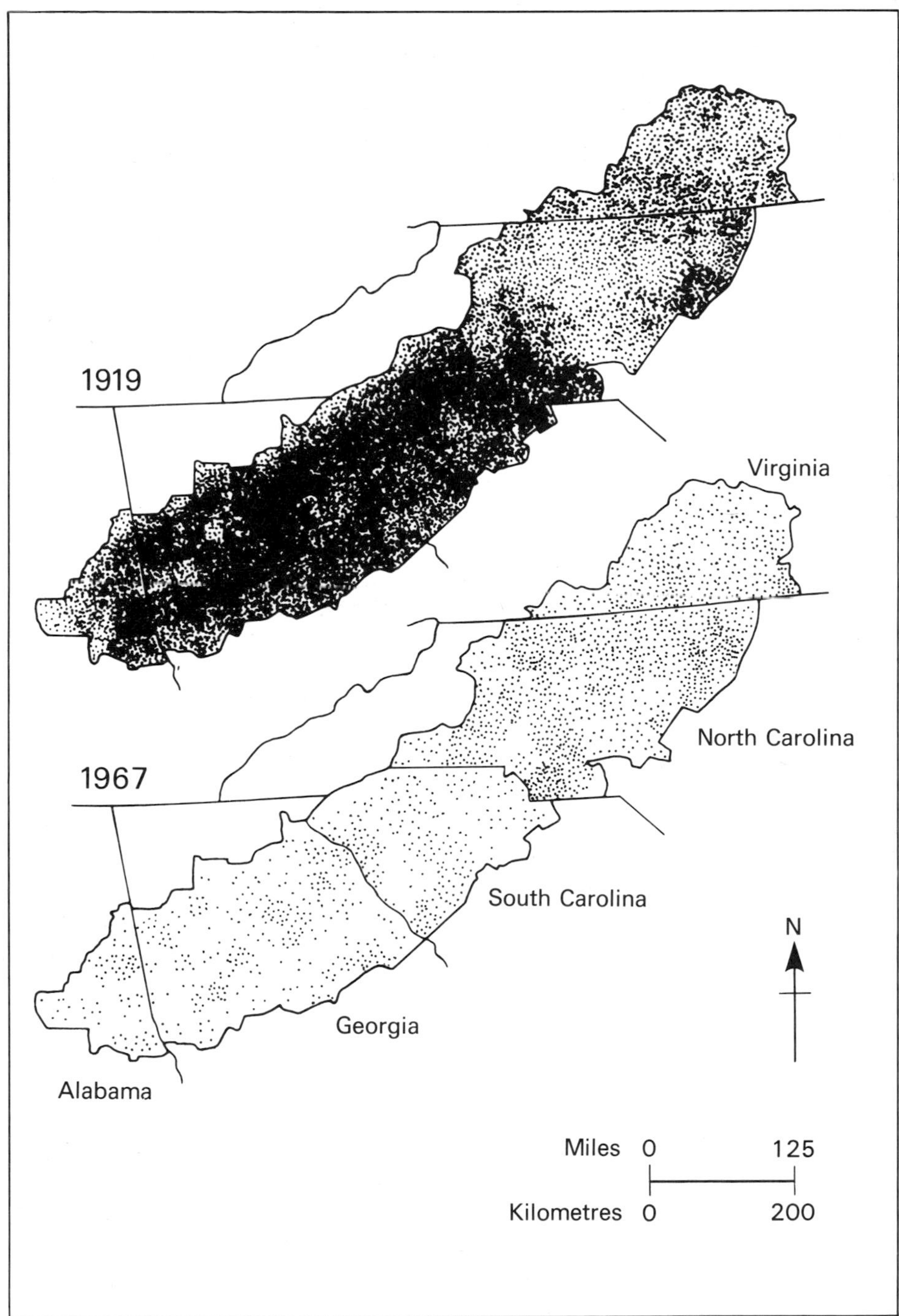

Figure 6.2. DECREASE IN ROW-CROP AGRICULTURE IN THE SOUTHERN PIEDMONT, 1919–1967. Each dot represents 1000 acres (405 ha). (Source: Meade and Trimble (1974))

major causes of the decline of certain ancient civilizations (e.g. Simkovitch 1916; Cowgill and Hutchinson, 1963). Olson and Hanfmann (1971) commented that:

> Important soil properties and their influence upon civilisations have too often been ignored. Soil studies now are being used with other resource inventories in planning for the future, but historical events too, can help predict more accurately and efficiently the future consequences of present-day decisions.

Vita-Finzi (1969), Butzer (1974) and others have discovered considerable archaeological and sedimentological evidence and also historical descriptions which relate land use to concomitant soil-erosion.

Nor should the effects of agriculture be neglected in those areas less vulnerable to erosion. The nature of crop-rotations can have important effects on soil-depletion and exhaustion. Glenn (1911) recorded that lack of rotation in tobacco cultivation in Appalachian America was a probable cause of soil-exhaustion and erosion in that area and Rockie (1939) has noted the beneficial effects of the change from monoculture wheat to wheat and fallow in the Palouse.

The use of fire as a technique of land management has also had considerable impact on vegetation and erosional processes in many parts of the world. Decline in the use of fire with the demise of Indian practices has been considered by Chavannes (1941), Humphrey (1958), Johnston (1963) and others as a major cause of the increase in woody plants in parts of the United States.

Other agricultural practices which may have environmental effects are the use of machinery which may compact the ground, the removal of hedges (common at present in Britain) and the use of fertilizers and pesticides. The last are probably the most important and their effects on water quality and ecology have been studied (e.g. Green, 1973, 1974). General management practices can also affect the environment and vegetation. Bahre and Bradbury (1978) have examined the effects of different practices of grazing control on either side of the Mexico/United States border by comparing old and recent photographs. Peterken (1969) and Rackham (1975) have demonstrated the effects of various woodland management practices.

Many changes ascribed to land-use alterations are similar to those which would result from changes in climate. A number of authors have considered these alternative hypotheses (e.g. Schumm and Lichty, 1963), and in one field, the study of arroyo development, alternative hypotheses have been exhaustively tested. Comprehensive reviews of theories on causes of arroyo formation, as well as evidence on the processes of development, are provided by Tuan (1966) and Cooke and Reeves (1976). Many of the arroyos which began to develop in the late nineteenth century have been studied. In the semi-arid south-west of the United States the development of arroyos appears to coincide very closely in time with the period of settlement and increasing numbers of cattle. An early hypothesis proposed that increased grazing was the cause of arroyo-cutting (Rich, 1911; Duce, 1918). However, Tuan (1966) and Denevan (1967), from their examination of historical sources, have suggested that cattle numbers were in fact high in the period well before the onset of nineteenth-century gullying. A period of arroyo-cutting in about the twelfth century AD has also been identified, for which the increasing cattle numbers explanation seems inadequate (Bryan, 1928b; Antevs, 1952). Bryan (1928b) suggested that climatic change was the primary cause

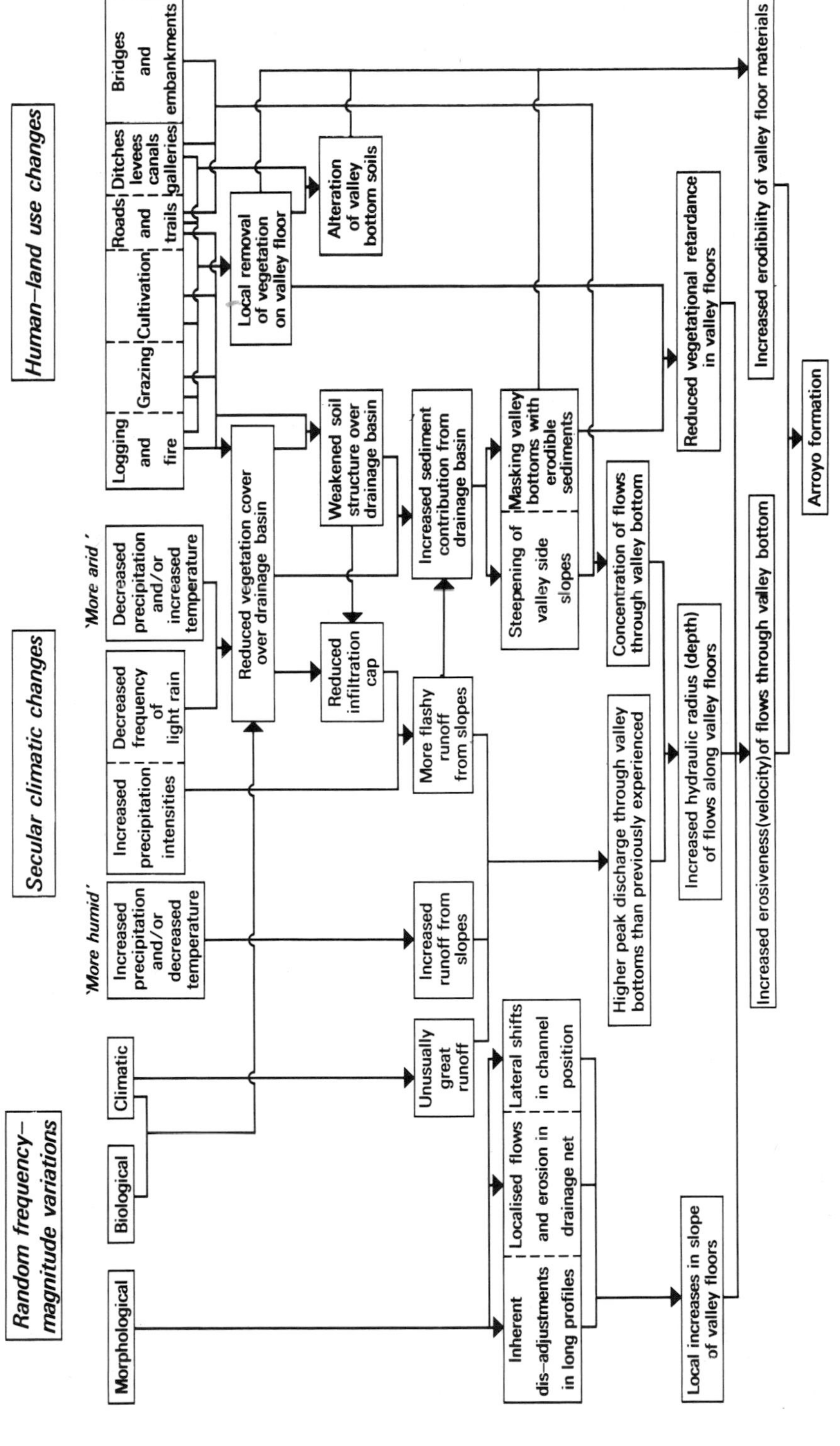

Figure 6.3. A MODEL OF ARROYO-FORMATION.
(Source: Cooke and Reeves (1976))

but that increased grazing may have provided a trigger in the nineteenth century. Others (e.g. Bull, 1964) have correlated the periods of cutting and filling with variations in rainfall. Whether cutting occurs in periods of low rainfall or during more intense summer storms is still a matter of controversy (Tuan, 1966). Possible causes of arroyo formation are set out in Figure 6.3.

Urbanization

The number of studies of the effects of urbanization on runoff and sediment-yields has increased over the past two decades and there are several reviews and collections of papers (Leopold, 1968; Detwyler and Marcus, 1972; Coates, 1974; Hollis, 1979). The type and characteristics of urbanization can influence the environmental effects that it induces. The location of sewage and stormwater outflows, the phase of urbanization (Figure 6.4) and the

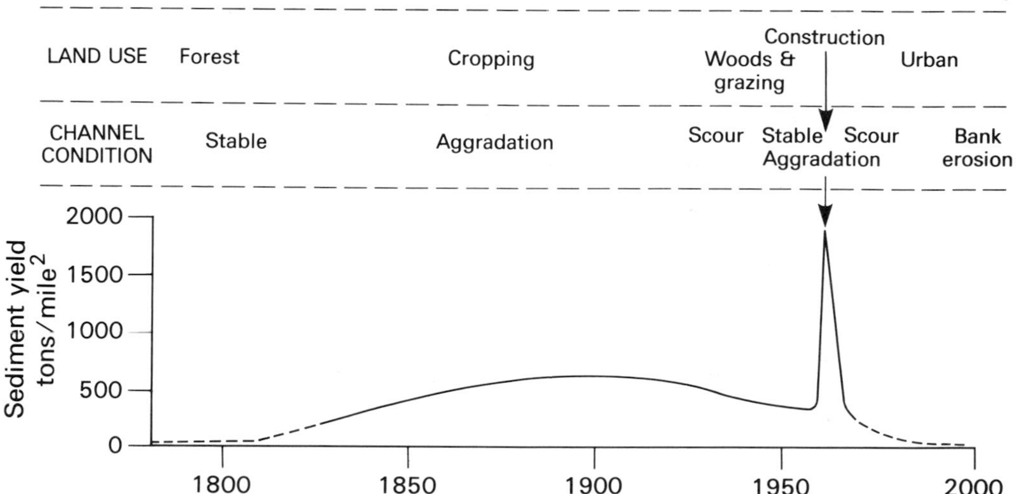

Figure 6.4. SEDIMENT-YIELD IN RELATION TO LAND USE IN THE PIEDMONT AREA OF THE UNITED STATES, *C.* A D 1800–2000. (Source: Wolman (1967))

type and density of housing have all to be taken into account (Hammer, 1972). The effects also vary with the climatic, geological, morphological and hydrological characteristics of an area. In general, if drainage remains within the basin then runoff, particularly peak flows, increases and channel-enlargement may take place. Sediment-yield also increases markedly in the construction phase (Wolman and Schick, 1967; Guy, 1970; Walling, 1979). Urbanization can directly alter the natural drainage and may obscure it when flow is put underground; Nunn (1979) has reconstructed the original network of streams in London partly from map and documentary evidence. The construction and surfacing of roads can cause increased and localized erosion (Gregory and Park, 1976); a general review of these effects is provided by Parizek (1971). Urbanization can also have a significant effect on groundwater level and quality and this may also have repercussions on water resources and agriculture and may cause saline intrusion on coasts and subsidence of land. Slope stability can be affected directly by construction on slopes and indirectly by alteration of groundwater and moisture levels (Coates, 1974).

Urbanization can alter climate by causing increases in temperature and air-turbulence and in precipitation due to air-pollution (Atkinson, 1979). There has been much investigation of air and water quality in recent years but most of the work uses 'technical' rather than historical sources. However, it should not be assumed that pollution is only a modern problem. Brimblecombe (1975) has compiled evidence of air-pollution created by the burning of coal in the thirteenth century in Britain. Prior to this, the major fuel has been charcoal; in 1257, Queen Eleanor visited Nottingham Castle and complained of the air being full of the 'stench of sea coal smoke'. Brimblecombe (1975) considered that lime kilns were a main source of pollution and used documents such as manorial rolls to assess the amounts of coal being burnt. By 1285 the problem was serious enough to require the establishment of a Commission on Pollution. Brimblecombe and Ogden (1977) discuss the extent to which art and literature can be used to assess pollution levels and the many difficulties caused by the emotive nature of this source material.

Water-pollution and changes in water quality are perhaps even more difficult to investigate for past periods since much pollution is invisible in the form of solutes or microscopic organisms. There is evidence in historical sources of water becoming unfit for use; such evidence mainly concerns bacteria, organisms and waterborne disease and has been little used in water-quality studies but more often in histories of public health. A study incorporating historical evidence of water quality is Gilmour's (1974) investigation of the impact of settlement on the coast of Australia over the last 150 years. He used travellers' descriptions and maps showing sources of fresh water which later became polluted or saline, and has evidence of deteriorating water quality provoking demands for the construction of sewerage systems.

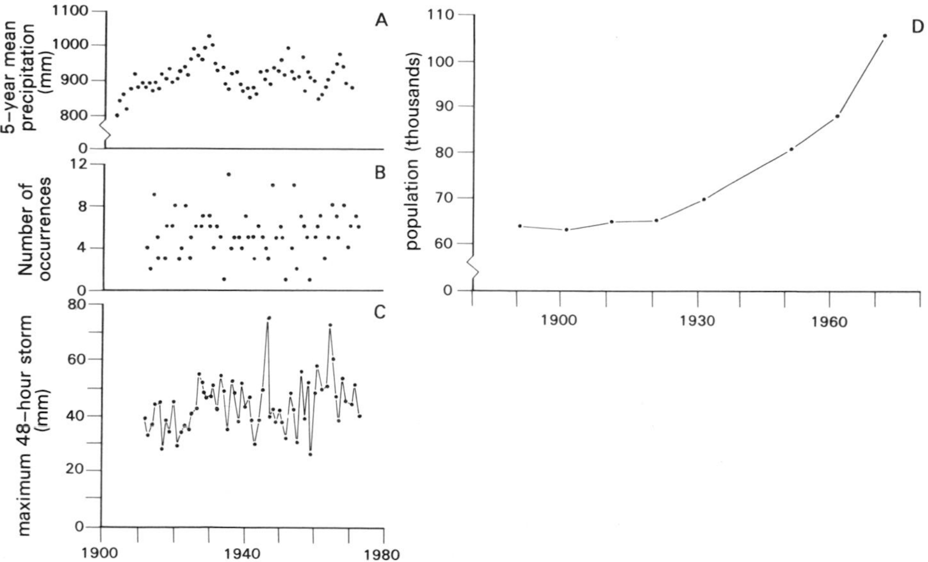

Figure 6.5. ANALYSIS OF CHANGES IN PRECIPITATION AND POPULATION IN THE RIVER BOLLIN BASIN, CHESHIRE, 1872–1973. A — Five-year moving means of total annual precipitation at Macclesfield, 1903–1970; B — Number of precipitation events greater than 25 mm in two days at Macclesfield, 1912–1972; C — Magnitude of heaviest two-day storm in each year at Macclesfield, 1912–1972; D — Population of four urban districts, 1891–1971. (Source: Mosley (1975))

A number of studies have also been made of the effects of urbanization on hydrological processes and river channels, many of which cover periods of up to thirty years by direct observation, repeat surveys and the use of readily available technical data (e.g. Wolman, 1967; Leopold, 1973; Warner and MacLean, 1977). Others use previous surveys and data acquired by official bodies for comparison with a current situation (Hollis and Luckett, 1976). Some purely historical studies have also been carried out. Mosley (1975) has investigated channel changes on the river Bollin-Dean in Cheshire, using maps of various dates to establish changes. He has found significant differences in type and rate of change between the two periods 1872–1935, and 1935–1973. Mosley investigated possible causes of this change, basing his enquiry on the premise, derived from theory and process studies, that flood-peaks and bed-material discharge had probably increased. He tested various explanatory hypotheses, including effects of changes in precipitation, land use, agricultural drainage and increased urbanization (Figure 6.5). Precipitation records were examined for changes in amounts and magnitude–frequency of events but no changes were found that would have produced the alterations in runoff. He suggested that changes in land use affecting evapotranspiration and thus runoff are unlikely to have been significant but that changes due to artificial drainage are more likely to have been important. Detailed information on tile drains in the catchment is not available but he examined national trends in agricultural drainage and concluded that this was unlikely to be a primary cause because of the lack of time coincidence. He has derived an index of urbanization from the population census and has suggested that the major period of expansion of impervious surfaces coincided with the change in behaviour of the Bollin. In view of the lack of evidence of any other major changes in land use in the watershed he concludes that urbanization in conjunction with some major storm events was the probable cause of the onset of disequilibrium in the behaviour of the river.

Drainage

The history of a number of major drainage schemes in Britain has been well documented. These include seminal studies of the Fens by Darby (1936), the Somerset levels by Williams (1963) (Figure 6.6) and the Hull valley by Sheppard (1958). Drainage works obviously have a direct effect on runoff and discharge in streams (Green, 1973) and on vegetation and soils, but they may also affect ground-water levels and sediment budgets. Drainage alters the ecology of an area, and, in conjunction with improved agriculture, can often exacerbate pollution from the use of fertilizers. The drainage of peat and marsh areas can also cause subsidence (e.g. Sharpe, 1938; Poland and Davis, 1971; Hutchinson, 1980).

Maps and written descriptions have been used to reconstruct pre-drainage conditions, supported by evidence of place-names. Legal documents, parliamentary and sewer court records can also be used to establish dates of drainage schemes. For more recent periods, maps, descriptions and agricultural records can show the effects on agriculture and land use. Evidence of the extent of pre-drainage flooding has been employed by Sheppard (1958) and Williams (1963). Ecclesiastical records can provide abundant information on some drained areas since monasteries and churches were major landowners. The petitions, arguments and disputes leading to legislation for drainage often give detailed accounts of flooding and its effects on soils and land use.

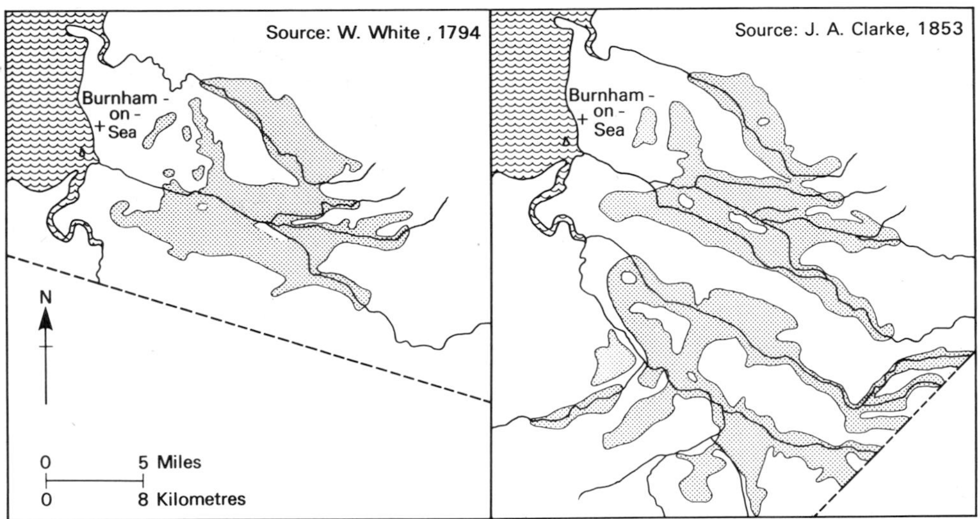

Figure 6.6. FLOODING IN THE SOMERSET LEVELS IN 1794 AND 1853.
White's map accompanied his proposal for more effective drainage of turf bogs and flooded lands near the rivers Brue and Axe. The extent of flooding in 1853 as recorded by Clark was typical of the previous three decades. (Source: Williams (1963))

The history of underdrainage in Britain has also been widely investigated (e.g. Sturgess, 1966; Collins and Jones, 1967; Phillips, 1969; Sheail, 1976) but it is only recently that study of its hydrological effects has been undertaken (e.g. Green, 1973, 1976, 1979). There is still a dearth of investigations of its historical impact in physical process terms. Green has summed up the possible consequences on streams and their interrelationships with other factors as follows:

> Flow in watercourses is changing, since the drainage activity causes change in the incidence of runoff in relation to rainfall. Much research is necessary here, to disentangle the different effects of different types of field under drainage—for instance, runoff is probably rendered more irregular by tile drainage unaccompanied by moling or sub-soiling, but may be smoothed where moling or sub-soiling has taken place. There must also be taken into account an apparent increase, noted also during the past ten years, of heavy rains in short periods; this probably has a similar effect to the tile-draining. A further consideration is that, although tile-draining may increase 'peakiness' of flow in the short-term, it may well prevent the build-up of enough water to cause severe floods.

The Drainage Acts of 1846 and 1849 provided means of obtaining money for land-drainage improvement and in the 1850s and 1860s pipe drainage was introduced under the supervision of government inspectors (Sturgess, 1966). The invention of a machine to make cheap clay pipes allowed further extension of the drained area, though a report of the House of Lords Committee on Improvement of Land in 1873 (Collins and Jones, 1967) remarked that investment in drainage in the period 1850–1873 was rather low. Sheail (1976) in his review of land improvement during the First World War stated that 'many water-courses and field ditches were so neglected that they flooded neighbouring farmland'. A Land Drainage Act was passed in 1918 to alleviate this. The main expansion of tile drainage is

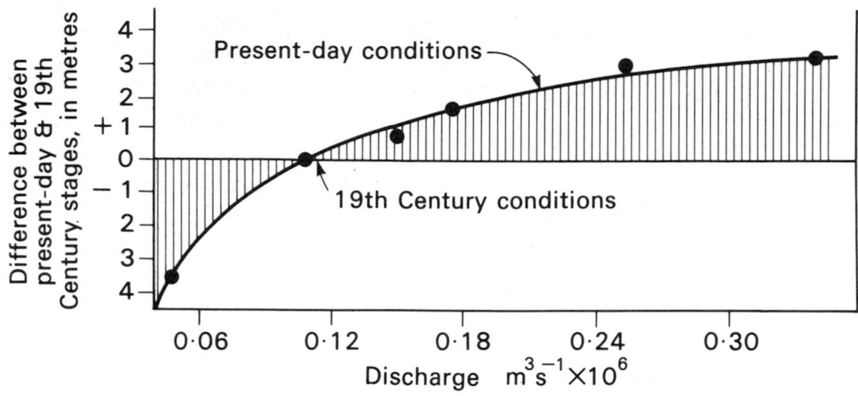

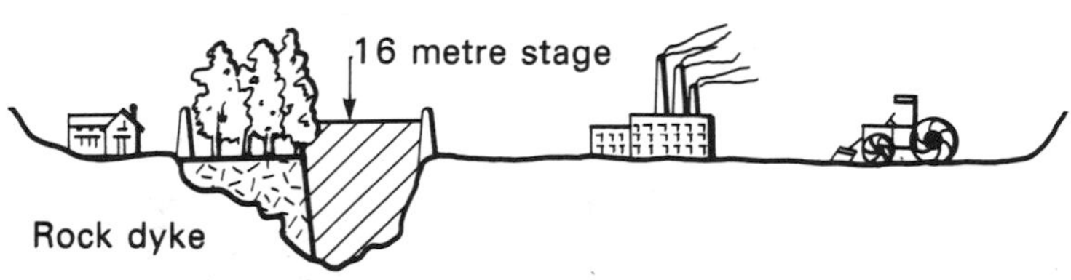

Figure 6.7. DIFFERENCE BETWEEN PRESENT-DAY AND PRE-DEVELOPMENT STAGES ON THE MISSISSIPPI AT ST LOUIS AND CROSS-SECTIONS AT ST LOUIS IN 1844 AND 1973.
(Source: Stevens *et al.* (1975))

much more recent; Green (1976) showed that the period since about 1965 has seen a considerable increase in the extent of field drainage in England and Wales as a whole.

Other activities directly and deliberately altering the flow of water in streams may be considered under the general heading of drainage. The effects of reservoir construction on streams have been studied (e.g. Stanley, 1951; Gregory and Park, 1974; Petts and Lewin, 1979). Such studies indicate the resulting changes in sediment-loads and channel-morphometry downstream. Meade and Trimble (1974) considered reservoir construction an alternative to land use change as a cause of decrease in sediment-loads of streams in the period 1919–1967 (Figure 6.1).

Some reservoirs and dams in the United States and elsewhere were constructed at the end of the nineteenth century and have produced clearly discernible hydrological effects. A number of authors have warned of the consequences of such human interference (e.g. Sonderreger, 1935), while Bergman and Sullivan (1963) cited the indirect effect of channel-reservoir seepage on channel vegetation. Historical studies include those of Schumm (1971)

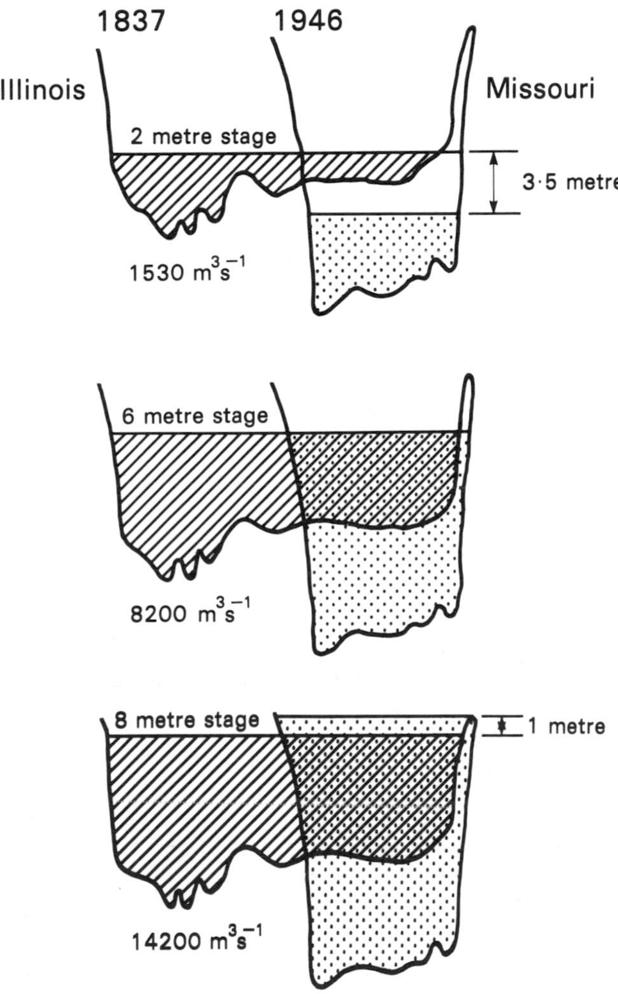

Figure 6.8. CROSS-SECTIONS OF THE MISSISSIPPI AT ST LOUIS SHOWING THE DIFFERENCES IN FLOW AREA BETWEEN 1837 AND 1946 FOR THREE DIFFERENCE FLOW STAGES.
(Source: Stevens *et al.* (1975)

and Piasecka (1974), in which river regulation is considered the cause of river-channel changes.

Abstraction of water from streams may similarly have an effect on sediment and morphology (Beckinsale, 1972). Dury (1973) has drawn attention to the effects of diverting streams for mills and Meyer (1927) reviewed some of the documents useful for reconstructing mill history in Britain.

Another activity with a direct effect on streams is channel regulation and modification; this includes straightening, embankment, diversion and use of other structures to control floods and erosion and to maintain flow. This affects discharge régime and sediment budgets and has implications for surrounding flood plains as well as the channel. Alterations in one part of the channel may also have repercussions elsewhere, as Winkley (1972) has shown for the Mississippi. This river has been intensively studied in respect of channel control because of the scale of human modifications (Matthes, 1947; Belt, 1975; Stevens *et al.*, 1975) (Figures 6.7 and 6.8). Laczay (1977) and others have shown the effects of channelization and a number of studies have demonstrated the changes which can take place after straightening of streams (Daniels, 1960; Noble and Palmquist, 1968).

Engineering construction

The effects of man-made structures can be ascertained by direct field observations, by comparison of affected areas before and after construction or by experimental simulation or theoretical calculation. Direct observations and comparison of areas have been used widely, especially in an historical context. Engineers usually make calculations or models of a structure before construction but the full effects cannot always be predicted in this way.

The construction of harbours, jetties, groynes and embankments has been shown to alter sediment movement along coasts causing depletion in one place and sedimentation elsewhere (e.g. Shepard and Wanless, 1971; Rawls, 1972; Seelig and Sorenson, 1973) (Figure 6.9). Bird (1979) has reviewed these effects. In a study of historical changes on the coast at Aberystwyth, So (1974) has shown how work undertaken to prevent beach-erosion and encourage beach-growth has only exacerbated erosion. Brookfield (1952), Burrows (1972) and others have demonstrated the role played by man-made structures in the silting of harbours on the coast of south-east England. The construction of a railway and associated embankments at Dawlish Warren in Devon was blamed by Martin (1876) for changes to the Dawlish sand spit.

Embankments and reinforcements have been used for erosion control and regulation of rivers but, as Winkley (1972) and others have indicated, partial control often moves the focus of erosion elsewhere. Encroachment on flood plains can alter flood and stream dynamics, and Johnson and Paynter (1967) considered that channel restriction by industrial works contributed to the occurrence of a cut-off on the river Irk in Oldham. Bridges and weirs usually have localized effects, the former often causing scour around piers. Poor maintenance and collapse of weirs will reduce channel control and Hooke (1977b) has shown how this may allow meandering to recommence in formerly straightened streams.

The addition or removal of structures may directly affect slopes by altering conditions of stability and indirectly by altering hydrological and vegetation conditions (Wayne, 1969; Zaruba and Mencl, 1969). There are many instances where the construction of railways, roads and the excavation of cuttings has caused landslides or aggravated slope instability. Justly famous is the Folkestone Warren landslip where the effects of the construction of the railway and of Folkestone Harbour pier have been analysed from documentary evidence (Toms, 1953; Wood, 1955; Hutchinson, 1969; Hutchinson *et al.*, 1980). Hutchinson and Gostelow (1976) have suggested that human interference on a slope may have been the trigger for a nineteenth-century landslide at Hadleigh, Essex. Walls and buildings can influence drainage and porewater pressures in slopes and thereby affect stability (e.g. Skempton and Hutchinson, 1969; Skempton, 1971). It is, therefore, often useful to examine the history of building works to understand slope dynamics.

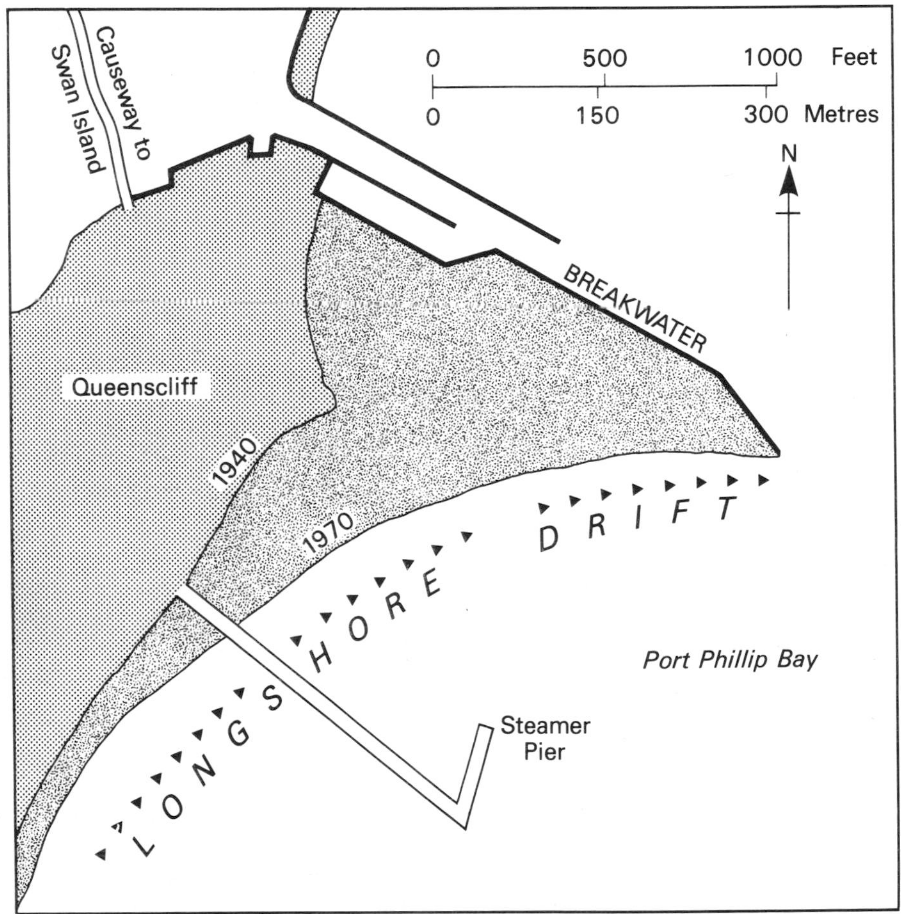

Figure 6.9. THE COAST AT QUEENSCLIFF, VICTORIA, SHOWING EXTENT OF SAND-ACCUMULATION AGAINST THE HARBOUR BREAKWATER.
(Source: Bird (1972))

Buildings, earthworks and other artificial structures of known date can protect the soils they bury from further pedogenic processes and thus preserve a record of past soil characteristics which can be subjected to modern scientific analysis. On Exmoor in

Somerset, for example, spoil banks constructed when the Pinkery 'Canal' was cut in 1833 preserve a soil untouched by later nineteenth-century reclamation (Crabtree and Maltby, 1976). Differentiation of an organic rich horizon at the surface of the soil enabled these investigators to examine the degree of soil development over a known period of time. It appears that soil developed at a rate of 8 cm per 140 years, surprisingly quickly, in their opinion, considering the harsh upland environment of Exmoor.

In short, most studies which examine the physical effects of man-made structures attempt to establish the situation prior to building and to identify subsequent changes. The date of construction can often be ascertained approximately from maps and large-scale plans as well as from newspaper descriptions, local histories and eye-witness accounts. Some surveys and documents associated with planning, design and construction may also be available from the organization or authority which commissioned the work. Illustrations may depict structures and provide evidence of the situation at a particular time.

Industry

Included under this head are the environmental consequences of a variety of activities such as dredging, quarrying, mining and the disposal of waste (Marsh, 1864; Sherlock, 1922; Jennings, 1965). For example, the removal of gravel offshore from the Devon coast is considered to have caused the destruction of the village of Hallsands in 1917 (Figure 6.10).

Figure 6.10. THE REMAINS OF THE VILLAGE OF HALLSANDS, DEVON, DESTROYED BY MARINE ACTION IN 1917 AS A CONSEQUENCE OF OFF-SHORE DREDGING

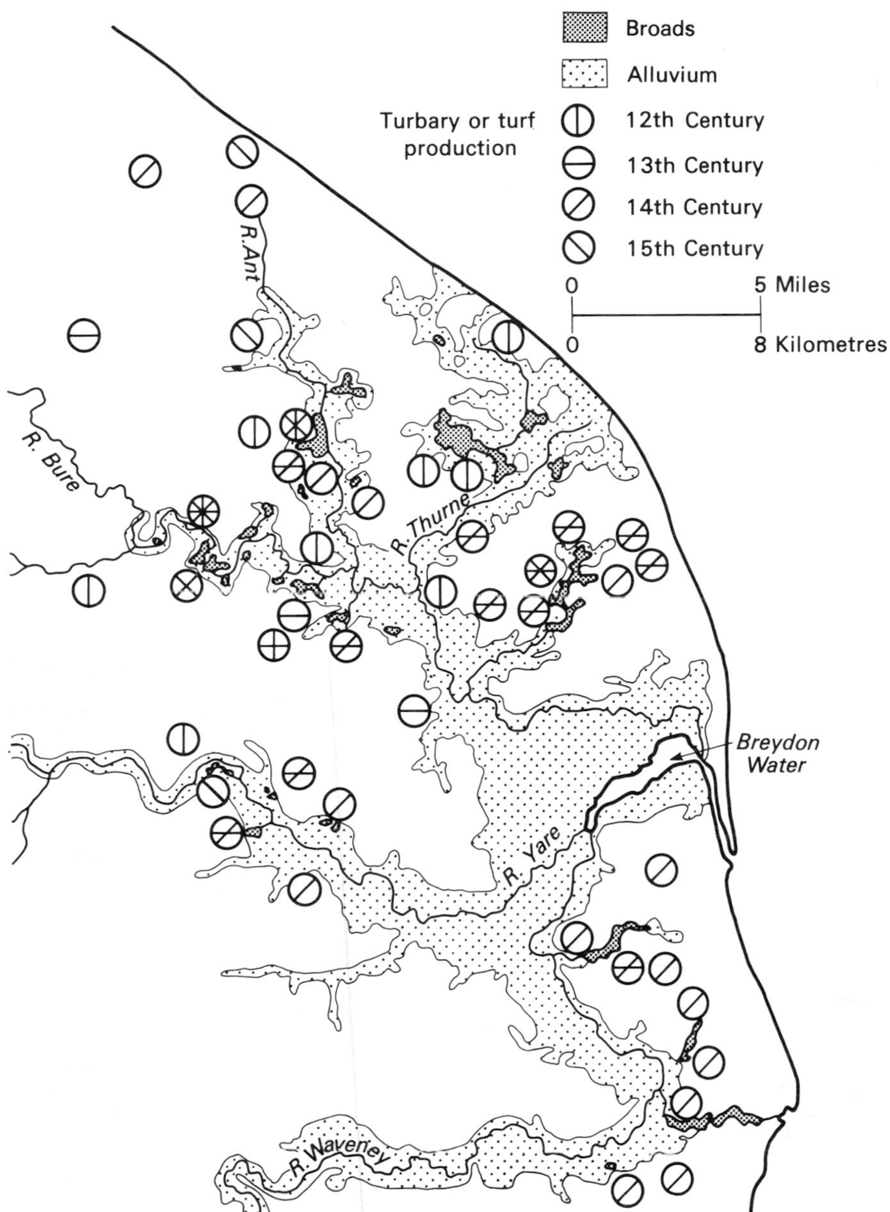

Figure 6.11. MEDIEVAL TURBARY IN EAST NORFOLK AND NORTH-EAST SUFFOLK.
(Source: Lambert *et al.* (1960)

The extraction of gravel from river channels can also have considerable side-effects, and these have been studied by Bull and Scott (1974).

An example of a study in which historical sources have been used to ascribe an 'industrial' origin for a landscape feature is the detailed investigation of the Norfolk Broads in England by Lambert *et al.* (1960). Stratigraphical and archaeological evidence were also used but, as Smith in this study commented:

> The peculiar character of the problems posed by the stratigraphy suggested that an alliance between the methods of the historical geographer and the geomorphologist may prove to be a valuable one.

Using, tithe, enclosure and other maps he highlighted the close relationship between field patterns and the shape of the water areas which implied a human origin. Much of the study is concerned with establishing from the documents the extent, importance and amount of peat-cutting, since physical and other evidence indicated that this was a probable cause of formation (Figure 6.11).

Mining can affect the physical environment in a number of ways. It may directly affect morphology by removal of material or dumping of spoil heaps while waste material contributed to streams can markedly alter sediment budgets and channel processes (Beckinsale, 1972). Mine-drainage can affect the quantity and quality of water (Biesecker and George, 1966) and mining may also cause land subsidence. The actual effects depend very much on the type and methods of mining, the nature of the materials and the amount of waste (Doerr and Guernsey, 1956). Hydraulic gold-mining in the Sierra Nevada of California in the last half of the nineteenth century discharged a great deal of waste material into streams and out to estuaries and harbours. Gilbert (1917) compiled evidence of the pre-mining (mid-nineteenth century) conditions in the San Francisco Bay and Sierra Nevada areas and, from comparative surveys, photographs and technical reports, he was able to calculate the amounts of deposition and sediment-transport and assess the impact of the debris. Graf (1979b) also focused on the effects of mining on streams in an historical study of gold- and silver-mining in mid-nineteenth-century Colorado. His sources were old photographs (Figure 6.12), prints, maps and written records, which show that mining activities were coincident with the initiation of stream entrenchment and the development of arroyos. Also in North America, Ireland *et al.* (1939) have suggested that the digging of clay pits may have been one cause of gullying in parts of the south-west United States. Wallwork (1956) has demonstrated by time and space association that brine pumping to extract salt in Cheshire has resulted in subsidence (Figure 6.13). Poland and Davis (1971) have also studied subsidence caused by the extraction of fluids, particularly in California.

As well as affecting rivers, the deliberate dumping of waste also affects landforms. Hutchinson *et al.* (1973) showed, again by time and space coincidence, that tipping of waste at Bury Hill near Wolverhampton, Staffordshire, appeared to be the cause of slope instability. Recent examples of waste disposal demonstrate the impact which human interference can have. At Denver, Colorado, waste was injected deep into the ground and within a month of the start of injection, earthquakes were reported (Healy *et al.*, 1968; Evans, 1973). No fewer than 710 were experienced within three and a half years yet none had been recorded between 1882 and the initiation of waste-injection. Numbers markedly decreased after pumping ceased. Similarly, Emiliani *et al.* (1973) have demonstrated by statistical analysis of frequencies that there is a link between nuclear explosions and earthquakes.

The investigation of past mining and other industrial activities can do more than assist with the explanation of landscape features or changes. Graf (1979b) commented that the

Figure 6.12. CULTURAL AND GEOMORPHOLOGICAL CHANGE AT NEVADEVILLE, COLORADO. Evidenced by photographs taken from approximately the same point in 1864 (top, p. 181), 1879 (bottom, p. 181) and 1898 (above). (Photographs by courtesy of the Denver Public Library, Western History Department (F 33820, F 5860 and F 10102 respectively))

results of an historical approach can provide 'insights into potential impacts of mining developments in other areas'. An applied use of historical sources is in areas where old mine-workings cause difficulties for land-use planning and construction. Historical documents, together with geophysical techniques, have been used to trace old workings (Maxwell, 1976; Dearman *et al.*, 1977).

6.3 CASE STUDY: Devon rivers

The following example is presented to provide some synthesis of the various factors discussed in this chapter. On a number of Devon rivers, maps indicate considerable changes in river-channel courses over the period 1840–1975 and field study has highlighted annual changes in the period 1974–1976. The rates of change are summarized in Table 6.1. Many stream sections show increasing rates of erosion and increased sinuosity. Two sets of potential causes—natural and anthropogenic—were considered and these are listed in Table 6.2, which also summarizes their expected/theoretical effects and their observed activity. Possible natural causes include changes in external factors such as rainfall and the pattern of flood events, and internal variations such as state of development of bends, loci

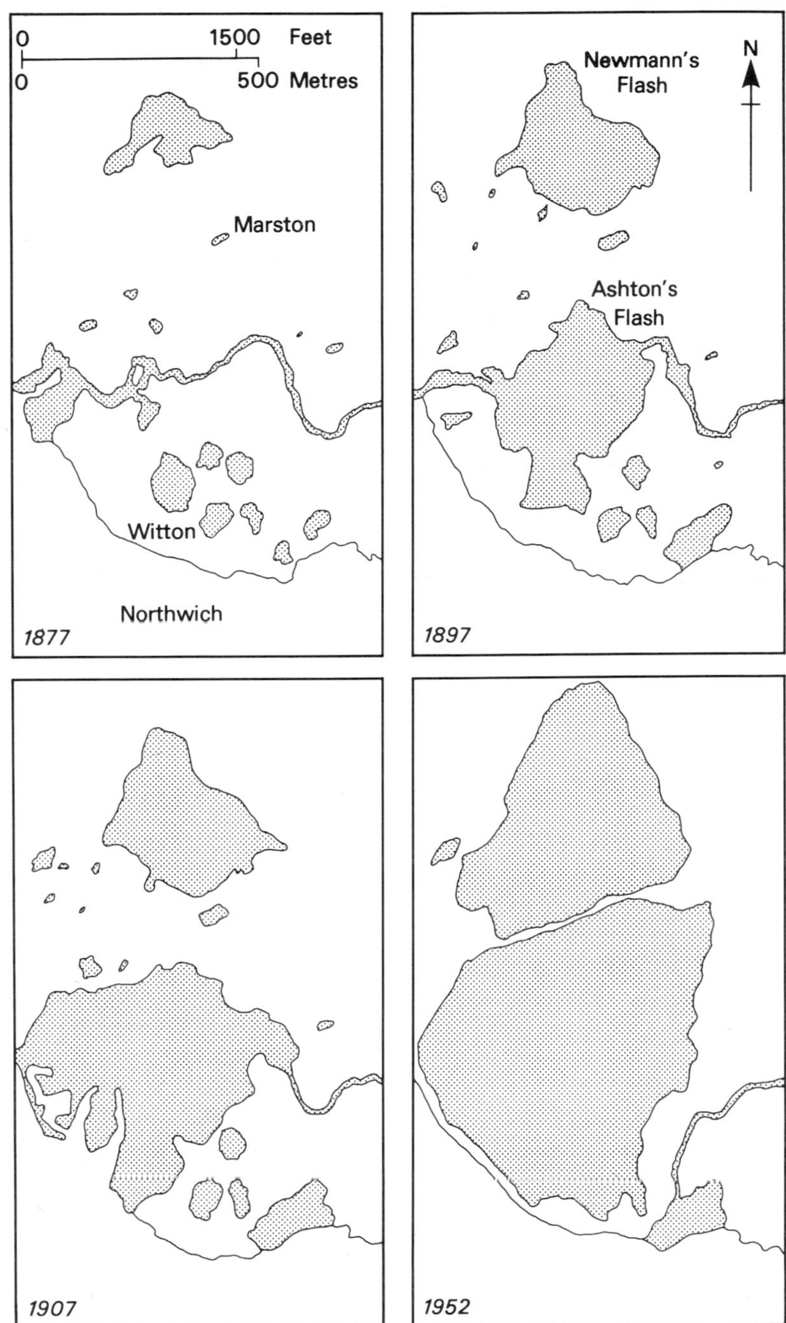

Figure 6.13. THE DEVELOPMENT OF SUBSIDENCE AT NORTHWICH, CHESHIRE, 1877–1952.
(Source: Wallwork (1956))

Table 6.1
RATES OF EROSION ON DEVON RIVERS, 1842–1975

Site	1842–1903	1903–N.G.	Map rates m/yr N.G.–1975	Mean	Field rates m/yr Max.
Exe A	0.68	0.92	1.79	0.63	0.71
Exe B	0.36	1.35	1.17	—	—
Exe C	0.11	0.05	1.08	0.62	1.10
Exe D	0.20	0.38	1.21	1.18	2.58
Exe E	Little change		0.92	1.03	2.40
Creedy	0	0.12	—	0.26	0.52
Culm—Lower	0.35	0.27	1.07	0.24	0.51
Culm—Upper	0.09	0.72	0.46	0.18	0.63
Axe 1	0.13	0.39	0.50	0.15	0.49
Axe 2	Alteration of direction of flow			0.42	1.16
Axe 3	0.06	0.09	0.50	0.29	0.60
Axe 4	0.05	0.30	0.58	0.46	1.00
Axe 5	0.13	0.15	0.69	0.33	0.69
Yarty Upper Bend	Alteration of flow		0.67	0.70	1.38
Yarty Lower Bend	0.18	0.30	0.92		
Coly	Straightened	0.33	Straightened	0.08	0.48
Hookamoor Brook	Change not detectable			0.08	0.19

of erosion and sediment variations. Rainfall records including manuscript records of an Exeter recording station were examined but no alteration in totals could be detected.

The longest river-gauge records in the area date from 1956, but some historical data have been collected for the river Exe (NERC, 1975). Major floods did occur in 1952, 1960 and 1968 which may have had a significant effect on the channels. Critical process thresholds may have been crossed at these times but no specific effects attributable to these floods are detectable in the field. Rates of erosion vary as meander-bends shift and as material of differing resistance is encountered, but this effect would be variable or random over time and would not exhibit a consistent pattern on different rivers.

A number of changes in human activities, both direct and indirect, were also examined. The general pattern of land use and its changes were determined from nineteenth-century agricultural essays, from agricultural statistics and from the Land Utilisation Survey of the 1930s, but no significant change corresponding in direction and time with the fluvial changes can be detected. It is difficult to obtain detailed records of agricultural drainage, especially for the last century, but from general knowledge and figures, oral evidence from agricultural advisors and such farm records as are extant it appears that the main period of drainage was in the late nineteenth century with an increase over the past ten years or so. The removal of hedges is another land management practice which has been marked in recent years but its effects are not known in detail. Much road-metalling has also occurred in the period since 1840, but it is probable that its effects are very localized. The growth of urban areas was investigated by examining census-returns but no significant increase in population upstream of the sites was found in any of the river basins. Nor are actively eroding sections related to the location of bridges, though there are some isolated cases of scour and, where weirs have collapsed, there has been increased meandering of streams. Abstraction of water for mill leats was formerly very important on these Devon rivers and it

Table 6.2
POSSIBLE CAUSES OF RIVER CHANNEL CHANGES IN DEVON, 1845–1970

Cause	Theoretical effect	Timing	Influence
(a) Natural			
Flood events	Increased erosion; alteration of morphology if threshold crossed	Largest events—1952, 1960, 1968	Possibly important
Change in composition and resistance of bank	Localized in particular sections of bank, alters erosion rates	No temporal control	Likely to be significant at many sites
Equilibrium-form relations	Controls rates of erosion	Depends on stage of development	General effects at most sites though difficult to identify
Change in locus of erosion	Zone of maximum erosion shifts	Changes through time with development of bend	Some effect at most sites
Rainfall variations	Change in amount of runoff and peakedness	General trends and fluctuations, 1840–1976	Difficult to distinguish
(b) Due to human activities			
Increase in field drainage	Depends on type, increase or decrease of peakedness	Main period 1850–1888, increase in last decade	Widespread drainage—probably important
Removal of hedges	Increased peakedness of run-off	Throughout 1840–1976, increase in last decade	Possibly important
Changes in land use	Increase in arable—more peaked run-off and soil erosion; decrease v.v.	Decline in arable 1880–1945, increase since 1945	Little effect prior to last decade?
Metalling of roads	Increased peakedness of run-off	Early twentieth century	Probably not important
Urban growth	Usually increased rate of run-off	Little change 1840–1976	Probably not important
Bridges	Stabilization	Destruction and construction	Some modification of movement
Weirs	Stabilization	Mostly collapse post-1880	Increase of meandering
Mill leats	Silting up—more flow in main streams	Silting up since late nineteenth century	More water in main channel—possibly enlargement

is possible that decline in mill-working and silting up of leats has increased discharge in the natural channels.

Thus it can be seen from this specific case study that many possible causes can be considered unlikely since either their timing and direction is incompatible with the river-channel changes or there has been no detectable change in the factor in the relevant period. Historical analysis suggests that fluvial changes in Devon in this period are due mainly to natural, autogenic processes and not to anthropogenic factors (Hooke, 1977b) but it is possible that human alterations at an earlier period are still affecting these streams.

7
Summary and Prospect

7.1 THE SOURCES

Our review of applications of historical sources in the earth sciences has highlighted those commonly used, those of limited value and some which are of potential value but as yet not fully exploited. Early maps, both published and manuscript, are the most widely employed sources in environmental investigations. Accurate, large-scale maps can provide a basis for measurement and quantitative analysis, although they only show features in two dimensions and at particular dates. Other graphical sources such as drawings, paintings and engravings have provided valuable evidence in some fields of study, notably glaciology, and might be more widely used in investigations of other landscape phenomena.

A wide variety of written sources is available to the earth scientist (Chapter 1) which range from items such as nineteenth-century technical articles containing much precise information to sources like manorial or estate papers with useful nuggets amongst much that is redundant. With these last, judgement must be made as to whether a search is worth the time it will take; often the efficient use of such sources requires some prior knowledge of dates or some indication that relevant data may be present. But we consider it a useful general rule not to underestimate the value of written sources which may initially appear fragmentary, vague or irrelevant.

Most statistical series and quantitative records date from modern times, and because they usually resemble modern records, we have not discussed them as fully as other sources. There are, however, some early records of weather and land use in manuscript form which require some pre-processing and interpretation, and these are reviewed briefly in Chapter 1.

The documents available in New World countries settled by Europeans during the last few centuries received separate treatment in Chapter 2. These have proved very valuable because they monitor the rapid environmental changes produced by the recent yet profound impact of human activities in these environments. In particular, they provide a basis for examining the effects of land clearance and the introduction of European agricultural methods.

Throughout the book we have stressed the importance of assessing the accuracy of sources with respect to the questions asked of them. In early studies assessment of accuracy was often ignored; nowadays careful testing is more usual. Conversely, some documents have been unfairly dismissed as being unreliable without proper testing.

7.2 TYPES OF APPLICATION

Historical records have at least two major roles in environmental research. In some applications they are perhaps the only available source of data. Thus in the analysis of the break-up of ice in Hudson's Bay made by Moodie and Catchpole (1975) documentary sources constitute the essential data-base. In other situations they can provide additional data to complete a picture or at least make it more comprehensive. We have discussed the nature and importance of the links between historical sources and field-derived data in Chapter 2. Historical sources can often provide critically important absolute dates, especially in the modern period for which radiometric techniques are at present less applicable.

In Chapters 4, 5 and 6 respectively we have reviewed the use of historical sources for making reconstructions at particular times, for providing data on processes and events and for explaining changes in the historical period. Cross-sectional reconstructions at particular dates involve few conceptual problems but rely for their success on the availability of a body of suitable source material. At its simplest, the analysis of change involves the comparison of two such cross-sections, but such an exercise can be fraught with danger; great care must always be taken when making inferences of continuous change from evidence at isolated dates. It may not always be the case, for example, that a landscape feature has changed continuously from the state depicted on one map to that on the next. The propriety of such interpolation must be judged by reference to known processes, field evidence and also contemporary descriptions. Eye-witness reports of the more catastrophic changes may be extant, particularly where the event affected lives or human activities and these can be used to help interpret the features produced and to calculate frequencies and likelihood of recurrence.

The identification of causes of change can require many types of information from a variety of sources corresponding to the multiplicity of factors operating in the physical environment. We have distinguished two main groups of factors in Chapter 6—natural and anthropogenic. Natural causes have to do with inherent changes within and interactions between, different parts of physical systems. Thus changes in climate, for example, can bring about changes in landforms and so it may be necessary for a geomorphologist to study historical climate sources. A wide range of human activities have an effect on the environment, and we have identified the most important as land use and agriculture, urbanization, drainage, engineering construction and industrial activities.

7.3 AREAS OF APPLICATION

A number of the major branches of earth science—meteorology and climatology, geomorphology, hydrology and biogeography—have been examined in this book. In all of these, historical sources are useful not only in relation to the reconstruction of past situations but also they can help with applied problems. As so many applied problems depend for their solution on the projection of past through present to future, it follows that historical investigations can have a major practical role to play (Table 7.1).

Climatology is the field in which fluctuations in the historical period have been most

Table 7.1
EXAMPLES OF THE POTENTIAL USE OF HISTORICAL SOURCES IN APPLIED PROBLEMS

Field of application	Information from sources	Use
Engineering		
Slopes	Stability of land Types of movement Location and frequency of movements Conditions and causes of instability Origin of features	Prediction of effects of construction and disturbance, e.g. cuttings embankments, drainage, buildings
Rivers and hydrology	Channel stability Type, extent and rate of change Effects of past interference Flow magnitudes and frequencies	Prediction of effects of alterations, e.g. channelization, design of bridges and other structures. Flood alleviation. Flow prediction
Coasts	Type, rate and distribution of changes Sediment transport patterns and régimes Effects of structures and practices	Prediction of effects of structures and interference, e.g. jetties, walls, groynes. Maintenance of shipping lanes
Climate and meteorology	Magnitude of fluctuations Trends and temporal patterns Associated conditions and effects	Climatic prediction
Rural and urban planning	Types and distribution of land use and their effects Magnitude, frequency, location and effects of natural hazards Distribution and effects of industrial activities, e.g. mining, waste disposal	Prediction of effects of alteration of land use. Design of warning schemes and alleviation procedures, land-use zoning. Prediction of effects of activities, location of old mine shafts and dumps

closely examined, in which the widest variety of historical sources has been used and in which the methods and techniques of using historical sources have been most highly developed. The general pattern of climate fluctuations has now been established and much work is continuing to provide detailed, quantitative information on the climatic characteristics of particular times, on producing homogeneous and continuous records as far back as possible and in analysing the frequency of certain types of weather. Workers in other fields could learn much from climatologists about how to use historical sources, especially on how to bring a range of material to bear on a question and how to obtain useful and often quantitative information from apparently vague and fragmentary records.

Coastal geomorphologists have made much use of historical sources and approaches largely because of the nature of the many practical problems they have been concerned with caused by rapidly eroding cliffs, depleting beaches, silting estuaries and shifting sand bars. The systematic study of historical changes in fluvial geomorphology has also burgeoned in the last decade or so. It is based mainly on map evidence and a result of this work has been the growing realization that fluvial changes are much more rapid than formerly thought. In studies of slopes, historical sources have been mostly used to date and interpret slides and rapid mass movements. The value of such historical investigation is being increasingly realized by engineers.

Historical sources have been used extensively and very successfully in the study of glaciers. Initial work emanated from Greenland, Iceland and Scandinavia, where a wealth of historical material dating back many centuries has permitted detailed reconstructions. Variations in Alpine and North American glaciers have also now been extensively studied, both in order to understand their régimes and processes and because of their value as climatic indicators.

In hydrology, the main applications of historical sources have been in the retrospective extension of flow records and in analysis of floods. This has included calculation of frequency and recurrence-intervals of flows and investigation of the characteristics of individual flood events.

Two main types of vegetation analysis have been conducted with historical sources. One relates to the transformation of vegetation under the influence of European settlement in the New World and in this, sources such as explorers' records and land surveys have provided useful evidence. The other type of analysis relates to vegetation changes under the predominant influence of man and agriculture. A wide variety of maps and written sources such as estate records have been used for this, often combined with pollen and other field-derived evidence.

7.4 PROSPECT

The emphasis of this book on the historical period should not be taken to imply that historical sources will supply all the answers to environmental questions. Historical sources provide one more form of evidence and are an additional tool for the environmental scientist. Although their wide applicability and value has been demonstrated in Chapters 4, 5 and 6, it is suspected that earth scientists will continue to view historical sources with the healthy circumspection that is quite right and proper. Indeed, as Ingram and Underhill (1979) remarked in a review of studies of the history of climate:

> undoubtedly some work produced earlier has been to some extent marred by the methodological weaknesses which are bound to afflict pioneering efforts. But a greater awareness of the problems involved in the scientific use of documentary sources is now current and the future prospects of this field of study seem excellent.

We hope that by bringing together a range of applications and discussing techniques in some detail this will engender a wider awareness of the potential as well as the limitations of historical sources in all branches of the earth sciences. There are certainly signs of an increasing use of historical sources, particularly for placing the results of modern process studies based on direct measurement into a longer time perspective. As the complexity of contemporary processes is revealed and man-made change accelerates, there is ever more need to examine the past both for interpreting the present environment and for predicting the effects of natural processes and human activities in the future.

Bibliography

Aakjaer, S. (1929). Villages, cadastres et plans parcellaires au Danemark. *Annales d'histoire économique et sociale* **1,** 562–575

Abbey, J. R. (1952). *Scenery of Great Britain and Ireland in Aquatint and Lithography, 1770–1860.* London, Curwen Press

Adams, I. H. (1966, 1970). *Scottish Record Office: Descriptive List of Plans,* 2 vols. Edinburgh

Adams, I. H. (1967). Large-scale manuscript plans in Scotland. *Journal of the Society of Archivists* **3,** 286–290

Aghassy, J. (1973). Man-induced badlands topography. In *Environmental Geomorphology and Landscape Conservation,* ed. D. R. Coates, Vol. 3, pp. 124–136. Stroudsberg; Dowden, Hutchinson and Ross

Ahlmann, H. W. (1953). *Glacier Variations and Climatic Fluctuations.* Bowman Memorial Lecture, American Geographical Society. New York

Akeroyd, A. V. (1972). Archaeological and historical evidence for subsidence in southern Britain. *Philosophical Transactions of the Royal Society* Series A, **272,** 151–169

Alestalo, J. (1971). Dendrochronological interpretation of geomorphic processes. *Fennia* **105,** 1–140

Alexander, C. S. and Nunnally, N. R. (1972). Channel stability on the Lower Ohio River. *Annals of the Association of American Geographers* **62,** 411–417

Alexander, C. S. and Prior, J. C. (1971). Holocene sedimentation rates in overbank deposits in the Black Bottom of the Lower Ohio River, Southern Illinois. *American Journal of Science* **270,** 361–372

Altmann, J. G. (1751). *Versuch einer historischen Beschreibung der helvetsichen Eisbergen.* Zurich

Ambraseys, N. N. (1979). A test case of historical seismicity: Isfahan and Chahar Mahal, Iran. *Geographical Journal* **145,** 56–71

Anderson, J. P. (1881, reprinted 1976). *The Book of British Topography—A Classified Catalogue of the Topographical Works in the Library of the British Museum Relating to Great Britain and Ireland.* Introduction by J. Simmons. Wakefield, E. P. Publishing

Anderson, M. G. and Calver, A. (1977). On the persistence of landscape features formed by a large flood. *Transactions of the Institute of British Geographers* New Series **2,** 243–254

Andrews, J. T. and Miller, G. H. (1980). Dating Quaternary deposits more than 10,000 years old. In *Timescales in Geomorphology,* eds R. A. Cullingford, D. A. Davidson and J. Lewin, pp. 263–287. Chichester, Wiley

Andrews, W. (1887). *Famous Frosts and Frost Fairs in Great Britain.* London, Redway

Antevs, E. (1952). Arroyo-cutting and filling. *Journal of Geology* **60,** 375–385

Arakawa, H. (1956). Dates of first or earliest snow covering for Tokyo since 1632. *Quarterly Journal of the Royal Meteorological Society* **82,** 222–226

Arakawa, H. (1957). Climatic change as revealed by the data from the Far East. *Weather* **12,** 46–51

Aranuvachapun, S. and Brimblecombe, P. (1978). Extreme ebbs in the Thames. *Weather* **33,** 126–131

Arber, M. A. (1941). The coastal landslips of West Dorset. *Proceedings of the Geological Association* **52,** 273–283

Arrowsmith, A. (1809). *Memoir Relative to the Construction of a Map of Scotland.* London

Ashby, C. M. (1967). *Preliminary Inventory of the Cartographic Records of the Forest Service.* Washington DC, The National Archives

Athearn, W. D. and Ronne, C. (1963). Shoreline changes at Cape Hatteras. *U.S. Naval Research Review* **16,** 17–24

Atkinson, B. (1979). Urban influences on precipitation in London. In *Man's Impact on the Hydrological Cycle in the United Kingdom,* ed. G. E. Hollis, pp. 123–133. Norwich, Geo Abstracts

Azimi, C. and Desvarreux, P. (1974). A study of one special type of mudflow in the French Alps. *Quarterly Journal of Engineering Geology* **7,** 329–338

Bahre, C. J. and Bradbury, D. E. (1978). Vegetation change along the Arizona–Sonora boundary. *Annals of the Association of American Geographers* **68,** 145–165

Bailey, E. B. (1952). *Geological Survey of Great Britain.* London, Thomas Murby

Bailey, R. G. (1973). Forest land use implications. In *Environmental Geomorphology and Landscape Conservation,* ed. D. R. Coates, Vol. 3, pp. 388–413. Stroudsberg: Dowden, Hutchinson and Ross

Bailey, R. W. (1935). Epicycles of erosion in the valleys of the Colorado Plateau Province. *Journal of Geology* **43,** 337–355

Baker, A. R. H. (1962a). Local history in early estate maps. *Amateur Historian* **5,** 66–71

Baker, A. R. H. (1962b). Some early Kentish estate maps and a note on their portrayal of field boundaries. *Archaeologia Cantiana* **72,** 177–184

Baker, A. R. H. (1965). Field patterns in seventeenth-century Kent. *Geography* **50,** 18–30

Baker, J. N. L. (1932). The climate of England in the seventeenth century. *Quarterly Journal of the Royal Meteorological Society* **58,** 421–439

Barnes, F. A. and King, C. A. M. (1957). The spit at Gibraltar Point, Lincolnshire. *East Midland Geographer* **8,** 22–31

Barrow, G. and Wills, L. H. (1913). *Records of London Wells.* London, Geological Survey

Barry, R. G. (1978). Climatic fluctuations during the periods of historical and instrumental record. In *Climatic Change and Variability,* eds A. B. Pittock, L. A. Frakes, D. Jenssen, J. A. Peterson and J. W. Zilliman, pp. 150–166. Cambridge, Cambridge University Press

Barth, T. F. W. (1942). Craters and fissure eruptions at Myvatn in Iceland. *Norsk Geografisk Tidsskrift* **9,** 58–81

Bateman, P. C. (1961). Willard D. Johnson and the strike-slip component of fault movement in the Owens Valley, California, earthquake of 1872. *Bulletin of the Seismological Society of America* **51,** 483–493

Beardmore, N. (1862). *Manual of Hydrology.* London, Waterlow and Sons

Beaumont, P. and Oberlander, T. M. (1973). Litter as a geomorphological aid—Death Valley, California. *Geography* **58,** 136–141

Beche, H. T. de la (1839). *Report on the Geology of Cornwall, Devon and West Somerset.* London, Geological Survey

Beckinsale, R. P. (1972). The effect upon river channels of sudden changes in sediment load. *Acta Geographica Debrecina* **10,** 181–186

Bell, B. (1970). The oldest records of the Nile floods. *Geographical Journal* **136,** 569–573

Bell, W. T. and Ogilvie, A. E. (1978). Weather compilations as a source of data for the reconstruction of European climate during the medieval period. *Climatic Change* **1,** 331–348

Belt, C. B. (1975). The 1973 flood and man's constriction of the Mississippi River. *Science* **189** (4204), 681–684

Bennett, H. H. (1931). Cultural changes in soils from the standpoint of erosion. *American Society of Agronomy Journal* **23,** 434–454

Benson, M. A. (1962). Factors influencing the occurrence of floods in a humid region of diverse terrain. *U.S. Geological Survey Water-Supply Paper* 1580–B

Bergman, D. L. and Sullivan, C. W. (1963). Channel changes on Sandstone Creek near Cheyenne, Oklahoma. *U.S. Geological Survey Professional Paper* 475C, 145–148

Bergsten, K. E. (1961). Sweden. In *A Geography of Norden,* ed. A. Somme, pp. 293–349. London, Heinemann

Best, R. H. and Coppock, J. T. (1962). *The Changing Use of Land in Britain.* London, Faber and Faber

Beveridge, W. H. (1921). Weather and harvest cycles. *Economic Journal* **31,** 429–452

Biesecker, J. E. and George, J. R. (1966). Stream quality in Appalachia as related to coal-mine drainage, 1965. *U.S. Geological Survey Circular* 526, 1–27

Binns, W. O. (1979). The hydrological impact of afforestation in Great Britain. In *Man's Impact on the Hydrological Cycle in the United Kingdom,* ed. G. E. Hollis, pp. 55–69. Norwich, Geo Abstracts

Bird, E. C. F. (1972). Our changing coastline. *Victoria's Resources* Sept.–Nov., 6–10

Bird, E. C. F. (1974). Assessing man's impact on coastal environments in Australia. In *Report on Symposium on the Impact of Human Activities on Coastal Zones.* Australian Unesco Committee for Man and the Biosphere, Publication No. 1, pp. 66–75. Australian Government Publishing Service

Bird, E. C. F. (1979). Coastal processes. In *Man and Environmental Processes,* eds K. J. Gregory and D. E. Walling, pp. 82–101. Folkestone, Dawson

Biswas, A. K. (1970). *History of Hydrology.* London, North-Holland Publishing Company

Bjerrum, L. (1954). Stability of natural slopes in quick clay. *Géotechnique* **5,** 101–119

Bjerrum, L. (1971). Quick clay slides. *Norwegian Geotechnical Institute* Publication No. 89.

Blakemore, M. J. and Harley, J. B. (1980). Concepts in the history of cartography: a review and perspective. *Cartographica* **17,** Monograph **26,** 1–120

Bloch, M. (1929). Le plan parcellaire document historique. *Annales d'histoire économique et sociale* **1,** 60–70, 390–8

Bluck, B. J. (1971) . Sedimentation in the meandering River Endrick. *Scottish Journal of Geology* **7,** 93–138

Bluhm, R. K. (1975). *Bibliography of British Newspapers: Wiltshire.* London, Library Association

Boer, G. de (1964). Spurn Head: Its history and evolution. *Transactions of the Institute of British Geographers* **34,** 71–89

Boer, G. de (1969). The historical variations of Spurn Point; the evidence of early maps. *Geographical Journal* **135,** 17–27

Bond, M. F. (1959–1960). Materials for transport history among the records of Parliament. *Journal of Transport History* **4**

Bond, M. F. (1971). *Guide to the Records of Parliament.* London, HMSO

Bowman, I. (1923). An American boundary dispute. *Geographical Review* **13,** 161–189

Bradley, R. S. (1976). *Precipitation History of the Rocky Mountain States.* Boulder, Colorado, Westview Press

Brandon, P. F. (1971a). Agriculture and the effects of floods and weather at Barnhorne during the late middle ages. *Sussex Archaeological Collections* **109,** 69–93

Brandon, P. F. (1971b). Late medieval weather in Sussex and its agricultural significance. *Transactions of the Institute of British Geographers* **54,** 1–17

Braund, M. (1980). An analysis of the effects of temperature and humidity on a variety of drawing media. *Bulletin of the Society of University Cartographers* **14,** 25–35

Bray, J. R. (1971). Vegetational distribution, tree growth and crop success in relation to recent climatic change. *Advances in Ecological Research* **7,** 177–233

Bray, J. R. and Struik, G. J. (1963). Forest growth and glacial chronology in eastern British Columbia and their relation to recent climatic trends. *Canadian Journal of Botany* **41,** 1245–1271

Briault, E. W. H. (1942). *Sussex (East and West).* The Land of Britain, the Report of the Land Utilisation Survey of Britain, parts 83–84

Brice, J. C. (1974). Evolution of meander loops. *Bulletin of the Geological Society of America* **85,** 581–586

Brimblecombe, P. (1975). Industrial air pollution in thirteenth-century Britain. *Weather* **30,** 388–396

Brimblecombe, P. and Ogden, C. (1977). Air pollution in art and literature. *Weather* **32,** 285–291

British Library, Science Reference Library, 22nd edition 1976. *Guide to Government Department and Other Libraries and Information Bureaux.* London, The British Library

British Museum (1844; reprinted 1962). *Catalogue of the Manuscript Maps, Charts and Plans and of the Topographical Drawings in the British Museum,* 3 vols. London

British Museum (1967, 1978). *Catalogue of Printed Maps, Charts and Plans,* 15 vols, and *Ten-year Supplement 1965–74.* London

Britton, C. E. (1937). A meteorological chronology to A.D. 1450. *Meteorological Office Geophysical Memoirs* No. 70. London, HMSO

Bromhead, E. N., (1978). Large landslides in London Clay at Herne Bay, Kent. *Quarterly Journal of Engineering Geology* **11,** 291–304

Brookfield, H. C. (1952). The estuary of the Adur. *Sussex Archaeological Collections* **90,** 153–163

Brooks, C. E. P. (1926). *Climate through the Ages,* 1st edition. London, Ernest Benn

Brooks, C. E. P. (1928). Periodicities in the Nile floods. *Memoirs of the Royal Meteorological Society* **2,** No. 12, 9–26

Brooks, C. E. P. and Glasspoole, J. (1928). *British Floods and Droughts.* London, Ernest Benn

Brooks, C. E. P. and Hunt, T. M. (1933). Variations of wind direction in the British Isles since 1341. *Quarterly Journal of the Royal Meteorological Society* **59,** 375–388

Brown, C. W. (1939). Hurricanes and shoreline changes in Rhode Island. *Geographical Review* **29,** 416–430

Brunsden, D. and Jones, D. K. C. (1976). The evolution of landslide slopes in Dorset. *Philosophical Transactions of the Royal Society* Series A, **283,** 605–631

Brunsden, D. and Kesel, R. H. (1973). Slope development on a Mississippi river bluff in historic time. *Journal of Geology* **81,** 576–597

Brunt, D. (1937). Climatic cycles. *Geographical Journal* **89,** 214–238

Bryan, K. (1923). Erosion and sedimentation in the Papago country, Arizona. *U.S. Geological Survey Bulletin* 730B, 19–90

Bryan, K. (1925). Date of channel trenching (arroyo cutting) in the arid southwest. *Science* **62,** 338–344

Bryan, K. (1928a). Change in plant associations by change in ground water level. *Ecology* **9,** 474–478

Bryan, K. (1928b). Historic evidence on changes in the channel of Rio Puerco, a tributary of the Rio Grande in New Mexico. *Journal of Geology* **36,** 265–282

Bryson, R. A. and Julian, P. R. (eds) (1962). Proceedings of conference on the climates of the eleventh and sixteenth centuries, Aspen, Colorado. *National Center for Atmospheric Research Technical Notes* 63–1. Boulder

Buchinsky, I. E. (1963). Climatic fluctuations in the arid zone of the Ukraine. In *UNESCO, Arid Zone Research 20,* 91–95

Buckinghamshire Record Office (1962). *Handlist of Buckinghamshire Estate Maps*

Buckinghamshire Record Office (1964). *Buckinghamshire Estate Maps*

Buffington, L. C. and Herbel, C. H. (1965). Vegetational changes on a semi-desert grassland range from 1858 to 1963. *Ecological Monographs* **35,** 139–164

Bull, W. B. (1964). History and causes of channel trenching in Western Fresno County, California. *American Journal of Science* **262,** 249–258

Bull, W. B. and Scott, K. M. (1974). Impact of mining gravel from urban stream beds in the south-western United States. *Geology* **2,** 171–174

Bullard, F. M. (1976). *Volcanoes of the Earth.* Austin, University of Texas Press

Burcham, L. T. (1956). Historical backgrounds of range land use in California. *Journal of Range Management* **9,** 81–86

Burkham, D. E. (1972). Channel changes of the Gila river in Safford Valley, Arizona, 1846–1970. *U.S. Geological Survey Professional Paper* 655–G

Burkham, D. E. (1976). Effects of changes in an alluvial channel on the timing, magnitude and transformation of flood waves, south-eastern Arizona. *U.S. Geological Survey Professional Paper* 655–K

Burroughs, W. J. (1981). Some observations on the climate in Britain during the 13th and 14th centuries. *Weather* **36,** 140–144

Burrows, C. J. (1976). Icebergs in the southern ocean. *New Zealand Geographer* **32,** 127–138

Burrows, M. (1972). The Cinque Ports. In *Environmental Geomorphology and Landscape Conservation,* ed. D. R. Coates, Vol. 1, pp. 350–369. Stroudsberg: Dowden, Hutchinson and Ross

Butzer, K. W. (1965). *Environment and Archaeology: an Introduction to Pleistocene Geography.* London, Methuen

Butzer, K. W. (1974). Accelerated soil erosion: a problem of man–land relationships. In *Perspectives on Environment,* eds I. R. Manners and M. W. Mikesell, pp. 57–78. Washington, Association of American Geographers

Byerly, P. (1951). History of earthquakes in the San Francisco Bay area. *California Division of Mines Bulletin* **154,** 151–160

Caird, J. (1852). *English Agriculture in 1850–51.* London, Longman, Brown, Green, and Longmans

Cameron, L. G. (1941). *Hertfordshire.* The Land of Britain, the Report of the Land Utilisation Survey of Britain, Part 80

Carey, W. C. (1963). The mechanisms of turns in alluvial streams. *Military Engineer* **363,** 14–16

Carey, W. C. (1969). Formation of flood plain lands. *Proceedings of the American Society of Civil Engineers, Journal of Hydraulics Division* 95 HY3, 981–994

Carlson, W. S. (1939). Movement of some Greenland glaciers. *Bulletin of the Geological Society of America* **50,** 239–256

Carney, T. F. (1972). *Content Analysis: a Technique for Systematic Inference from Communications.* London, Batsford

Carr, A. P. (1962). Cartographic record and historical accuracy. *Geography* **47,** 135–145

Carr, A. P. (1969). The growth of Orford Spit: cartographic and historical evidence from the sixteenth century. *Geographical Journal* **135,** 28–39

Carr, A. P. (1980). The significance of cartographic sources in determining coastal change. In *Timescales in Geomorphology,* eds R. A. Cullingford, D. A. Davidson and J. Lewin, pp. 69–78. Chichester, Wiley

Carrara, P. E. (1979). The determination of snow avalanche frequency through tree-ring analysis and historical records at Ophir, Colorado. *Bulletin of the Geological Society of America* **90,** 773–780

Carrara, P. E. and McGimsey, R. G. (1981). The late neo-glacial histories of the Agassiz and Jackson glaciers, Glacier National Park, Montana. *Arctic and Alpine Research* **13,** 183–196

Carter, K (compiler) and Pitman, E. V. (editor) (1974). *Dorset: a Catalogue of the Books and Other Printed Materials on the History, Topography, Geology, Archaeology, Natural History and Biography of Dorset in Dorset County Library.* Dorchester, Dorset County Council

Catchpole, A. J. W. and Moodie, D. W. (1978). Archives and the environmental scientist. *Archivaria* **6,** 113–136

Catchpole, A. J. W., Moodie, D. W. and Kaye, B. (1970). Content analysis: a method for the identification of dates of first freezing and first breaking from descriptive accounts. *Professional Geographer* **22,** 252–257

Catchpole, A. J. W., Moodie, D. W. and Milton, D. (1976). Freeze-up and break-up of estuaries on Hudson Bay in the eighteenth and nineteenth centuries. *Canadian Geographer* **20,** 279–297

Chambers, J. D. and Mingay, G. E. (1966). *The Agricultural Revolution 1750–1880.* London, Batsford

Chandler, R. J. (1970a). The degradation of lias clay slopes in an area of the east Midlands. *Quarterly Journal of Engineering Geology* **2,** 161–181

Chandler, R. J. (1970b). Solifluction on low-angled slopes in Northamptonshire. *Quarterly Journal of Engineering Geology* **3,** 65–69

Chandler, R. J. (1971). Landsliding on the Jurassic escarpment near Rockingham, Northamptonshire. In Institute of British Geographers Special Publication No. 3. *Slopes: Form and Process,* ed. D. Brunsden, pp. 111–128

Chavannes, E. (1941). Written records of forest succession. *Scientific Monthly* **53,** 76–80

Chu Ko-Chen (1973). A preliminary study on the climatic fluctuations during the last 5000 years in China. *Scientia Sinica* **16,** 226–256

Church, M. (1980). Records of recent geomorphological events. In *Timescales in Geomorphology,* eds R. A. Cullingford, D. A. Davidson and J. Lewin, pp. 13–30. Chichester, Wiley

Clark, A. H. (1956). The impact of exotic invasion on the remaining New World mid-latitude grasslands. In *Man's Role in Changing the Face of the Earth,* ed. W. L. Thomas, pp. 737–762. Chicago, University of Chicago Press

Clark, K. (1949). *Landscape into Art.* London, John Murray

Clarke, R. T. and McCulloch, J. S. G. (1979). The effect of land use on the hydrology of small upland catchments. In *Man's Impact on the Hydrological Cycle in the United Kingdom,* ed. G. E. Hollis, pp. 71–78. Norwich, Geo Abstracts

Cloet, R. L. (1954). Hydrographic analysis of Goodwin Sands and the Brake Bank. *Geographical Journal* **120,** 203–215

Cloet, R. L. (1963). Hydrographic analysis of the sandbanks in the approaches to Lowestoft Harbour. *Admiralty Marine Science Publication No. 6*

Clough, H. W. (1933). The 11-year sunspot period, secular periods of solar activity and synchronous variations in terrestrial phenomena. *Monthly Weather Review* **60,** 99–108

Clout, H. D. and Sutton, K. (1969). The cadastre as a source for French rural studies. *Agricultural History* **43,** 215–223

Coates D. R. (ed.) (1972, 1973, 1974). *Environmental Geomorphology and Landscape Conservation.* Vol. 1 (1972) *Prior to 1900*; Vol. 2 (1974) *Urban Areas,* Vol. 3 (1973) *Non-urban.* Stroudsberg: Dowden, Hutchinson and Ross

Coates, D. R. and Vitek, J. D. (eds) (1980). *Thresholds in Geomorphology.* London, George Allen and Unwin

Cocco, E. (1976). The Italian North Ionian coast; tendency towards erosion. *Marine Geology* **21,** 49–57

Coleman, J. M. (1969). Brahmaputra River: channel processes and sedimentation. *Sedimentary Geology* **3,** 129–239

Collin, A. (1846, reprinted 1956). *Landslides in Clay.* University of Toronto Press

Collins E. J. T. and Jones, E. L. (1967). Sectoral advance in English agriculture, 1850–80. *The Agricultural History Review* **15,** 65–81

Colon, J. A. (1953). A study of hurricane tracks for forecasting purposes. *Monthly Weather Review* **81,** 53–66

Cooke, R. U. and Reeves, R. W. (1976). *Arroyos and Environmental Change in the American South-West.* Oxford, Oxford University Press

Cooper, W. S. (1937). The problem of Glacier Bay, Alaska. *Geographical Review* **27,** 37–62

Cooperider, C. K. and Hendricks, B. A. (1937). Soil erosion and stream flow on range and forest land of the Upper Rio Grande watershed and its relation to land resources and human welfare. *U.S. Department of Agriculture Technical Bulletin 567*

Coppock, J. T. (1954). Land-use change in the Chilterns, 1931–51. *Transactions of the Institute of British Geographers* **20,** 113–140

Coppock, J. T. (1956). The statistical assessment of British agriculture. *The Agricultural History Review* **4,** 4–21, 66–79

Coppock, J. T. (1957). The changing arable in the Chilterns 1875–1957. *Geography* **42,** 217–229

Coppock, J. T. (1958). Changes in farm and field boundaries in the nineteenth century. *Amateur Historian* **3,** 292–298

Coppock, J. T. (1968). Maps as sources for the study of land use in the past. *Imago Mundi* **22,** 37–49

Costa, J. E. (1975). Effects of agriculture on erosion and sedimentation in the Piedmont Province, Maryland. *Bulletin of the Geological Society of America* **86,** 1281–1286

Cottam, W. P. and Stewart, G. (1940). Plant succession as a result of grazing and of meadow desiccation by erosion since settlement in 1862. *Journal of Forestry* **31,** 613–626

Cowgill, U. M. and Hutchinson, G. E. (1963). Ecological and geochemical archaeology in the southern Maya lowlands. *South west Journal of Anthropology* **19,** 267–286

Cox, E. G. (1949). *Reference Guide to the Literature of Travel. Vol. 3 Great Britain.* Seattle, University of Washington

Cozens-Hardy, B. (1924–1929). Cley-next-the-Sea and its marshes. *Transactions of the Norfolk and Norwich Naturalists Society* **12,** 354–373

Crabtree, K. and Maltby, E. (1976). Soil and land use change on Exmoor; significance of a buried profile on Exmoor. *The Proceedings of the Somersetshire Archaeological and Natural History Society* **119,** 38–43

Craddock, J. M. (1976). Annual rainfall in England since 1725. *Quarterly Journal of the Royal Meteorological Society* **102,** 823–840

Craddock, J. M. (1977). A homogenous record of monthly rainfall totals for Norwich for the years 1836 to 1976. *Meteorological Magazine* **106,** 267–278

Craddock, J. M. and Craddock, E. (1977). Rainfall at Oxford from 1767 to 1814, estimated from the records of Dr. Thomas Hornsby and others. *Meteorological Magazine* **106,** 361–372

Craddock, J. M. and Wales-Smith, B. G. (1977). Monthly rainfall totals representing the East Midlands for the years 1726 to 1975. *Meteorological Magazine* **106,** 97–111

Craddock, J. M. and Weller, M. J. (1975). Reflections on some unusual years. *Meteorological Magazine* **104,** 61–69

Crandell, D. R. and Mullineaux, D. R. (1967). Volcanic hazards at Mount Rainier, Washington. *U.S. Geological Survey Bulletin 1238,* 1–26

Crandell, D. R. and Waldron, H. H. (1973). Volcanic hazards in the Cascade Range. In *Focus on Environmental Geology,* ed. R. W. Tank, pp. 19–28. Oxford, Oxford University Press

Crane, R. S., Kaye, F. B. and Prior, M. E. (1927, reprinted 1966). *A Census of British Newspapers and Periodicals, 1620–1800.* London, Holland Press

Crone, G. R. (1966). *Maps and their Makers, an Introduction to the History of Cartography.* London, Hutchinson

Cruff, R. W. and Rantz, S. E. (1965). A comparison of methods used in flood frequency studies for coastal basins in California. *U.S. Geological Survey Water Supply Paper 1580E*

Cullingford, R. A., Davidson, D. A. and Lewin, J. (eds) (1980). *Timescales in Geomorphology.* Chichester, Wiley

Cumberland, K. B. (1941). A century's change: natural to cultural vegetation in New Zealand. *Geographical Review* **31,** 529–554

Curry, R. R. (1969). Holocene climatic and glacial history of the central Sierra Nevada, California. *Geological Society of America Special Paper 123,* eds. S. A. Schumm and W. C. Bradley, pp. 1–47

Curry, R. R. (1971). Soil destruction associated with forest management and prospects for recovery in geologic time. *Association of South-eastern Biologists Bulletin* **18,** 117–128

Daniels, R. B. (1960). Entrenchment of the Willow Drainage Ditch, Harrison County, Iowa. *American Journal of Science* **258,** 161–176

Darby, H. C. (1933). The agrarian contribution to surveying in England. *Geographical Journal* **82,** 529–535

Darby, H. C. (ed.) (1936). *An Historical Geography of England before AD 1800.* Cambridge, Cambridge University Press

Darby, H. C. (1951). The changing English landscape. *Geographical Journal* **117,** 377–398

Darby, H. C. (1954). Some early ideas on the agricultural regions of England. *The Agricultural History Review* **2,** 30–47

Darby, H. C. (1956). The clearing of the woodland in Europe. In *Man's Role in Changing the Face of the Earth,* ed. W. L. Thomas, pp. 183–216. Chicago, University of Chicago Press

Darby, H. C. (ed.) (1973). *A New Historical Geography of England.* Cambridge, Cambridge University Press

Davies, B. E. and Lewin, J. (1974). Chronosequences in alluvial soils with special reference to historic lead pollution in Cardiganshire, Wales. *Environmental Pollution* **6,** 49–57

Davis, M. B. (1976). Erosion rates and land use history in southern Michigan. *Environmental Conservation* **3,** 139–148

Davison, C. (1924). *A History of British Earthquakes.* Cambridge, Cambridge University Press

Day, A. (1967). *The Admiralty Hydrographic Service 1795–1919.* London, HMSO

Dearman, W. R., Baynes, F. J. and Pearson, R. (1977). Geophysical detection of disused mineshafts in the Newcastle-upon-Tyne area, noth-east England. *Quarterly Journal of Engineering Geology* **10,** 257–269

Deneven, W. M. (1967). Livestock numbers in nineteenth century New Mexico and the problem of gullying in the south-west. *Annals of the Association of American Geographers* **57,** 691–703

Denton, G. H. and Karlén, W. (1973). Holocene climatic variations—their pattern and possible cause. *Quaternary Research* **3,** 155–205

Derville, M. T. (1936). *The Level and the Liberty of Romney Marsh.* London, Headley Brothers

Detwyler, T. R. (1971) (ed.). *Man's Impact on Environment.* New York, McGraw-Hill

Detwyler, T. R. and Marcus, M. G. (eds.) (1972). *Urbanisation and Environment.* California, Duxbury Press

Dickinson, P. G. M. (1968). *Maps in the County Record Office, Huntingdon.* St Ives

Dietz, R. A. (1952). The evolution of a gravel bar. *Annals of Missouri Botanical Garden* **39,** 249–254

Dightman, R. A. and Beatty, M. E. (1952). Recent Montana glacier and climate trends. *Monthly Weather Review* **80,** 77–81

Dixon, D. (1973). *Local Newspapers and Periodicals of the Nineteenth Century.* Victorian Studies Handlist No. 6. Leicester, University of Leicester Victorian Studies Centre

Doerr, A. and Guernsey, L. (1956). Man as a geomorphological agent: the example of coal mining. *Annals of the Association of American Geographers* **46,** 197–210

Dolan, R. (1973). Barrier islands: natural and controlled. In *Coastal Geomorphology,* ed. D. R. Coates, pp. 263–278. Binghampton, State University of New York

Dolan, R. and Bosserman, K. (1972). Shoreline erosion and the lost colony. *Annals of the Association of American Geographers* **62,** 424–426

Dopson, L. (1955). The John Rylands Library, Manchester. *Amateur Historian* **2,** 202–206

Douch, R. (1952, 1960). *A Handbook of Local History: Dorset, and a Supplement of Additions and Corrections to 1960.* Bristol, University of Bristol Department of Extra-Mural Studies

Duce, J. T. (1918). The effect of cattle on the erosion of canyon bottoms. *Science* **47,** 450–452

Dugdale, W. (1662). *The History of Imbanking and Drayning.* London, Alice Warren

Duncan, A. M. Chester, D. K. and Guest, J. E. (1981). Mount Etna volcano: environmental impact and problems of volcanic prediction. *Geographical Journal* **147,** 164–178

Duncanson, H. B. (1909). Observations on the shifting of the channel of the Missouri River since 1883. *Science* **29,** 752, 869–871

Dunn, G. E. and Miller, B. I. (1960). *Atlantic Hurricanes.* Baton Rouge, Louisiana State University Press

Dury, G. H. (1973). Magnitude frequency analysis and channel morphometry. In *Fluvial Geomorphology,* ed. M. Morisawa, pp. 91–121. Binghampton, State University of New York

Eardley, A. J. (1938). Yukon channel shifting. *Bulletin of the Geological Society of America* **49,** 343–358

Edgar, D. E. and Melhorn, W. N. (1974). Drainage basin response: documented historical change and theoretical considerations. Purdue University, *Water Resources Research Centre Technical Report No. 52*

Ehrenberg, R. E. (1975). Bibliography to resources on historical geography in the National Archives. In *Pattern and Process. Research in Historical Geography,* ed. R. E. Ehrenberg, pp. 315–349. Washington DC, Harvard University Press

Eide, O. and Bjerrum, L. (1954). The slide at Bekkelaget. *Géotechnique* **5,** 88–100

El-Ashry, M. T. and Wanless, H. R. (1968). Photo interpretation of shoreline changes between Cape Hatteras and Fear (North Carolina). *Marine Geology* **6,** 347–379

Emery, K. O. (1960). Weathering of the Great Pyramid. *Journal of Sedimentary Petrology* **30,** 140–143

Emery, F. V. and Smith, C. G. (1976). A weather record from Snowdonia, 1697–98. *Weather* **31,** 142–150

Emiliani, C., Harrison, C. and Swanson, M. (1973). Underground nuclear explosions and the control of earthquakes. In *Focus on Environmental Geology,* ed. R. W. Tank, pp. 88–90. New York, Oxford University Press

Emmison, F. G. (1947). *Catalogue of Maps in the Essex Record Office 1566–1855.* Supplements 1952, 1964, 1968 and 2nd edition 1969. Chelmsford, Essex County Council

Emmison, F. G. (1963). Estate maps and surveys. *History* **48,** 34–37

Emmison, F. G. (1966). *Archives and Local History.* London, Methuen

Emmison, F. G. and Fowler, G. H. (1930). *Catalogue of Maps in the Bedfordshire County Muniments.* Bedford, Bedford County Council

Evans, D. M. (1973). Man-made earthquakes in Denver. In *Focus on Environmental Geology,* ed. R. W. Tank, pp. 76–87. New York, Oxford University Press

Evans, E. J. (1976). *The Contentious Tithe. The Tithe Problem and English Agriculture 1750–1850.* London, Routledge and Kegan Paul

Evans, E. J. (1978). *Tithes and the Tithe Commutation Act 1836.* London, Bedford Square Press

Evenari, M., Shanon, L., Tadmor, N. and Aharoni, Y. (1961). Ancient agriculture in the Negev. *Science* **133,** 979–996

Everitt, B. L. (1968). Use of the cottonwood in an investigation of the recent history of a flood plain. *American Journal of Science* **266,** 417–439

Eyles, R. J. (1977a). Birchams Creek: the transition from a chain of ponds to a gully. *Australian Geographical Studies* **15,** 146–157

Eyles, R. J. (1977b). Changes in drainage networks since 1820, Southern Tablelands, New South Wales. *Australian Geographer* **13,** 377–386

Fairhall, D. (1978). Pastor power on ice. *The Guardian,* 24 August

Fels, R. (1967). *Der wirtschaftende Mensch als Gestalter der Erde,* 2nd edition. Stuttgart

Field, W. O. (1932). The glaciers of the northern part of Prince William Sound, Alaska. *Geographical Review* **22,** 361–388

Fisk, H. N. (1944). *Geological Investigations of the Alluvial Valley of the Lower Mississippi Valley.* Vicksburg, Mississippi River Commission

Fisk, H. N., McFarlan, E., Kolb, C. R. and Wilbert, L. J. (1954). Sedimentary framework of the modern Mississippi delta. *Journal of Sedimentary Petrology* **24,** 76–99

Flett, J. S. (1937). *The First Hundred Years of the Geological Survey of Great Britain.* London, HMSO

Flohn, H. (1967). Die Klimaschwankungen in historischer Zeit. In *Die Schwankungen und*

Pendelungen des Klimas in Europa seit dem Beginn der regelmassigen Instrumenten-Beobachtungen 1670, H.V. Rudloff. Braunschweig

Folland, C. K. and Wales-Smith, B. G. (1977). Richard Towneley and 300 years of regular rainfall measurement. *Weather* **32,** 438–445

French, R. A. (1972). Historical geography in the USSR. In *Progress in Historical Geography*, ed. A. R. H. Baker, pp. 111–128. Newton Abbot, David and Charles

Friis, H. R. (1958). Highlights in the first hundred years of surveying and mapping and geographical exploration of the United States by the Federal Government 1775–1880. *Surveying and Mapping* **18,** 186–206

Friis, H. R. (1965). A brief review of the development and status of geographical and cartographical activities of the United States Governments 1776–1818. *Imago Mundi* **19,** 68–80

Friis, H. R. (1975). The role of the United States topographical engineers in compiling a cartographic image of the Plains Region. In *Images of the Plains. The Role of Human Nature in Settlement,* eds B. W. Blouet and M. P. Lawson, pp. 59–74. Lincoln, University of Nebraska Press

Fritts, H. C. (1976). *Tree Rings and Climate.* New York, Academic Press

Fritts, H. C., Lofgren, G. F. and Gordon, G. A. (1979). Variations in climate since 1602 as reconstructed from tree rings. *Quaternary Research* **12,** 18–46

Fussell, G. E. (1935). *The Exploration of England: a Select Bibliography of Travel and Topography, 1570–1815.* London, Mitre Press

Gabert, J. (1965). Le Pointe du Chay. *Norois* **47,** 353–357

Galway, J. G. (1977). Some climatological aspects of tornado outbreaks. *Monthly Weather Review* **105,** 477–484

Garrad, G. H. (1954). *A Survey of the Agriculture of Kent.* London, Royal Agricultural Society of England

Gedymin, A. V. (1960, trans. 1968). The use of old Russian land survey data in geographic research for agricultural purposes, *Soviet Geography: Review and Translation* **9,** 602–624

Gilbert, G. K. (1917). Hydraulic mining debris in the Sierra Nevada. *U.S. Geological Survey Professional Paper* 105

Giles, A. W. (1927). Tornadoes in Virginia, 1814–1925. *Monthly Weather Review* **55,** 169–175

Gilmour, A. J. (1974). the impact of man on coastal systems in Victoria. In *Report on Symposium on the Impact of Human Activities on Coastal Zones.* Australian UNESCO Committee for Man and the Biosphere, Publication No. 1, pp. 127–146. Australian Government Publishing Service

Glässer, E. (1969). Zur Frage der anthropogen bedingten Vegetation, vor allem in Mitteleuropa. *Die Erde* **100**, 37–45

Glasspoole, J. (1933). The rainfall over the British Isles of each of the eleven decades during the period 1820 to 1929. *Quarterly Journal of the Royal Meteorological Society* **59,** 253–260

Glenn, L. C. (1911). Denudation and erosion in the southern Appalachian region and the Monongahela Basin. *U.S. Geological Survey Professional Paper* 72. Abridged in

Environmental Geomorphology and Landscape Conservation, ed. D. R. Coates, Vol. 3, pp. 36–56

Gole, C. V. and Chitale, S. V. (1966). Inland delta building activity of Kosi River. *Proceedings of the American Society of Civil Engineers, Journal of Hydraulics Division* **92,** HY2, 111–126

Gottschalk, L. C. (1945). Effects of soil erosion on navigation in Upper Chesapeake Bay. *Geographical Review* **35,** 219–238

Goudie, A. S. (ed.) (1981). *Geomorphological Techniques.* London, George Allen and Unwin

Graf, W. L. (1979a). Catastrophe theory as a model for change in fluvial systems. In *Adjustments of the Fluvial System,* eds D. D. Rhodes and G. P. Williams, pp. 13–32. Dubuque, Kendall Hunt Publishing Company

Graf, W. L. (1979b). Mining and channel response. *Annals of the Association of American Geographers* **69,** 262–275

Grant, P. J. (1965). Major régime changes of the Tukituki River, Hawke's Bay since about 1650 AD. *Journal of Hydrology New Zealand* **4,** 17–30

Green, C. (1961). East Anglian coastline levels since Roman times. *Antiquity* **35,** 21–28

Green, F. H. W. (1970). Weather notes from an Aberdeenshire diary 1783 to 1792. *Weather* **25,** 553–554

Green, F. H. W. (1973). Aspects of the changing environment: some factors affecting the aquatic environment in recent years. *Journal of Environmental Management* **1,** 377–391

Green, F. H. W. (1974). Changes in artificial drainage, fertilisers and climate in Scotland. *Journal of Environmental Management* **2,** 107–121

Green, F. H. W. (1976). Recent changes in land use and treatment. *Geographical Journal* **142,** 12–26

Green, F. H. W. (1979). Field underdrainage and the hydrological cycle. In *Man's Impact on the Hydrological Cycle in the United Kingdom,* ed. G. E. Hollis, pp. 9–17. Norwich, Geo Abstracts

Green, F. H. W. and Moon, H. P. (1940). *Hampshire.* The Land of Britain, the Report of the Land Utilisation Survey of Britain, Parts 89–90

Gregory, K. J. (1976). Lichens and the determination of river channel capacity. *Earth Surface Processes* **1,** 273–285

Gregory, K. J. (ed). (1977). *River Channel Changes.* Chichester, Wiley

Gregory, K. J. and Park, C. C. (1974). Adjustment of river channel capacity downstream from a reservoir. *Water Resources Research* **10,** 870–873

Gregory, K. J. and Park, C. C. (1976). The development of a Devon gully and man. *Geography* **61,** 77–82

Gregory, K. J. and Walling, D. E. (eds) (1979). *Man and Environmental Processes.* Folkestone, Dawson

Gregory, K. J. and Williams, R. F. (1981). Physical geography from the newspaper. *Geography* **66,** 42–52

Grieve, H. (1959). *Examples of English Handwriting, 1150–1750,* 2nd edition. Chelmsford, Essex Record Office Publications

Grigg, D. B. (1967). The changing agricultural geography of England: a commentary on the sources available for the reconstruction of the agricultural geography of England,

1770–1850. *Transactions of the Institute of British Geographers* **41,** 73–96

Grove, J. M. (1966). The Little Ice Age in the massif of Mont Blanc. *Transactions of the Institute of British Geographers* **40,** 129–143

Grove, J. M. (1972). The incidence of landslides, avalanches and floods in western Norway during the Little Ice Age. *Arctic and Alpine Research* **4,** 131–138

Guy, H. P. (1970). Sediment problems in urban areas. *U.S. Geological Survey Circular* 601-E, 1–8

Hadley, R. F. (1977). Some concepts of erosional processes and sediment yield in a semi-arid environment. In *Erosion, Research Techniques, Erodibility and Sediment Delivery,* ed. T. J. Toy, pp. 73–82. Norwich, Geo Abstracts

Haggard, H. R. (1902). *Rural England,* 2 vols. London, Longmans, Green

Hammer, T. R. (1972). Stream channel enlargement due to urbanisation. *Water Resources Research* **8,** 1530–1540

Handy, R. L. (1972). Alluvial cut-off dating from subsequent growth of a meander. *Bulletin of the Geological Society of America* **83,** 475–480

Happ, S. C. (1944). Effect of sedimentation on floods in the Kickapoo Valley, Wisconsin. *Journal of Geology* **52,** 53–68

Happ, S. C., Rittenhouse, G. and Dobson, G. C. (1940). Some principles of accelerated stream and valley sedimentation. *U.S. Department of Agriculture Technical Bulletin* 695

Harcup, S. E. (1968). *Historical, Archaeological and Kindred Societies in the British Isles, a List.* London, University of London Institute of Historical Research

Hardie, M. (1966–1968). *Water Colour Painting in Britain,* 3 vols. London, Batsford

Hardman, G. and Venstrom, C. (1941). A 100-year record of Truckee River runoff estimated from changes in levels and volumes of Pyramid and Winnemucca Lakes. *Transactions of the American Geophysical Union* **22,** 71–90

Hardy, J. R. (1966). An ebb-flood channel system and coastal changes near Winterton, Norfolk. *East Midland Geographer* **4,** 24–30

Harley, J. B. (1965). The re-mapping of England, 1750–1800. *Imago Mundi* **19,** 56–67

Harley, J. B. (1968a). The evaluation of early maps: towards a methodology. *Imago Mundi* **22,** 62–80

Harley, J. B. (1968b). Error and revision in early Ordnance Survey maps. *Cartographic Journal* **5,** 115–124

Harley, J. B. (1972). *Maps for the Local Historian, a Guide to the British Sources.* London, National Council of Social Service

Harley, J. B. (1975). *Ordnance Survey Maps, a Descriptive Manual.* Southampton, Ordnance Survey

Harley, J. B. (1979). *The Ordnance Survey and Land-use Mapping: Parish Books of Reference and the County Series 1:2500 maps, 1855–1918.* Historical Geography Research Series No. 2

Harley, J. B. and Phillips, C. W. (1964). *The Historian's Guide to Ordnance Survey Maps.* London, National Council of Social Service

Harrison, A. E. (1956). Fluctuations of the Nisqually Glacier, Mount Rainier, Washington, since 1750. *Journal of Glaciology* **2,** 675–683

Harvey, D. (1969). *Explanation in Geography.* London, Edward Arnold

Hastings, J. R. (1959). Vegetation change and arroyo cutting in south-eastern Arizona. *Arizona Academy of Science Journal* **1,** 60–67

Hastings, J. R. and Turner, R. M. (1965). *The Changing Mile.* Tucson, University of Arizona Press

Hatori, T. (1976). Documents of tsunami and coastal deformation in Tokai District associated with the Ansei earthquake of December 23, 1854. *Bulletin of the Earthquake Research Institute,* University of Tokyo **51,** 13–28

Hatori, T. (1977). Field investigation of the Tokai tsunamis in 1707 and 1854 along the Shizuoka coast. *Bulletin of the Earthquake Research Institute,* University of Tokyo **52,** 407–439

Hatori, T. (1978). Monuments of the Nankaido tsunamis of 1605, 1707 and 1854 in the Shikoku district. *Bulletin of the Earthquake Research Institute,* University of Tokyo **53,** 423–445

Healy, J. H., Rubey, W. W., Griggs, D. T. and Raleigh, C. B. (1968). The Denver earthquakes. *Science* **161,** 1301–1310

Heathcote, R. L. (1972). The artist as geographer: landscape painting as a source for geographical research. *Proceedings of the Royal Geographical Society of Australasia* **73,** 1–21

Hector, L. C. (1980). *The Handwriting of English Documents.* Dorking, Kohler and Coombes

Heim, A. (1882). Der Bergsturz von Elm. *Deutschen Geologischen Gesellschaft Zeitschrift* **34,** 74–115

Helley, E. J. and LaMarche, V. C. (1973). Historic flood information for northern California streams from geological and botanical evidence. *U.S. Geological Survey Professional paper* 485–E

Henderson, H. C. K. (1936). Our changing agriculture: the distribution of arable land in the Adur Basin, Sussex, from 1780 to 1931. *Journal of the Ministry of Agriculture* **43,** 625–633

Henderson, H. C. K. (1952). Agriculture in England and Wales in 1801. *Geographical Journal* **118,** 338–345

Hendry, G. W. (1931). The adobe brick as a historical source. *Agricultural History* **5,** 110–127

Herbin, R. and Pebereau, A. (1953). *Le cadastre français.* Paris, Francis Lefèbvre

Heusser, C. J. (1956). Postglacial environments in the Canadian Rocky Mountains. *Ecological Monographs* **26,** 263–302

Heusser, C. J. (1957). Variations of Blue, Hoh and White glaciers during recent centuries. *Arctic* **10,** 139–150

HMSO. *Surface Water Yearbooks*

Hollis, G. E. (ed.) (1979). *Man's Impact on the Hydrological Cycle in the United Kingdom.* Norwich, Geo Abstracts

Hollis, G. E. and Luckett, J. K. (1976). The response of natural river channels to urbanisation: two case studies from south-east England. *Journal of Hydrology* **30,** 351–363

Holsti, O. R. (1969). *Content Analysis for the Social Sciences.* Ontario, Don Mills

Homan, W. M. (1938). The marshes between Hythe and Pett. *Sussex Archaeological Collections* **79,** 199–223

Hooke, J. M. (1977a). An analysis of changes in river channel patterns. Unpublished Ph.D. thesis, University of Exeter

Hooke, J. M. (1977b). The distribution and nature of changes in river channel patterns. In *River Channel Changes,* ed. K. J. Gregory, pp. 265–280. Chichester, Wiley

Hooke, J. M. and Perry, R. A. (1976). The planimetric accuracy of tithe maps. *Cartographic Journal* **13,** 177–183

Horn, P. (1980). *The Rural World 1780–1850. Social Change in the English Countryside.* London, Hutchinson

Horner, L. (1847). On the origin of Monte Nuovo. *Quarterly Journal of the Geological Society of London* **3,** 19–22

Howard, A. D. (1939). Hurricane modification of the offshore bar of Long Island, New York. *Geographical Review* **29,** 400–415

Howard, P. (1979). The use of art works in landscape studies. *Landscape Research* **4,** 10–12

Howe, E. (1909). Landslides in the San Juan Mountains, Colorado. *U.S. Geological Survey Professional Paper* 67

Howe, G. M., Slaymaker, H. O. and Harding, D. M. (1967). Some aspects of the flood hydrology of the upper catchments of the Severn and Wye. *Transactions of the Institute of British Geographers* **41,** 33–58

Howell, E. J. (1941). *Shropshire.* The Land of Britain, the Report of the Land Utilisation Survey of Britain, Part 66

Hoyt, J. B. (1958). The cold summer of 1816. *Annals of the Association of American Geographers* **48,** 118–131

Hsu, K. J. (1975). Catastrophic debris streams (sturzstroms) generated by rockfalls. *Bulletin of the Geological Society of America* **86,** 129–140

Huber, W. L. (1930). San Francisco earthquakes of 1865 and 1868. *Bulletin of the Seismological Society of America* **20,** 261–272

Hull, F. (ed.) (1973). *Catalogue of Estate Maps 1590–1840 in the Kent County Archives Office.* Maidstone, Kent County Council

Humphrey, R. R. (1958). The desert grassland: a history of vegetational change and an analysis of causes. *Botanical Review* **24,** 193–252

Hussey, K. M. and Zimmerman, H. L. (1953). Rate of meander development as exhibited by two streams in Story County, Iowa. *Iowa Academy of Science Proceedings* **60,** 390–392

Hutchinson, J. N. (1961). A landslide on a thin layer of quick clay at Furre, central Norway. *Géotechnique* **11,** 69–94

Hutchinson, J. N. (1965a). The landslide of February 1959 at Vibstad in Namdalen. *Norwegian Geotechnical Institute* **61,** 1–16

Hutchinson, J. N. (1965b). A survey of the coastal landslides of Kent. *Building Research Station, Note No. EN 35/65*

Hutchinson, J. N. (1969). A reconsideration of the coastal landslides at Folkestone Warren, Kent. *Géotêchnique* **19,** 6–38

Hutchinson, J. N. (1980). The record of peat wastage in the East Anglian fenlands at Holme Post, 1848–1978 AD. *Journal of Ecology* **68,** 229–249

Hutchinson, J. N., Bromhead, E. N. and Lupini, J. F. (1980). Additional observations on the Folkestone Warren landslides. *Quarterly Journal of Engineering Geology* **13,** 1–31

Hutchinson, J. N. and Gostelow, T. P. (1976). The development of an abandoned cliff in

London clay at Hadleigh, Essex. *Philosophical Transactions of the Royal Society* Series A **283,** 557–604

Hutchinson, J. N., Somerville, S. H. and Petley, D. J. (1973). A landslide in periglacially disturbed Etruria marl at Bury Hill, Staffordshire. *Quarterly Journal of Engineering Geology* **6,** 377–404

Imeson, A. C., Kwaad, F. J. P. M. and Mücher, H. J. (1980). Hillslope processes and deposits in forested areas of Luxembourg. In *Timescales in Geomorphology,* eds R. A. Cullingford, D. A. Davidson and J. Lewin, pp. 31–42. Chichester, Wiley

Ineson, J. (1966). Groundwater: principles of network design. *International Association for Scientific Hydrology, Publication No. 68,* 476–83. Quebec.

Ingram, M. J. and Underhill, D. J. (1979). The use of documentary sources for the study of past climates. Paper at *International Conference on Climate and History,* University of East Anglia Climatic Research Unit, 59–90

Ingram, M. J., Underhill, D. J. and Wigley, T. M. L. (1978). Historical climatology. *Nature* **276,** 329–334

Innes, J. L. (1981). A manual for lichenometry—comment. *Area* **13,** 237–241

Ireland, H. A., Sharpe, C. F. S. and Eargle, D. H. (1939). Principles of gully erosion in South Carolina. *U.S. Department of Agriculture Technical Bulletin 633*

Jacobsen, T. and Adams, R. M. (1958). Salt and silt in ancient Mesopotamian culture. *Science* **128,** 1251–1258

Jeans, D. N. (1966). The breakdown of Australia's first rectangular grid survey. *Australian Geographical Studies* **4,** 119–128

Jeans, D. N. (1966–1967). Territorial divisions and the locations of towns in New South Wales, 1826–1842. *Australian Geographer* **10,** 243–255

Jeans, D. N. (1978). Use of historical evidence for vegetation mapping in NSW. *Australian Geographer* **14,** 93–97

Jennings, J. N. (1965). Man as a geological agent. *Australian Journal of Science* **28,** 150–156

Johnson, H. B. (1975). The United States Land Survey as a principle of order. In *Pattern and Process. Research in Historical Geography,* ed. R. E. Ehrenberg, pp. 114–130. Washington DC, Harvard University

Johnson, R. H. and Paynter, J. (1967). The development of a cut-off in the River Irk at Chadderton, Lancashire. *Geography* **52,** 41–49

Johnston, M. C. (1963). Past and present grasslands of southern Texas and north-eastern Mexico. *Ecology* **44,** 456–466

Jones E. L. (1964). *Seasons and Prices. The Role of the Weather in English Agricultural History.* London, Allen and Unwin

Jones, M. (1977). *Finland, Daughter of the Sea.* Folkestone, Dawson

Jones, R. C. (1975). Assessment of records and use of historic flood records. *Institution of Civil Engineers, Flood Studies Conference,* pp. 39–42. London

Kain, R. J. P. (1974). The tithe commutation surveys. *Archaeologia Cantiana* **89,** 101–118

Kain, R. J. P. (1975). R. K. Dawson's proposal in 1836 for a cadastral survey of England

and Wales. *Cartographic Journal* **12,** 81–88

Kain, R. J. P. (1977) Tithe surveys and the rural landscape of England and Wales. *Society of University Cartographers Bulletin* **11,** 1–3

Kain, R. J. P. (1979). The London Topographical Society's edition of Thomas Milne's map of London. *Society of University Cartographers Bulletin* **13,** 16–22

Karlén, W. (1973). Holocene glacier and climatic variations, Kebnekaise Mountains, Swedish Lapland. *Geografiska Annaler* **55A,** 29–63

Kaye, C. A. (1976). The geology and early history of the Boston area of Massachusetts: A Bicentennial approach. *U.S. Geological Survey Bulletin 1476*

Kelsey, J. (1972). Geodetic aspects concerning possible subsidence in south-eastern England. *Philosophical Transactions of the Royal Society* Series A **272,** 141–149

Kemp, D. D. (1976). Winter weather in West Fife in the 18th century according to the 'Annals of Dunfermline'. *Weather* **31,** 400–404

Kenoyer, L. A. (1933). Forest distribution in south-western Michigan as interpreted from the original Land Survey (1826–32). *Papers of the Michigan Academy of Arts, Science and Letters* **19,** 107–111

Kerridge, E. (1966). The manorial survey as an historical source. *Amateur Historian* **7,** 2–7

Kestner, F. J. (1962). The old coastline of the Wash. *Geographical Journal* **128,** 457–478

Kestner, F. J. (1970). Cyclic changes in Morecambe Bay. *Geographical Journal* **136,** 85–97

Kidson, C. (1950). Dawlish Warren: a study of the evolution of the sand spits across the mouth of the River Exe in Devon. *Transactions of the Institute of British Geographers* **16,** 69–80

Kiely, E. R. (1947, reprinted 1979). *Surveying Instruments: Their History.* Columbus, Ohio; Carben Surveying Book Reprints

Kington, J. A. (1975a). An analysis of the seasonal characteristics of 1781–4 and 1968–71 using the P.S.C.M. indices. *Weather* **30,** 109–114

Kington, J. A. (1975b). A comparison of British Isles weather type frequencies in the climatic record from 1781 to 1971. *Weather* **30,** 21–24

Kington, J. A. (1976a). An examination of monthly and seasonal extremes using historical weather maps from 1781: October 1781. *Weather* **31,** 151–158

Kington, J. A. (1976b). An introduction to an examination of monthly and seasonal extremes in the climatic record of England and Wales using historical daily synoptic weather maps from 1781 onward. *Weather* **31,** 72–78

Kirikov, S. V. (1960). Les changements dans la distribution des oiseaux de la partie européene de l'Union Soviétique aux XVIIe-XIXe siècles. *Proceedings of the XIIth International Ornithological Congress, Helsinki 1958* Helsinki, 404–421

Kishimoto, H. (1968). *Cartometric Measurements.* Zurich; Geographisches Institut der Universtat, Zurich

Klimek, K. and Trafas, K. (1972). Young–Holocene changes in the course of the Dunajec River in the Beskid Sadecki Mountains. *Studia Geomorphologica Carpatho–Balcanica* **6,** 85–92

Knox, J. C. (1972). Valley alluviation in south-western Wisconsin. *Annals of the Association of American Geographers* **62,** 401–410

Knox, J. C. (1977). Human impacts on Wisconsin stream channels. *Annals of the Association of American Geographers* **67,** 323–342

Knox, J. C., Bartlein, P. J., Hirschboeck, K. K. and Muckenheim, R. J. (1976). The response of floods and sediment yields to climate variation and land use in the Upper Mississippi Valley. *Institute of Environmental Studies* Report No. 52. Center for Geographical Analysis, University of Wisconsin, Madison

Koc, L. (1972). 19th and 20th centuries changes in Vistula channel between Plock and Torun. *Przeglad Geograficzny* **44,** 703–719

Koch, L. (1940). Survey of north Greenland. *Meddelelser om Grønland* **130,** Part 1, 1–364

Koch, L. (1945). The east Greenland ice. *Meddelelser om Grønland* **130,** Part 3, 1–374

Kondrat'yev N.Ye and Popov, I. V. (1967). Methodological prerequisites for conducting network observations of the channel process. *Soviet Hydrology* **3,** 273–297

Kraus, E. B. (1955a). Secular changes of east-coast rainfall regimes. *Quarterly Journal of the Royal Meteorological Society* **81,** 430–439

Kraus, E. B. (1955b). Secular changes of tropical rainfall regimes. *Quarterly Journal of the Royal Meteorological Society* **81,** 198–210

Laczay, I. A. (1977). Channel pattern changes of Hungarian rivers: the example of the Hernad river. In *River Channel Changes,* ed. K. J. Gregory, pp. 185–192. Chichester, Wiley

Ladurie, E. Le Roy (1972). *Times of Feast, Times of Famine.* London, George Allen and Unwin

Lamb, H. H. (1963). What can we find out about the trend of our climate? *Weather* **18,** 194–216

Lamb, H. H. (1964). Britain's climate in the past. Lecture to British Association for Advancement of Science. Published in Lamb, H. H., 1966, *The Changing Climate,* London, Methuen, pp. 170–195

Lamb, H. H. (1965). The early medieval warm epoch and its sequel. *Palaeogeography, Palaeoclimatology, Palaeoecology* **1,** 13–37

Lamb, H. H. (1966). *The Changing Climate: Selected Papers.* London, Methuen

Lamb, H . H. (1970). Volcanic dust in the atmosphere; with a chronology and assessment of its meteorological significance. *Philosophical Transactions of the Royal Society* Series A **266,** 425–533

Lamb, H. H. (1977). *Climate Present, Past and Future. Vol.II Climatic History and the Future.* London, Methuen

Lamb, H. H. and Johnson, A. I. (1959). Climatic variation and observed changes in the general circulation. *Geografiska Annaler* **41,** 94–134

Lambert, J. M., Jennings, J. N., Smith, C. T., Green, C. and Hutchinson, J. N. (1960). The making of the Broads. *Royal Geographical Society Research Series No. 3*

Landsberg, H. E. (1967). Two centuries of New England climate. *Weatherwise* **20,** 52–57

Lavergne, L. de (1855). *The Rural Economy of England, Scotland and Ireland.* Edinburgh and London, William Blackwood

Lawrence, D. B. (1950). Glacier fluctuation for six centuries in south-eastern Alaska and its relation to solar activity. *Geographical Review* **40,** 191–223

Lawrence, D. B. and Elson, J. A. (1953). Periodicity of deglaciation in North America since the late Wisconsin Maximum. *Geografiska Annaler* **35,** 83–104

Lawrence, D. B. and Lawrence, E. G. (1961). Response of enclosed lakes to current glaciopluvial climatic conditions in middle latitude western North America. *New York*

Academy of Science Annals **95,** 341–350

Lawson, M. P., Reiss, A., Phillips, R. and Livingstone, K. (1971). Nebraska droughts. *Department of Geography, University of Nebraska, Occasional Paper No. 1*

Laxton, P. (1976). The geodetic and topographical evaluation of English County Maps, 1740–1840. *Cartographic Journal* **13,** 37–54

Leblanc, G. (1981). A closer look at the September 16, 1732, Montreal earthquake. *Canadian Journal of Earth Sciences* **18,** 539–550

Legget, R. F. and Lasalle, P. (1978). Soil studies at Shipshaw, Quebec: 1941 and 1969. *Canadian Geotechnical Journal* **15,** 556–564

Leopold, L. B. (1951a). Rainfall frequency: an aspect of climatic variation. *American Geophysical Union Transactions* **32,** 347–357

Leopold, L. B. (1951b). Vegetation of south-western watersheds in the nineteenth century. *Geographical Review* **41,** 295–316

Leopold, L. B. (1968). Hydrology for urban land planning—A guidebook on the hydrologic effects of urban land use. *U.S. Geological Survey Circular* 554, 1–18

Leopold, L. B. (1973). River channel change with time: an example. *Bulletin of the Geological Society of America* **84,** 1845–1860

Leopold, L. B. and Snyder, C. T. (1951). Alluvial fills near Gallup, New Mexico. *U.S. Geological Survey Water Supply Paper* 110–A, 1–17

Lewin, J. (1977). Channel pattern changes. In *River Channel Changes* ed. K. J. Gregory, pp. 167–184. Chichester, Wiley

Lewin, J. and Brindle, B. J. (1977). Confined meanders. In *River Channel Changes,* ed. K. J. Gregory, pp. 221–233. Chichester, Wiley

Lewin, J. and Hughes, D. (1976). Assessing channel change on Welsh rivers. *Cambria* **3,** 1–10

Lewin, J. and Weir, M. J. C. (1977). Morphology and recent history of the Lower Spey. *Scottish Geographical Magazine* **93,** 45–51

Lewis, G. M. (1962). Changing emphases in the description of the natural environment of the American Great Plains area. *Transactions of the Institute of British Geographers* **30,** 75–90

Lewis, G. M. (1965a). Early American exploration and the Cis-Rocky Mountain desert, 1803–23. *Great Plains Journal* **5,** 1–11

Lewis, G. M. (1965b). Three centuries of desert concepts in the Cis-Rocky Mountain west. *Journal of the West* **4,** 457–468

Lewis, G. M. (1966). Regional ideas and reality in the Cis-Rocky Mountain west. *Transactions of the Institute of British Geographers* **38,** 135–150

Lewis, W. V. (1932). The formation of Dungeness Foreland. *Geographical Journal* **80,** 309–324

List and Index Society, 1971, 1972. *Inland Revenue Tithe Maps and Apportionments,* Vols 68 and 83. London, Swift (P.& D.)

Locke, W. W., Andrews, J. T. and Webber, P. J. (1979). A manual for lichenometry. *British Geomorphological Research Group Technical Bulletin* **26**

Lockwood, J. G. (1979). Water balance of Britain, 50 000 yr B.P. to the present day. *Quaternary Research,* **12,** 297–310

Login, T. (1872). Geological changes of rivers in N. India. *Geological Society Quarterly Journal* **28,** 186–200

Louderback, G. D. (1947). Central California earthquakes of the 1830s. *Bulletin of the Seismological Society of America* **37,** 33–74

Lovegrove, H. (1953). Old shore lines near Camber Castle. *Geographical Journal* **119,** 200–207

Lowdermilk, W. C. (1927). Erosion control in Japan. *Journal of the Association of Chinese American Engineers* **8,** 3–13

Lowe, E. J. (1870). *Natural Phenomena and Chronology of the Seasons.* London, Bell and Daldy

Lyell, C. (1867). *Principles of Geology.* 10th edition, 2 vols, 1st edition 1830. London, John Murray

Lynam, E. W. O'F. (1953). *The Mapmaker's Art. Essays on the History of Maps.* London, Batchworth Press

McGregor, O. R. (1961). The historiography of English farming. In *English Farming Past and Present,* Lord Ernle, pp. 79–145

Maejima, I. and Koike, Y. (1976). An attempt at reconstructing the historical weather situations in Japan. *Geographical Report of Tokyo Metropolitan University* **11,** 1–12

Malin, J. C. (1946). Dust storms in Kansas: Part one, 1850–1860, Part two 1861–1880, Part three, 1881–1900. *Kansas Historical Quarterly* **14,** 129–144, 265–296, 391–413

Malin, J. C. (1953). Soil, animal and plant relations of the grassland, historically reconsidered. *Scientific Monthly* **76,** 207–220

Maling, D. H. (1968). How long is a piece of string? *Cartographic Journal* **5,** 147–156

Manley, G. (1946). Temperature trend in Lancashire 1753–1945. *Quarterly Journal of the Royal Meteorological Society* **72,** 1–31

Manley, G. (1953). The mean temperature of central England, 1698–1952. *Quarterly Journal of the Royal Meteorological Society* **79,** 242–261

Manley, G. (1959). Temperature trends in England, 1698–1957. *Archiv für Meteorologie Geophysik und Bioklimatologie* Serie B **9,** 413–433

Manley, G. (1961). Late and postglacial climatic fluctuations and their relationship to those shown by the instrumental record of the past 300 years. *New York Academy of Sciences Annals* **95,** 162–172

Manley, G. (1975). 1684: The coldest winter in the English instrumental record. *Weather* **30,** 382–388

Marschner, F. J. (1959). *Land Use and its Patterns in the United States.* US Department of Agriculture Handbook No. 153

Marsh, G. P. (1864). *Man and Nature; or, Physical Geography as Modified by Human Action.* London, Sampson Low

Marshall, W. (1817). *A Review and Complete Abstract of the Reports to the Board of Agriculture on the Several Counties of England*

Martin, J. M. (1872). Exmouth Warren and its threatened destruction. *Report and Transactions of the Devonshire Association* **5,** 84–89

Martin, J. M. (1876). The changes of Exmouth Warren. *Report and Transactions of the Devonshire Association* **8,** 453–460

Martin, J. M. (1893). Some further notes on Exmouth Warren. *Report and Transactions of the Devonshire Association* **25,** 406–415

Mason, L. (1963). Using historical records to determine climax vegetation. *Journal of Soil and Water Conservation* **18,** 190–194

Matthes, G. H. (1947). Mississippi River cut-offs. *Transactions of the American Society of Civil Engineers* **113,** 1–39

Matthews, W. (1950). *British Diaries. An Annotated Bibliography of British Diaries Written between 1442 and 1942.* Berkeley and Los Angeles, University of California Press

Mathews, W. H. (1951). Historic and prehistoric fluctuations of Alpine glaciers in the Mount Garibaldi map-area, south-western British Columbia. *Journal of Geology* **59,** 357–380

Maunder, E. W. (1922). The sun and sunspots 1820–1920. *Monthly Notices of Royal Astronomical Society* **82,** 534–543

Maxwell, G. M. (1976). Old mine shafts and their location by geophysical surveying. *Quarterly Journal of Engineering Geology* **9,** 283–290

Mayewski, P. A. and Jeschke, P. A. (1979) Himalayan and trans-Himalayan glacier fluctuations since AD 1812. *Arctic and Alpine Research* **11,** 267–287

Meade, R. H. and Trimble, S. W. (1974). Changes in sediment loads in rivers of the Atlantic drainage of the United States since 1900. *International Association for Scientific Hydrology Bulletin* **113,** 99–104

Meaden, G. T. (1975). Tornadoes at Scarborough AD 1975 and 1165. *Weather* **30,** 303–305

Meaden, G. T. (1976). Late summer weather in Kent, 55 BC. *Weather* **31,** 264–270

Meaden, G. T. (1977). Summer weather in south-east England, 54 BC. *Weather* **32,** 33–35

Mellor, G. R. (1955). History from newspapers. *Amateur Historian* **2,** 97–107

Merian, M. (1644). *Topographia Helvetiae, Rhaetiae et Valesiae etc.* Amsterdam

Messerli, B., Messerli, P., Pfister, C. and Zumbuhl, H. J. (1978). Fluctuations of climate and glaciers in the Bernese Oberland, Switzerland and their geoecological significance, 1600 to 1975. *Arctic and Alpine Research* **10,** 247–260

Meyer, G. M. (1927). Early water mills in relation to changes in the rainfall of east Kent. *Quarterly Journal of the Royal Meteorological Society* **53,** 407–419

Mikesell, M. W. (1969). The deforestation of Mount Lebanon. *Geographical Review* **59,** 1–28

Milford, R. T. and Sutherland, D. M. (1936). *Catalogue of English Newspapers and Periodicals in the Bodleian Library, 1622–1800.* Oxford, Oxford Bibliographical Society

Miller, J. P. and Wendorf, F. (1958). Alluvial chronology of the Tesuque Valley, New Mexico. *Journal of Geology* **66,** 177–194

Minchinton, W. E. (1953). Agricultural returns and the Government during the Napoleonic Wars. *The Agricultural History Review* **1,** 29–43

Moir, D. G. (ed.) (1973). *The Early Maps of Scotland to 1850.* Edinburgh, Royal Scottish Geographical Society

Monmonier, M. S. (1980). The hopeless pursuit of purification in cartographic communication: a comparison of graphic-arts and perceptual distortions of graytone symbols. *Cartographica* **17,** 24–39

Moodie, D. W. (1971). Content analysis: a method for historical geography. *Area* **3,** 146–149

Moodie, D. W. (1977). The Hudson's Bay Company's archives: a resource for historical geography. *Canadian Geographer* **21,** 268–274

Moodie, D. W. and Catchpole, A. J. W. (1975). *Environmental Data from Historical Documents by Content Analysis: Freeze-up and Break-up of Estuaries on Hudson Bay 1714–1871.* Manitoba Geographical Studies, 5

Moon, K. (1969). Perception and appraisal of the south Australian landscape 1836–1850. *Proceedings of the Royal Geographical Society of Australasia* **70,** 41–64

Moore, T. R. (1979). Land use and erosion in the Machakos Hills. *Annals of the Association of American Geographers* **69,** 419–431

Moore, D. P. and Matthews, W. H. (1978). The Rubble Creek landslide, south-western British Columbia. *Canadian Journal of Earth Sciences* **15,** 1039–1052

Moore, J. G., Nakamura, K. and Alcaraz, A. (1973). The 1965 eruption of Taal Volcano. In *Focus on Environmental Geography,* ed. R. W. Tank, pp. 11–18. New York, Oxford University Press

Morgan, R. P. C. (1980). Soil erosion and conservation in Britain. *Progress in Physical Geography* **4,** 24–47

Morse, E. S. (1869). On the landslides in the vicinity of Portland, Maine. *Proceedings of the Boston Society of Natural History* **12,** 235–244

Mosby, J. E. G. (1938). *Norfolk.* The Land of Britain, the Report of the Land Utilisation Survey of Britain, Part 70

Mosley, M. P. (1975). Channel changes on the River Bollin, Cheshire, 1872–1973. *East Midland Geographer* **6,** 185–199

Mottershead, D. N. (1980). Lichenometry—some recent applications. In *Timescales in Geomorphology,* eds R. A. Cullingford, D. A. Davidson and J. Lewin, pp. 95–108. Chichester, Wiley

Mullins, E. L. C. (1958). *Texts and Calendars. An Analytical Guide to Serial Publications.* London, Royal Historical Society

Muntz, A. P. (1969). Federal cartographic archives: a profile. *Prologue: The Journal of the National Archives* **1,** 3–7

Murphy, J. (1978). Measures of map accuracy and some early Ulster maps. *Irish Geography* **11,** 88–101

Murton, B. J. (1968). Mapping the immediate pre-European vegetation on the east coast of the North Island of New Zealand. *Professional Geographer* **20,** 262–264

Natural Environment Research Council (1975). *Flood Studies Report.* London, HMSO

Nelson, J. G. (1966). Man and geomorphic process in the Chemung River Valley, New York and Pennsylvania. *Annals of the Association of American Geographers* **56,** 24–32

Newland, D. H. (1916). Landslides in unconsolidated sediments: with a description of some occurrences in the Hudson Valley. *New York State Museum and Science Service Bulletin* **187,** 79–105

Nicholls, P. H. (1972). On the evolution of a forest landscape. *Transactions of the Institute of British Geographers* **56,** 57–76

Nilsson, G. and Martvall, S. (1972). The Ore River and its meanders. *Uppsala Universitet Naturgeografiska Institutionen Report No. 19*

Noble, C. and Palmquist, R. C. (1968). Meander growth in artificially straightened streams. *Iowa Academy of Science Proceedings* **75,** 234–242

Norden, J. (1966). *Orford Ness: a Selection of Maps by John Norden presented to J. A.*

Steers. Cambridge, Cambridge University Press

Nordseth, K. (1973). Floodplain construction on a braided river. The islands of Koppang öyene on the River Glomma. *Norsk Geografisk Tidsskrift* **27,** 109–126

Nossin, J. J. (1965). The geomorphic history of the northern Pahang delta. *Journal of Tropical Geography* **20,** 54–64

Nunn, G. W. A. (ed.) (1952). *British Sources of Photographs and Pictures.* London, Cassell

Nunn, P. D. (1979). Post-Devensian drainage evolution of the Thames in central London. Unpublished paper presented to the Institute of British Geographers Annual Conference, Manchester

Nuttli, O. W. (1973). The Mississippi Valley earthquakes of 1811 and 1812. *Bulletin of the Seismological Society of America* **63,** 227–248

Oakeshott, G. B. (1976). *Volcanoes and Earthquakes.* San Francisco, McGraw-Hill

Oldfield, F. (1963). Pollen analysis and man's role in the ecological history of the south-east Lake District. *Geografiska Annaler* **45,** 23–40

Oldfield, F. (1969). Pollen analysis and the history of land use. *Advancement of Science* **25,** 298–311

Oldfield, F. (1970). The ecological history of Blelham Bog National Nature Reserve. In *Studies in the Vegetational History of the British Isles,* eds D. Walker and R. G. West, pp. 141–57. Cambridge, Cambridge University Press

Oldfield, F. (1981). Peats and lake sediments: formation, stratigraphy, description and nomenclature. In *Geomorphological Techniques,* ed. A. S. Goudie, pp. 306–326. London, George Allen and Unwin

Oldfield, F., Rummery, T., Bloemendal, J., Dearing, J. and Thompson, R. (1979). The persistence of fire-induced magnetic oxides in soils and lake sediments. *Annales de Géophysique* **35,** 103–107

Oliver, J. (1958). William Bulkeley's record of the weather of Anglesey, 1734–1743, 1747–1760. *Quarterly Journal of the Royal Meteorological Society* **84,** 126–133

Oliver, J. and Kington, J. A. (1970). The usefulness of ships' log-books in the synoptic analysis of past climates. *Weather* **25,** 520–528

O'Loughlin, C. L. (1969). Streambed investigations in a small mountain catchment. *New Zealand Journal of Geology and Geophysics* **12,** 684–706

Olson, G. W. and Hanfmann, G. M. A. (1971). Some implications of soils for civilisations. *New York Food and Life Science Quarterly* **4,** 11–14

Orwin, C. S. and Whetham, E. H. (1964). *History of British Agriculture 1846–1914* London, Longmans

Osborne, B. S. and Reimer, D. L. (1973). Content analysis and historical geography: a note on evaluative assertion analysis. *Area* **5,** 96–100

Ovenden, J. C. and Gregory, K. J. (1980). The permanence of stream networks in Britain. *Earth Surface Processes* **5,** 47–60

Owen, A. E. B. (1952). Coastal erosion in east Lincolnshire. *Lincolnshire Historian,* Spring, 330–341

Parizek, R. R. (1971). Impact of highways on the hydrogeologic environment. In *Environmental Geomorphology,* ed. D. R. Coates, pp. 151–199. Binghampton, State

University of New York

Parry, M. L. (1975). County maps as historical sources. A sequence of surveys in south-east Scotland. *Scottish Studies* **19,** 15–26

Parry, M. L. (1978). *Climatic Change, Agriculture and Settlement.* Folkestone, Dawson

Pattison, W. D. (1956). Use of the U.S. Public Land Survey plats and notes as descriptive sources. *The Professional Geographer* **8,** 10–14

Pattison, W. D. (1957, reprinted 1964). *Beginnings of the American Rectangular Land Survey System,* 1784–1800. Chicago, Department of Geography, University of Chicago Research Paper No. 50

Pattison, W. D. (1975). Reflections on the American rectangular land survey system. In *Pattern and Process. Research in Historical Geography,* ed. R. E. Ehrenberg, pp. 131–138. Washington DC, Harvard University Press

Pearson, M. G. (1973). Snowstorms in Scotland, 1782 to 1786. *Weather* **28,** 195–201

Pearson, M. G. (1976). Snowstorms in Scotland 1729 to 1830. *Weather* **31,** 390–393

Pearson, M. G. (1978). Snowstorms in Scotland 1831 to 1861. *Weather* **33,** 392–399

Perpillou, A. (1935). Les documents cadastraux dans les études de géographie économique (exemple du Limousin). *Bulletin de l'Association de Géographes Français* **84,** 10–18

Perry, P. (1969). Twenty five years of New Zealand historical geography. *New Zealand Geographer* **25,** 93–105

Peterken, G. F. (1969). Development of vegetation in Staverton Park, Suffolk. *Field Studies* **3,** 1–39

Peterssen, O. (1914). Climatic variations in historic and prehistoric time. *Uv Svenska Hydrografik—Biologiska Kommissionens Skrifter* 1–25

Petts, G. E. and Lewin, J. (1979). Physical effects of reservoirs on river systems. In *Man's Impact on the Hydrological Cycle in the United Kingdom,* ed, G. E. Hollis, pp. 79–91. Norwich, Geo Abstracts

Pfister, C. (1979). The reconstruction of past climate: the example of the Swiss historical weather documentation project (16th to early 19th century). *International Conference on Climate and History,* University of East Anglia, Climatic Research Unit, 128–147

Pfister, C. (1980). The Little Ice Age: Thermal and wetness indices for central Europe. *Journal of Interdisciplinary History* **10,** 665–696

Phillips, A. D. M. (1969). Underdraining and the English claylands, 1850–1880: a review. *The Agricultural History Review* **17,** 44–55

Phillips, A. D. M. (1980). The seventeenth century maps and surveys of William Fowler. *Cartographic Journal* **17,** 100–110

Phillips, W. S. (1963). *Vegetational Changes in Northern Great Plains.* Tucson, University of Arizona

Piasecka, J. E. (1974). Hydrographic changes in Warta Valley during the last 200 years. *Czasopismo Geograficzne* **45,** 229–238

Pickup, G. (1976). Geomorphic effects of changes in river runoff, Cumberland Basin, New South Wales. *Australian Geographer* **13,** 188–193

Pickwell, R. (1878). The encroachments of the sea from Spurn Point to Flamborough Head and the works executed to prevent the loss of land. *Proceedings of the Institution of Civil Engineers* **51,** 191–212

Piest, R. F., Beer, C. E. and Spomer, R. G. (1976). Entrenchment of drainage in western Iowa and north-western Missouri. *Proceedings of the 3rd Inter-Agency Sedimentation*

Conference, Water Resources Council, 548–560

Piest, R.F., Elliott, L. S. and Spomer, R. G. (1977). Erosion of the Tarkio drainage system, 1845–1976. *Transactions of the American Society of Agricultural Engineers* **20,** 485–488

Pinkett, H. T. (1970). Forest service records as research material. *Forest History* **13,** 18–29

Pitts, J. (1973). The Bindon landslip of 1839. *Proceedings of the Dorset Natural History and Archaeological Society* **95,** 18–29

Poland, J. F. and Davis, G. H. (1971). Land subsidence due to withdrawal of fluids. In *Man's Impact on Environment,* ed. T. R. Detwyler, pp. 370–382. New York, McGraw-Hill

Pollard, M. (1968). Floods according to an ancient Egyptian scientific chronicle. *Journal of Geophysical Research* **73,** 7158–7159

Pool, I. de S. (ed.) (1959). *Trends in Content Analysis.* Urbana, University of Illinois Press Press

Potter, H. R. (1978). *The Use of Historic Records for the Augmentation of Hydrological Data.* Wallingford, Institute of Hydrology Report 46

Potter, N. (1969). Tree-ring dating of snow avalanche tracks and the geomorphic activity of avalanches, Northern Absaroka Mountains, Wyoming. *Geological Society of America Special Paper* **123,** eds S. A. Schumm and W. C. Bradley, 141–165

Powell, J. M. (1970). The Victorian survey system, 1837–1860. *New Zealand Geographer* **26,** 50–69

Powell, J. M. (1972). The records of the New Zealand Lands and Survey Department. *New Zealand Geographer* **28,** 72–77

Prentiss, L. W. (1952). Gulf hurricanes and their effects on the Texas coast. *Proceedings of 2nd Conference of Coastal Engineers* 208–216

Price, D. J. (1955). Medieval land surveying and topographical maps. *Geographical Journal* **121,** 1–10

Price, R. J. (1980). Rates of geomorphological change in proglacial areas. In *Timescales in Geomorphology,* eds R. A. Cullingford, D. A. Davidson and J. Lewin, pp. 79–93. Chichester, Wiley

Prince, H. (1959). The tithe surveys of the mid-nineteenth century. *The Agricultural History Review* **7,** 14–26

Prince, H. C. (1971). Real, imagined and abstract worlds of the past. *Progress in Geography* **3,** 1–86

Public Record Office (1963). *Guide to the Contents of the Public Record Office.* London, HMSO

Public Record Office (1967). *Maps and Plans in the Public Record Office. I British Isles, c. 1410–1860.* London, HMSO

Pugh, R. B. (1967). The structure and aims of the Victoria History of the Counties of England. *Bulletin of the Institute of Historical Research* **40,** 65–73

Rackham, O. (1975). *Hayley Wood: its History and Ecology.* Cambridgeshire and Isle of Ely Naturalists' Trust Ltd

Rackham, O. (1980). *Ancient Woodland: its History, Vegetation and Uses in England.* London, Edward Arnold

Randall, J. (1880). *History of Madeley.* Madeley, The Wrekin Echo. Republished 1975, Shrewsbury, Shopshire County Library

Randhawa, M. S. (1952). Progressive dessication of northern India in historical times. *Indian Forester* **78,** 497–505

Ratcliffe, R. A. S. (1978). The story of the Royal Meteorological Society. *Weather* **33,** 261–268

Ravenhill, W. L. D. (1971). The missing maps from John Norden's survey of Cornwall. In *Exeter Essays in Geography,* eds K. J. Gregory and W. L. D. Ravenhill, pp. 93–104. Exeter, University of Exeter

Ravenhill, W. L. D. (1972). *John Norden's Maps of Cornwall and its Nine Hundreds.* Exeter, University of Exeter

Ravenhill, W. L. D. (1973). The mapping of Great Haseley and Latchford: an episode in the surveying career of Joel Gascoyne. *Cartographic Journal* **10,** 105–111

Ravenhill, W. L. D. and Gilg, A. (1974). The accuracy of early maps? Towards a computer aided method. *Cartographic Journal* **11,** 48–52

Rawls, O. G. (1972). A case history of shoreline effects of jetties and channel improvements at the mouth of St John's River. *Shore and Beach* **40,** 33–35

Reagan, A. B. (1924). Recent changes in the Plateau Region. *Science* **60,** 283–285

Redman, J. B. (1852). On the alluvial formations and the local changes of the south coast of England. *Minutes of the Proceedings of the Institution of Civil Engineers* **11,** 162–223

Redman, J. B. (1864). The east coast between the Thames and the Wash. *Minutes of the Proceedings of the Institution of Civil Engineers* **23,** 186–256

Redstone, L. J. and Steer, F. W. (1953). *Local Records, their Nature and Care.* London, G. Bell and Sons

Reid, H. F. (1914). The Lisbon earthquake of November 1, 1755. *Bulletin of the Seismological Society of America* **4,** 53–80

Reynolds, G. (1965). A history of raingauges. *Weather* **20,** 106–114

Rich, J. L. (1911). Recent stream trenching in the semi-arid portion of south-western New Mexico. *American Journal of Science* **32,** 237–245

Richeson, A. W. (1966). *English Land Measuring to 1800: Instruments and Practices.* Cambridge, Massachusetts; Society for the History of Technology

Rittmann, A. and Rittmann, L (1976). *Volcanoes.* London, Orbis Publishing

Roberts, B. K., Turner, J. and Ward, P. F. (1973). Recent forest history and land use in Weardale, northern England. In *Quaternary Plant Ecology,* eds H. J. B. Birks and R. G. West, pp. 207–221. Oxford, Blackwell

Robinson, A. H. W. (1951). The changing navigation routes of the Thames estuary. *Journal of the Institute of Navigation* **4,** 357–370

Robinson, A. H. W. (1955). The harbour entrances of Poole, Christchurch and Pagham. *Geographical Journal* **121,** 33–50

Robinson, A. H. W. (1960). Ebb-flood channel systems in sandy bays and estuaries. *Geography* **45,** 183–199

Robinson, A. H. W. (1962). *Marine Cartography in Britain. A History of the Sea Chart to 1855.* Leicester, Leicester University Press

Robinson, A. H. W (1964). The inshore waters, sediment supply and coastal changes of part of Lincolnshire. *East Midland Geographer* **3,** 307–321

Robinson, A. H. W. (1966). Residual currents in relation to shoreline evolution of the East Anglian coast. *Marine Geology* **4,** 57–84

Robinson, G. (1981). A statistical analysis of agriculture in the Vale of Evesham during the 'great agricultural depression'. *Journal of Historical Geography* **7**, 37–52

Rockie, W. A. (1939). Man's effects on the Palouse. *Geographical Review* **29**, 34–45

Rodger, E. M. (1972). *The Large Scale County Maps of the British Isles, 1596–1850: a Union List,* 2nd edition. Oxford, Bodleian Library

Royal Commission on Coast Erosion and Reclamation (1907). London, HMSO

Royal Commission on Historical Manuscripts (1979). *Record Repositories in Great Britain. A Geographical Directory,* 6th edition. London, HMSO

Rude, G. T. (1923). Shore changes at Cape Hatteras. *Annals of the Association of American Geographers* **12**, 87–95

Russell, R. J. (1954). Alluvial morphology of Anatolian rivers. *Annals of the Association of American Geographers* **44**, 363–391

Schattner, I. (1962). *The Lower Jordan Valley. Jerusalem, Magness Press, Publications of Hebrew University*

Schell, I. I. (1961). The ice off Iceland and the climates during the last 1200 years, approximately. *Geografiska Annaler* **43**, 354–362

Schmudde, T. H. (1963). Some aspects of landforms of the Lower Missouri River flood plain. *Annals of the Association of American Geographers* **53**, 60–73

Schneider, S. H. and Mass, C. (1975). Volcanic dust, sunspots, and temperature trends. *Science* **190** (4216), 741–746

Schove, D. J. (1961). Tree rings and climatic chronology. *New York Academy of Sciences Annals* **95**, 605–622

Schumm, S. A. (1971). Fluvial geomorphology: Channel adjustment and river metamorphosis. In *River Mechanics,* ed. H. W. Shen, 5.1–5.22. Fort Collins, Colorado State University

Schumm, S. A. (1973). Geomorphic thresholds and complex response of drainage systems. In *Fluvial Geomorphology,* ed. M. Morisawa, pp. 299–310. Binghampton, State University of New York, Publications in Geomorphology

Schumm, S. A. (1979). Geomorphic thresholds: the concept and its applications. *Transactions of the Institute of British Geographers* New Series 4, 485–515

Schumm, S. A. and Lichty, R. W. (1963). Channel widening and flood plain construction along the Cimarron River in south-western Kansas. *U.S. Geological Survey Professional Paper 352-D,* 71–88

Schumm, S. A. and Lichty, R. W. (1965). Time, space and causality in geomorphology. *American Journal of Science* **263**, 110–119

Sears, P. B. (1925). The natural vegetation of Ohio. *Ohio Journal of Science* **25**, 139–149

Sedgwick, W. (1914). Weather in the seventeenth century (Last Quarter) I Spring and summer to end of July. *Symon's Meteorological Magazine* **49**, 125–130

Sedgwick, W. (1914). Weather in the seventeenth century (Last Quarter) II Harvest and autumn. *Symon's Meteorological Magazine* **49**, 157–161

Seed, H. B. (1968). Landslides during earthquakes due to soil liquefaction. *Proceedings of the American Society of Civil Engineers, Journal of Soil Mechanics and Foundations Division* **94**, SM5, 1055–1122

Seelig, W. N. and Sorensen, R. M. (1973). Texas shoreline changes. *Shore and Beach* **41**, 23–25

Seth, S. K. (1963). A review of evidence concerning changes of climate in India during the protohistorical and historical periods. In *UNESCO, Arid Zone Research Series* **20,** 443–454

Seymour, W. A. (ed.) (1980). *A History of the Ordnance Survey.* Folkestone, Dawson

Shantz, H. L. and Turner, B. L. (1958). *Photographic Documentation of Vegetational Changes in Africa over a Third of a Century.* Tucson, University of Arizona College of Agriculture

Sharp, R. P. and Nobles, L. H. (1953). Mudflow of 1941 at Wrightwood, southern California. *Bulletin of the Geological Society of America* **64,** 547–560

Sharpe, C. F. S. (1938). *Landslides and Related Phenomena.* New York, Columbia Geomorphic Studies

Sheail, J. (1976). Land improvement and reclamation: the experiences of the First World War in England and Wales. *The Agricultural History Review* **24,** 110–125

Sheail, J. (1980). *Historical Ecology: the Documentary Evidence.* Cambridge, Institute of Terrestrial Ecology

Shepard, F. P. and Wanless, H. R. (1971). *Our Changing Coastlines.* New York, McGraw-Hill

Sheppard, J. A. (1957). The medieval meres of Holderness. *Transactions of the Institute of British Geographers* **23,** 75–86

Sheppard, J. A. (1958). The draining of the Hull Valley. *East Yorkshire Local History Society* No. 8

Sheppard, T. (1909). Changes on the east coast of England within the historical period. I Yorkshire. *Geographical Journal* **34,** 500–513

Sheppard, T. (1912). *Lost Towns of the Yorkshire Coast.* London, Brown and Sons Ltd

Sherlock, R. L. (1922). *Man as a Geologic Agent.* London, Willerby

Short, T. (1749). *A General Chronological History of the Air, Weather, Seasons, Meteors, etc.* London, Longman and Millar

Shvets, G. E. and Zaika, V. E. (1976). Multicentury dynamics of the Dnieper streamflow regime and its parameters. In *International Geography 1976,* ed. I. R. Gerasimov, Vol. 2, pp. 219–221. Moscow

Sigafoos, R. S. and Hendricks, E. L. (1961). Botanical evidence of the modern history of Nisqually Glacier, Washington. *U.S. Geological Survey Professional Paper 387-A,* A1–20

Simkovitch, V. G. (1916). Rome's fall reconsidered. *Political Science Quarterly* **31,** 201–243

Skelton, R. A. (1967). The military survey of Scotland 1747–1755. *Scottish Geographical Magazine* **83,** 5–16

Skelton, R. A. (1970). *County Atlases of the British Isles 1579–1850.* London, Carta Press

Skempton, A. W. (1964). Long-term stability of clay slopes. *Géotechnique* **14,** 77–102

Skempton, A. W. (1971). The Albion Mill foundations. *Géotechnique* **21,** 203–210

Skempton, A. W. and Hutchinson, J. N. (1969). Stability of natural slopes and embankment foundations. *7th International Conference of Soil Mechanics and Foundation Engineers,* Mexico, 291–340

Skempton, A. W. and Petley, D. J. (1967). The strength along structural discontinuities in stiff clays. *Proceedings of the Geotechnical Conference on Shear Strength Properties* Vol. 2, 29–46

Smith, C. D. and Parry, M. L. (eds) (1981). *Consequences of Climatic Change.* University of Nottingham, Department of Geography

Smith, L. (1973). *Devon Newspapers: a Finding List.* Exeter, Standing Conference on Devon History

Smith, W. and Stamp, L. D. (1941). *Lancashire.* The Land of Britain, the Report of the Land Utilisation Survey of Britain, Part 45

Smyth, W. J. (1976). Estate records and the making of the Irish landscape: an example from County Tipperary. *Irish Geography* **9,** 29–49

Snead, R. E. (1964). Active mud volcanoes of Baluchistan, West Pakistan. *Geographical Review* **54,** 546–560

So, C. L. (1971). Mass movements associated with the rainstorm of June 1966 in Hong Kong. *Transactions of the Institute of British Geographers* **53,** 55–65

So, C. L. (1974). Some coast changes around Aberystwyth and Tanybwlch, Wales. *Transactions of the Institute of British Geographers* **62,** 143–153

Somers-Cocks, J. V. (1977). *Devon Topographical Prints, 1660–1870—a Catalogue and Guide.* Exeter, Devon Library Services

Sonderegger, A. L. (1935). Modifying the physiographical balance by conservation measures. *Transactions of the American Society of Civil Engineers* **100,** 284–304

Somerville, R. (1951). *Handlist of Record Publications.* London, British Records Association

Stafford, D. B. and Langfelder, J. (1971). Air photo survey of coastal erosion. *Photogrammetric Engineering* **37,** 565–575

Stamp, L. D. (ed.) (1961). A history of land use in arid regions. *UNESCO, Arid Zone Research Series 17*

Stamp, L. D. (1962). *The Land of Britain, its Use and Misuse,* 3rd edition. London, Longmans, Green

Stamp, L. D. and Willatts, E. C. (1941). *Surrey.* The Land of Britain, the Report of the Land Utilisation Survey of Britain, Part 81

Stanford, J. L. (1975). A pictorial survey of Iowa tornadoes over three-quarters of a century. *Weather* **30,** 43–55

Stanley, J. W. (1951). Retrogression of the Lower Colorado River after 1935. *Transactions of the American Society of Civil Engineers* **116,** 943–957

Stearns, F. W. (1949). Ninety years change in a northern hardwood forest in Wisconsin. *Ecology* **30,** 350–358

Steer, F. W. (1962). *A Catalogue of Sussex Estate and Tithe Award Maps.* Sussex Record Society, Vol. 61

Steer, F. W. (1968). *A Catalogue of Sussex Maps.* Sussex Record Society, Vol. 66

Steers, J. A. (1926). Orford Ness: a study in coastal physiography. *Proceedings of the Geologists Association* **37,** 306–325

Steers, J. A. (1951). Recent changes on the marshland coast of North Norfolk. *Transactions of the Norfolk and Norwich Naturalists' Society* **17,** 206–213

Steers, J. A. (ed.) (1971). *Applied Coastal Geomorphology.* London, Macmillan

Steers, J. A. and Jensen, H. A. P. (1953). Winterton Ness. *Transactions of the Norfolk and Norwich Naturalists' Society* **17,** 259–274

Stephens, W. B. (1973). *Sources for English Local History.* Manchester, Manchester

University Press

Stephenson, J. and East, W. G. (1936). *Berkshire.* The Land of Britain, the Report of the Land Utilisation Survey of Britain, Part 78

Sternberg, H. O'R. (1968). Man and environmental change in South America. In *Biogeography and Ecology in South America,* ed. E. J. Fittkau *et al.,* Vol. 1, pp. 413–445. The Hague

Stevens, M. A., Simons, D. B. and Schumm, S. A. (1975). Man-induced changes of middle Mississippi River. *Proceedings of the American Society of Civil Engineers, Journal of Waterways Division* **101,** WW2, 119–133

Stoddart, D. R. (1971). Coastal reefs and islands and catastrophic storms. In *Applied Coastal Geomorphology,* ed. J. A. Steers, pp. 155–197. London, Macmillan

Stone, J. C. (1972). Techniques of scale measurement of historical maps. In *International Geography,* eds W. P. Adams and F. M. Helleiner, pp. 452–454. Toronto, University of Toronto Press

Stone, J. C. and Gemmell, A. M. D. (1977). An experiment in the comparative analysis of distortion on historical maps. *Cartographic Journal* **14,** 7–11

Stone, P. J. *et al.* (eds) (1966). *The General Enquirer: A Computer Approach to Content Analysis.* Cambridge, Massachusetts, MIT Press

Striffler, W. D. (1964). Sediment, streamflow and land use relationships in northern Lower Michigan. *U.S. Forest Service Research Paper LS-16*

Sturgess, R. W. (1966). The agricultural revolution on the English clays. *The Agricultural History Review* **14,** 104–121

Suklje, L. and Vidmar, S. (1961). A landslide due to long term creep. *Proceedings of 5th International Conference of Soil Mechanics and Foundation Engineers* Vol. 2, 727–735, Paris

Sundborg, A. (1956). The River Klarälven: a study of fluvial processes. *Geografiska Annaler* **38,** 127–316

Taher, M. (1974). Fluvial processes and geomorphology of the Brahmaputra Plain. *Geographical Review of India* **36,** 38–44

Tamura, T. (1978). An analysis of relationships between the areal distribution of earthquake-induced landslides and the earthquake magnitude. *Geographical Review of Japan* **51,** 662–672

Tank, R. W. (ed.) (1973). *Focus on Environmental Geology.* New York, Oxford University Press

Tannehill, I. R. (1938). *Hurricanes: their Nature and History.* Princeton University Press. 1969 reprint, New York, Greenwood Press

Tate, W. E. (1978). *A Domesday of English Enclosure Acts and Awards.* Edited with an introduction by M. E. Turner. Reading, The University of Reading

Taylor, E. G. R. (1947). The surveyor. *Economic History Review* **17,** 121–133

Taylor, E. G. R. (1954). *The Mathematical Practitioners of Tudor and Stuart England.* Cambridge, Cambridge University Press

Third, B. M. W. (1957). The significance of Scottish estate plans and associated documents, some local examples. *Scottish Studies* **1,** 39–64

Thirsk, J. (1955). The content and sources of English agrarian history after 1500. *The Agricultural History Review* **3,** 66–79

Thirsk, J. and Imray, J. (1958). *Suffolk Farming in the Nineteenth Century.* Suffolk Records Society, Vol. 1

Thom, B. G. (1974a). Coastal erosion in eastern Australia. *Search* **5,** 198–209

Thom, B. G. (1974b). Contemporary coastal erosion—geologic or historic. In *Impact of Human Activities on Coastal Zones,* Report of Symposium of Australian UNESCO Commission on Man and the Biosphere, pp. 58–65

Thomas, C. (1966). Estate surveys as sources in historical geography. *National Library of Wales Journal* **14,** 451–468

Thomas, D. (1958). The statistical and cartographic treatment of the acreage returns of 1807. *Geographical Studies* **5,** 15–25

Thomas, D. (1963). *Agriculture in Wales during the Napoleonic Wars; a Study in the Geographical Interpretation of Historical Sources.* Cardiff, University of Wales Press

Thomas, H. E. (1962). The meteorologic phenomenon of drought in the south-west. *U.S. Geological Survey Professional Paper 372-A*

Thomas, H. E. *et al.* (1963). Effects of drought in basins of interior drainage. *U.S. Geological Survey Professional Paper 372-E*

Thomas, W. L. (ed.) (1956). *Man's Role in Changing the Face of the Earth.* Chicago, University of Chicago

Thompson, P. (1978). *The Voice of the Past, Oral History.* Oxford, Oxford University Press

Thomson, D. H. (1938). A 100 years record of rainfall and water levels in the chalk at Chilgrove, West Sussex. *Transactions of the Institution of Water Engineers* **43,** 154–196

Thorarinsson, S. (1939). Hoffellsjökull, its movements and drainage. *Geografiska Annaler* **21,** 189–215

Thorarinsson, S. (1943). Oscillations of the Iceland glaciers in the last 250 years. *Geografiska Annaler* **35,** 1–54

Thorarinsson, S. (1970). The Lakagigar eruption of 1783. *Bulletin Volcanologique* **33,** 910–929

Thorarinsson, S., Einarsson, T. and Kjartansson, G. (1959). On the geology and geomorphology of Iceland. *Geografiska Annaler* **41,** 135–169

Thornes, J. B. and Brunsden, D. (1977). *Geomorphology and Time.* London, Methuen

Thornthwaite, C. W., Sharpe, C. F. S. and Dosch, E. F. (1942). Climate and accelerated erosion in the arid and semi-arid south-west with special reference to the Polacca Wash drainage basin, Arizona. *U.S. Department of Agriculture Technical Bulletin No. 808*

Thorpe, H. (1957). A special case of heath reclamation in the Alheden district of Jutland, 1700–1955. *Transactions of the Institute of British Geographers* **23,** 87–121

Thrower, N. J. W. (1966). *Original Survey and Land Subdivision. A Comparative Study of the Form and Effect of Contrasting Cadastral Surveys.* Chicago, Association of American Geographers and Rand McNally

Tinsley, H. (1976). Cultural influences on Pennine vegetation with particular reference to North Yorkshire. *Transactions of the Institute of British Geographers* New Series **1,** 310–322

Titow, J. Z. (1959–1960). The evidence of weather in the account rolls of the Bishopric of Winchester, 1209–1350. *Economic History Review* 2nd Series, **12,** 360–407

Titow, J. Z. (1970). Le climat à travers les rôles de comptabilité de l'évêché de Winchester (1350–1450). *Annales Economies Sociétés, Civilisations* **25,** 312–350

Toms, A. H. (1953). Recent research into the coastal landslides at Folkestone Warren, Kent, England. *Proceedings of the 3rd International Conference on Soil Mechanics,* Zurich **2,** 288–293

Towl, R. N. (1935). The behaviour history of the 'Big Muddy'. *Engineering News-Record* **115,** 262–264

Townley, S. D. and Allen, M. W. (1939). Descriptive catalog of earthquakes of the Pacific coast of the United States, 1769 to 1928. *Bulletin of the Seismological Society of America* **29,** 1–297

Trask, P. D. (1973). The Mexican volcano Paricutin. In *Focus on Environmental Geology,* ed. R. W. Tank, pp. 5–11. New York, Oxford University Press

Trewartha, G. T. (1940). The vegetal cover of the driftless cuestaform hill land. *Transactions of the Wisconsin Academy of Sciences, Arts and Letters* **32,** 361–382

Trimble, S. W. (1970). The Alcovy River swamps: the result of culturally accelerated sedimentation. *Bulletin of the Georgia Academy of Science* **28,** 131–141

Trimble, S. W. (1976). Sedimentation in Coon Creek Valley, Wisconsin. *Proceedings of the 3rd Inter-Agency Sedimentation Conference,* Water Resources Council, 5–100 to 5–112

Tuan, Y. F. (1966). New Mexican gullies: a critical review and some recent observations. *Annals of the Association of American Geographers* **56,** 573–597

Turner, H. H. (1925). Note on the 284-year cycle in Chinese earthquakes. *Monthly Notices of Royal Astronomical Society* **1,** 220–226

Turner, M. (1980). *English Parliamentary Enclosure, its Historical Geography and Economic History.* Folkestone, Dawson

UNESCO and International Council on Archives (1975). *International Directory of Archives, Archivum,* Vols 22–23 (1972–1973). Paris, Presses Universitaires de France

Upcott, W. (1818, reprinted 1978). *A Bibliographical Account of the Principal Works Relating to English Topography.* Wakefield, E.P. Publishing

Ursic, S. J. and Dendy, F. E. (1965). Sediment yields from small watersheds under various land uses and forest covers. *Proceedings of the Federal Inter-Agency Sedimentation Conference* 1963, **970,** 47–52. U.S. Department of Agriculture, Miscellaneous Publications

Usami, T. (1976). Map showing disaster areas of historical earthquakes in Japan. *Bulletin of the Earthquake Research Institute,* University of Tokyo **51,** 39–44

Usami, T. (1977). Study of the earthquake of March 11, 1853 based on newly collected old documents. *Bulletin of the Earthquake Research Institute,* University of Tokyo **52,** 333–342

Usami, T. (1978). Study of earthquakes in the Sanriku district during the Edo period. *Bulletin of the Earthquake Research Institute,* University of Tokyo **53,** 379–406

Valentin, H. (1953). Present vertical movements of the British Isles. *Geographical Journal* **119,** 299–305

Valentin, H. (1954). Land loss at Holderness, 1852–1952. In *Applied Coastal Geomorphology,* 1971, ed. J. A. Steers, pp. 116–137. London, Macmillan

Vita-Finzi, C. (1969)). *The Mediterranean Valleys: Geological Changes in Historical Times.* Cambridge, Cambridge University Press.

Vita-Finzi, C. (1978). *Archaeological Sites in their Setting.* London, Thames and Hudson

Vogel, W. (1929). Les plans parcellaires—Allemagne. *Annales d'histoire économique et sociale* **1,** 225–229

Vorsey, L. de (1973). Florida's seaward boundary: a problem in applied historical geography. *The Professional Geographer* **25,** 214–220

Vries, B. F. de (1966). *The Role of the Land Surveyor in the Development of New Zealand, 1840–76.* Victoria University of Wellington M.A. thesis

Wadell, H. (1935). Ice floods and volcanic eruptions in Vatnajökull. *Geographical Review* **25,** 131–136

Wahl, E. W. (1968). A comparison of the climate of the eastern United states during the 1830s with the current normals. *Monthly Weather Review* **96,** 73–82

Wales-Smith, B. G. (1971). Monthly and annual totals of rainfall representative of Kew, Surrey, from 1697 to 1970. *Meteorological Magazine* **100,** 345–357

Walker, M. F. and Taylor, J. A. (1976). Post-Neolithic vegetation changes in the western Rhinogau, Gwynedd, north-west Wales. *Transactions of the Institute of British Geographers* New Series **1,** 323–345

Wall, J. (ed.) (1977). *Directory of British Photographic Collections.* London, Heinemann

Walling, D. E (1979). The hydrological impact of building activity: a study near Exeter. In *Man's Impact on the Hydrological Cycle in the United Kingdom,* ed. G. E. Hollis, pp. 135–151. Norwich, Geo Abstracts

Wallis, H. (1981). The history of land use mapping. *Cartographic Journal* **18,** 45–48

Wallwork, K. L. (1956). Subsidence in the mid-Cheshire industrial area. *Geographical Journal* **122,** 40–53

Walne, P. (1969). *A Catalogue of Manuscript Maps in the Hertfordshire Record Office.* Hertford, Hertfordshire County Record Office

Walters, G. (1968). Themes in the large-scale mapping of Wales in the eighteenth century. *Cartographic Journal* **5,** 135–146

Warner, R. F., McLean, E. J. and Pickup, G. (1977). Changes in an urban water resource, an example from Sydney, Australia. *Earth Surface Processes* **2,** 29–38

Warren, K. F. (1965). Introduction to the map resources of the British Museum. *Professional Geographer* **17,** 1–17

Watson, J. W. (1967). *Mental Images and Geographical Reality in the Settlement of North America.* University of Nottingham, Cust Foundation Lecture

Watson, W. B. (1962). Summary report of the history section. In *Proceedings of Conference on the Climates of the Eleventh and Sixteenth Centuries,* eds R. A. Bryson and P. R. Julian, Aspen, Colorado. National Center for Atmospheric Research Technical Notes, 63–1. Boulder

Watts, F. B. (1960). The natural vegetation of the Southern Great Plains of Canada. *Geographical Bulletin, Canada Department of Mines Geographical Branch* **14,** 25–43

Wayne, W. J. (1969). Urban geology—A need and a challenge. *Proceedings of the Indiana Academy of Science* **78,** 49–63

Weidick, A. (1959). Glacial variations in West Greenland in historical time. *Meddelelser om Grønland* **158,** 1–96

Weil, G. E. (1978). *Répertoire des bibliothèques, collections, depôts de manuscrits et archives dans le monde.* Paris, Berger-Levrault

Welch, D. M. (1970). Substitution of space for time in a study of slope development. *Journal of Geology* **78**, 234–239

Werritty, A. and Ferguson, R. I. (1980). Pattern changes in a Scottish braided river over 1, 30 and 200 years. In *Timescales in Geomorphology,* eds R. A. Cullingford, D. A. Davidson and J. Lewin, pp. 53–68. Chichester, Wiley

West, J. (1962). *Village Records.* London, Macmillan

Whitaker, W. (1889). *The Geology of London and of Part of the Thames Valley.* London, Geological Survey

Whitaker, W. and Reid, C. (1899). *The Water Supply of Sussex from Underground Sources.* London, Geological Survey

White, G. (1977 reprint). The great frost of January 1776. *Weather* **32,** 106–108

Wightman, W. R. (1968). The pattern of vegetation in the Vale of Pickering area c. 1300 A.D. *Transactions of the Institute of British Geographers.* **45,** 125–142

Wilhelmy, H. (1966). Der 'Wandernde' Strom. *Erdkunde* **20,** 265–276

Willan, T. S. (1964). *River Navigation in England, 1600–1750.* London, Cass

Willatts, E. C. (1937). *Middlesex and the London Region.* The Land of Britain, the Report of the Land Utilisation Survey of Britain, Part 79

Williams, H. (1955). The great eruption of Goseguina, Nicaragua in 1835. *University of California Publications in Geological Sciences* **29,** 21–45

Williams, M. (1963). The draining and reclamation of the Somerset Levels, 1770–1833. *Transactions of the Institute of British Geographers* **33,** 163–179

Williams, W. W. (1956). An east coast survey—some recent changes in the coast of East Anglia. *Geographical Journal* **122,** 317–334

Williams, W. W. and Fryer, D. H. (1953). Benacre Ness: an east coast erosion problem. *Royal Institute of Chartered Surveyors Journal* **32,** 772–781

Winkley, B. R. (1972). River regulation with the aid of nature. *International Commission on Irrigation and Drainage* **29,** 433–457

Wise, S. M. (1980). Caesium-137 and lead-210: A review of the techniques and some applications in geomorphology. In *Timescales in Geomorphology,* eds R. A. Cullingford, D. A. Davidson and J. Lewin, pp. 109–127. Chichester, Wiley

Wolman, M. G. (1967). A cycle of sedimentation and erosion in urban river channels. *Geografiska Annaler* **49A,** 385–395

Wolman, M. G. and Leopold, L. B. (1957). River flood plains: some observations on their formation. *U.S. Geological Survey Professional Paper 282C,* 87–107

Wolman, M. G. and Schick, A. P. (1967). Effects of construction on fluvial sediment, urban and suburban areas of Maryland. *Water Resources Research* **3,** 451–464

Womack, W. R. and Schumm, S. A. (1977). Terraces of Douglas Creek, north-western Colorado: an example of episodic erosion. *Geology* **5,** 72–76

Wood, W. H. A. (1924). Rivers and man in the Indus-Ganges alluvial plain. *Scottish Geographical Magazine* **40,** 1–16

Wood, A. M. M. (1955). Folkestone Warren landslips: investigations 1948–1950. *Proceedings of the Institution of Civil Engineers* **4,** 410–428

Woolley, R. R. (1946). Cloudburst floods in Utah 1850–1938. *U.S. Geological Survey Water Supply Paper 994*

Worsley, P. (1981a). Lichenometry. In *Geomorphological Techniques,* ed. A. S. Goudie, pp. 302–305. London, George Allen and Unwin

Worsley, P. (1981b). Radiocarbon dating: principles, application and sample collection. In *Geomorphological Techniques,* ed. A. S. Goudie, pp. 277–283. London, George Allen and Unwin

Wright, C. E. (1957). Topographical drawings in the Department of Manuscripts, British Museum. *Archives* Vol. 3, 78–87

Wright, P. B. (1968). Wine harvests in Luxembourg and the biennial oscillation in European summers. *Weather* **23,** 300–304

Wynn, G. (1977). Discovering the antipodes, a review of historical geography in Australia and New Zealand, 1969–1975 with a bibliography. *Journal of Historical Geography* **3,** 251–265

Yates, E. M. (1964). Map of Over Haddon and Meadowplace, near Bakewell, Derbyshire c. 1528. *The Agricultural History Review* **12,** 121–124

York, J. C. and Dick-Peddie, W. A. (1969). Vegetation changes in southern New Mexico during the past hundred years. In *Arid Lands in Perspective,* eds W. G. McGinnies and B. J. Goldman, pp. 157–166. Tucson, University of Arizona Press

Youd, T. L. and Hoose, S. N. (1978). Historic ground failures in northern California triggered by earthquakes. *U.S. Geological Survey Professional paper 993*

Zaruba, Q. and Mencl, V. (1969). *Landslides and their Control.* Prague, Elsevier

Zhekulin, V. S. (1965, trans. 1968). Some thoughts on the subject of historical geography. *Soviet Geography: Review and Translation* **9,** 570

Zuidam, R. A. Van (1975). Geomorphology and archaeology: evidences of interrelation at historical sites in the Zaragoza region, Spain. *Zeitschrift für Geomorphologie* **19,** 319–328

Subject Index

Author Index